GW00949835

MASS OUTFLOW IN ACTIVE GALACTIC NUCLEI: NEW PERSPECTIVES

COVER ILLUSTRATION:

The lower left panel shows an optical image of the Seyfert 1 galaxy NGC 3783 from the Palomar Sky Survey. The upper right panel shows spectra of the active nucleus from the Space Telescope Imaging Spectrograph (STIS) on the *Hubble Space Telescope (HST)*, the *Far-Ultraviolet Spectroscopic Explorer (FUSE)*, and the *Chandra X-ray Observatory (CXO).* The specra are plotted in units of relative flux as a function of radial velocity with respect to the systemic velocity of the host galaxy. All three spectra show absorption lines at negative radial velocities (from the C IV doublet, Lyβ, and O VII, respectively) indicating mass outflow of gas from the nucleus. Four major kinematic components of absorption are labeled in the *HST*/STIS and *FUSE* spectra.

A SERIES OF BOOKS ON RECENT DEVELOPMENTS IN ASTRONOMY AND ASTROPHYSICS

Publisher

THE ASTRONOMICAL SOCIETY OF THE PACIFIC
390 Ashton Avenue, San Francisco, California, USA 94112-1722
Phone: (415) 337-1100 E-Mail: catalog@astrosociety.org
Fax: (415) 337-5205 Web Site: www.astrosociety.org

A listing of all other ASP Conference Series Volumes and IAU Volumes published by the ASP is cited at the back of this volume

**ASTRONOMICAL SOCIETY OF THE PACIFIC
CONFERENCE SERIES**

Volume 255

MASS OUTFLOW IN ACTIVE GALACTIC NUCLEI: NEW PERSPECTIVES

Proceedings of a workshop held at
The Catholic University of America
Washington, D. C., USA
8-10 March 2001

Edited by

D. Michael Crenshaw
Department of Physics and Astronomy, Georgia State University
Atlanta, Georgia, USA

Steven B. Kraemer
Department of Physics
The Catholic University of America, Washington, D.C., USA
Laboratory for Astronomy and Solar Physics
NASA/Goddard Space Flight Center, Greenbelt, Maryland, USA

and

Ian M. George
Joint Center for Astrophysics, Department of Physics
University of Maryland, Baltimore County, Baltimore, Maryland, USA
Laboratory for High Energy Astrophysics
NASA/Goddard Space Flight Center, Greenbelt, Maryland, USA

Library of Congress Cataloging in Publication Data
Main entry under title

Card Number: 2002100147
ISBN: 1-58381-095-1

ASP Conference Series - First Edition

Printed in United States of America by Sheridan Books, Chelsea, Michigan

Table of Contents

SESSION I: X-ray Absorption and Emission

Oral Presentations

Poster Presentations

SESSION II: Intrinsic UV Absorption

Oral Presentations

Poster Presentations

SESSION III: Broad Absorption-Line QSOs

Oral Presentations

Poster Presentations

SESSION IV: Narrow-Line Region, Broad-Line Region and Jets

Oral Presentations

Poster Presentation

SESSION VI: Discussion

Preface

Mass Outflow in Active Galactic Nuclei: New Perspectives

A Workshop Held at The Catholic University of America, 2001 March 8 – 10.

Over the past decade, a great amount of interest has been generated in the study of mass outflow in active galactic nuclei (AGN). X-ray spectra from the *Advanced Satellite for Cosmology and Astrophysics* (*ASCA*) confirmed the presence of a "warm absorber" (characterized by O VII and O VIII absorption edges) in least half of all Seyfert 1 galaxies. Ultraviolet spectra from the *Hubble Space Telescope* (*HST*) revealed that a similar fraction of Seyferts show absorption lines from moderately ionized species (C IV, N V); these lines are typically blueshifted with respect to the systemic velocity of the host galaxy, indicating outflow from the nucleus. Ground based observations revealed the presence of high-velocity clouds (up to 0.1c) ejected from quasars, and spectropolarimetry helped to constrain the geometry of the broad absorption lines (BALs) found in about 10% of all quasars. Much of the progress in this field was summarized in the proceedings of a workshop (ASP Conference Series Volume 128) on "Mass Ejection From AGN", which was held at the Carnegie Observatories on 1997 February 19 – 21.

Since the workshop in 1997, new and exciting results on mass outflow from AGN have become available through new space missions or instruments, including the Space Telescope Imaging Spectrograph (STIS) on board *HST*, the *Far-Ultraviolet Spectroscopic Explorer* (*FUSE*), the *Chandra X-ray Observatory* (*CXO*), and the *X-ray Multi-mirror Mission* (*XMM-Newton*), principally due to their increased light gathering power and spectral resolution compared to previous missions. New ground-based facilities and projects, including the Keck Observatory and the Sloan Digital Sky Survey (SDSS), have helped to greatly expand the sample of quasars with intrinsic absorption. Finally, dynamical models, such as those that invoke accretion disk winds driven by radiative and/or magnetic forces, are reaching a level of sophistication that yields testable predictions.

The rapid pace of discoveries in this field suggested to us that the time was ripe for another conference on mass outflow in AGN. A scientific organizing committee was formed, and it was decided that the conference should take the form of a workshop. "Mass Outflow in Active Galactic Nuclei: New Perspectives" was held on 2001 March 8 – 10 March at The Catholic University of America, in Washington, DC. The stated purpose of the workshop was to review the new observations and their implications, investigate the correspondence between the observations and models, and stimulate new directions for future collaborative efforts. Scientific topics included intrinsic UV and X-ray absorption, broad absorption-line quasars, evidence for radial outflow in the emission line regions (narrow-line region and broad-line region), photoionization and shock models of the emission-line and absorption regions, new observations and models of jets, and dynamical outflow models.

The format of the workshop was such that researchers were allowed "equal time" to present their observations and views. There were no invited talks (except for an invited review of the workshop), and each speaker was allowed 15 minutes (plus 5 minutes for questions). Posters were displayed all three days of the workshop, and ample time was provided for viewing and discussion of the posters. Time was provided at the end of each session for a group discussion of the relevant oral and poster presentations. A review of the workshop and general discussion were held on the afternoon of the last day. At the conclusion of the workshop, there was a general consensus that this format was very satisfactory, and particularly conducive to an open exchange of ideas.

These proceedings provide a record of the presentations at the workshop. The contributions are arranged by session, with papers based on oral presentations preceding those based on posters. We have edited the papers for consistency in style and presentation and for grammatical and typographical errors, but not for scientific content. All of the first authors were given a chance to review their edited papers prior to publication.

We thank the Scientific Organizing Committee for its role in shaping the focus and format of the workshop; members, in addition to the editors, were Fred Hamann, John Hutchings, Smita Mathur, Norm Murray, Hagai Netzer, and Joe Shields. We thank the following groups or institutions for providing facilities and/or financial support for the workshop: The Catholic University of America (CUA), NASA Headquarters, and the STIS Instrument Definition Team (IDT) at Goddard Space Flight Center. In particular, we gratefully acknowledge advice and support from STIS IDT members Bruce Woodgate (PI) and Ted Gull (Deputy PI). We thank John Convey (Provost, CUA) and the following individuals in the Department of Physics at CUA for guidance, advice, and/or assistance with the conference: Charles Montrose (chair), Fred Bruhweiler, Gail Hershey, Annie James, and Matt Austin. We are grateful to Westover Consultants, Inc. for providing logistical support for this meeting, and in particular to Gale Quilter, for a superb job of coordinating the effort.

Last, but certainly not least, we owe a great debt of gratitude to our invited reviewer, Ray Weymann. Ray sponsored the 1997 workshop on mass outflow, and has been a guiding force in this field for many years. We are pleased that we were able to coax Ray out of retirement to review the conference results and offer his own insightful comments. For those interested in a summary of the recent developments in this field, we recommend that they begin with his excellent review paper at the end of this volume.

2001 October 26
Mike Crenshaw
Steve Kraemer
Ian M George
(editors)

List of Participants

Aldcroft, Thomas L.,
Harvard-Smithsonian Center for Astrophysics, 60 Garden St., Cambridge, MA 02138, USA <taldcroft@cfa.harvard.edu>

Arav, Nahum
Astronomy Department, University of California Berkeley, 601 Campbell Hall, Berkeley, CA 94720, USA <arav@astro.berkeley.edu>

Baker, Joanne
Astronomy Department, University of California Berkeley, 601 Campbell Hall, Berkeley, CA 94720, USA <jcb@astro.berkeley.edu>

Ballantyne, David
Institute of Astronomy, University of Cambridge, Madingley Road, Cambridge CB3 OHA, UK <drb@ast.cam.ac.uk>

Behar, Ehud
Columbia University, 550 W. 120th Street, New York, NY 10027, USA <behar@astro.columbia.edu>

Bottorff, Mark
Department of Physics and Astronomy, University of Kentucky, 177 Chemistry-Physics Building, Lexington, KY 40506, USA <bottorff@pa.uky.edu>

Branduardi-Raymont, Graziella,
Mullard Space Science Laboratory, University College London, Holmbury St. Mary, Dorking, Surrey, RH5 6NT, UK <gbr@mssl.ucl.ac.uk>

Brotherton, Michael
National Optical Astronomy Observatories, 950 N. Cherry Avenue Tucson, AZ 85726, USA <mbrother@noao.edu>

Calvani, Humberto
Department of Physics & Astronomy, Johns Hopkins University, 3400 North Charles Street, Baltimore, MD 21218, USA <calvani@pha.jhu.edu>

Canalizo, Gabriela
Lawrence Livermore National Laboratory 7000 East Avenue, L-413, Livermore, CA 94550, USA <canalizo@igpp.ucllnl.org>

Cecil, Gerald
Department of Physics and Astronomy, University of North Carolina, Chapel Hill, NC 27599, USA <gerald@thececils.org>

Chambers, Ken
University of Hawaii, 2680 Woodlawn Drive, Honolulu, HI 96822, USA <chambers@ifa.hawaii.edu>

Chatzichristou, Eleni
NASA/Goddard Space Flight Center, Mail Code 681, Bldg 21, Greenbelt, MD 20770, USA <eleni@astarti.gsfc.nasa.gov>

Chelouche, Doron
Tel-Aviv University, PO Box 39040, Tel-Aviv 69978, Israel
`<doron@wise.tau.ac.il>`

Cohen, Marshall
Caltech, Caltech 102-24, Pasadena, CA 91125, USA
`<mhc@astro.caltech.edu>`

Collinge, Matthew
Pennsylvania State University, 525 Davey Lab, University Park, PA 16802, USA `<collinge@astro.psu.edu>`

Crenshaw, Mike
Department of Physics and Astronomy, Georgia State University, Atlanta, GA 30303, USA `<crenshaw@chara.gsu.edu>`

de Kool, Martijn
Mount Stromlo Observatory, Cotter Road, Weston, ACT 2611, Australia `<dekool@mso.anu.edu.au>`

Elvis, Martin,
Harvard-Smithsonian Center for Astrophysics, 60 Garden St., Cambridge, MA 02138, USA `<elvis@cfa.harvard.edu>`

Eracleous, Michael
Pennsylvania State University, 525 Davey Lab, University Park, PA 16802, USA `<mce@astro.psu.edu>`

Everett, John
University of Chicago, 5640 S Ellis Avenue, Chicago, IL 60637, USA
`<everett@apollo.uchicago.edu>`

Frank, Adam
Department of Physics and Astronomy, University of Rochester, Rochester, NY 14627, USA `<afrank@pas.rochester.edu>`

Gabel, Jack
The Catholic University of America, 620 Michigan Avenue, NE, Washington, DC 20064, USA `<gabel@iacs.gsfc.nasa.gov>`

Gallagher, Sarah
Pennsylvania State University, 525 Davey Lab, University Park, PA 16802, USA `<gallsc@astro.psu.edu>`

Galliano, Emmanuel
The European Southern Observatory Chile, Alonzo de Cordova 3107, Vitalura, Casilla 19001, Santiago 19 Chile `<egallian@eso.org>`

Ganguly, Rajib
Pennsylvania State University, 525 Davey Lab, University Park, PA 16802, USA `<ganguly@astro.psu.edu>`

Gaskell, Martin
Department of Physics and Astronomy, University of Nebraska, Lincoln, NE 68588, USA `<MGASKELL1@unl.edu>`

George, Ian M.,
Joint Center for Astrophysics, Department of Physics, University of Maryland, Baltimore County, 1000 Hilltop Circle, Baltimore, MD 21250, USA, *and* NASA/Goddard Space Flight Center, Mail Code 662, Greenbelt, MD 20771, USA `<ian.george@umbc.edu>`

Green, Paul J.,
Harvard-Smithsonian Center for Astrophysics, 60 Garden St., Cambridge, MA 02138, USA <green@cfa.harvard.edu>

Hall, Patrick
Princeton University, Peyton Hall, Ivy Lane, Princeton, NJ 08544, USA <pathall@astro.princeton.edu>

Hamann, Fred
University of Florida, 211 Bryant Space Science Center, Gainesville, FL 32611, USA <hamann@astro.ufl.edu>

Heap, Sara
NASA/Goddard Space Flight Center, Code 681, Greenbelt, MD 20771, USA <heap@srh.gsfc.nasa.gov>

Hines, Dean
Steward Observatory, 933 N. Cherry Avenue, Tucson, AZ 85721, USA <dhines@as.arizona.edu>

Hutchings, John
DAO, NRC of Canada, 5071 W. Saanich Road, Victoria, BC V9E 2E7, Canada <john.hutchings@nrc.ca>

Hutsemékers, Damien
The European Southern Observatory Chile, Alonso de Cordoba 3107, Vitalura, Casilla 19001, Santiago 19, Chile <dhutseme@eso.org>

Junkkarinen, Vesa
University of California, San Diego, CASS 0424, 9500 Gilman Drive, La Jolla, CA 92093, USA <vesa@ucsd.edu>

Kaastra, Jelle S.,
SRON, Sorbonnelaan 2, 3584 CA Utrecht, The Netherlands <j.kaastra@sron.nl>

Kaiser, Mary Beth
Dept of Physics and Astronomy, Johns Hopkins University, 3400 N. Charles Street, Baltimore, MD 21029, USA <kaiser@pha.jhu.edu>

Kallman, Timothy
NASA/Goddard Space Flight Center, Code 662, Greenbelt, MD 20771, USA <tim@xstar.gsfc.nasa.gov>

Kaspi, Shai,
Department of Astronomy and Astrophysics, The Pennsylvania State University, University Park, PA 16802, USA <shai@astro.psu.edu>

Kazanas, Demosthenes
NASA/Goddard Space Flight Center, Code 661, Greenbelt, MD 20771, USA <kazanas@milkyway.gsfc.nasa.gov>

Koide, Shinji
Toyama University, 3190 Gofuku, Toyama, Toyama 930-8555, Japan <koidesin@ecs.toyama-u.ac.jp>

Kollatschny, Wolfram
Uni-Sternwarte, Geismarlandstr. 11, 27083 Goettingen, Germany <wkollat@uni-sw.gwdg.de>

Korista, Kirk
Depatment of Physics, Western Michigan University, 1120 Everett Tower, Kalamazoo, MI 49008, USA <korista@wmich.edu>

Kraemer, Steve
Department of Physics, The Catholic University of America 200 Hannan Hall, Washington, DC 20064, USA, *and* NASA/Goddard Space Flight Center, Mail Code 681, Greenbelt, MD 20771, USA <kraemer@yancey.gsfc.nasa.gov>

Kriss, Gerard
Space Telescope Science Institute, 3700 San Martin Drive, Baltimore, MD 21218, USA <gak@stsci.edu>

Kukula, Marek
Institute for Astronomy, Royal Observatory Edinburgh, Blackford Hill, Edinburgh EH9 3HJ, UK <mjk@roe.ac.uk>

Laor, Ari
Technion, Haita 32000, Israel <laor@phobus.technion.ac.il>

Leighly, Karen
The University of Oklahoma, 440 W. Brooks Street, Norman, OK 73019, USA <leighly@ou.edu>

Mathur, Smita,
The Ohio State University, 140 West 18th Avenue, Columbus, OH 43210-1173, USA <smita@astronomy.ohio-state.edu>

Max, Claire,
Lawrence Livermore National Laboratory, PO Box 808, #L-413, Livermore, CA 94551, USA <max1@llnl.gov>

Murray, Norman,
CITA, University of Toronto, 60 St. George Street, Toronto, ON M4S 1B4, Canada <murray@cita.utoronto.ca>

Netzer, Hagai,
Tel Aviv University, Ramat Aviv, Tel Aviv, Israel 69978, Israel <netzer@wise.tau.ac.il>

Ogle, Patrick M.,
University of California Santa Barbara, Physics Department, Santa Barbara, CA 93106, USA <pmo@xmmom.physics.ucsb.edu>

Perlman, Eric,
Joint Center for Astrophysics, Department of Physics, University of Maryland, Baltimore County, 1000 Hilltop Circle, Baltimore, MD 21250, USA, <perlman@jca.umbc.edu>

Poludnenko, Alexei,
Department of Physics and Astronomy, University of Rochester, Rochester, NY 14627, USA <wma@pas.rochester.edu>

Proga, Daniel,
NASA/Goddard Space Flight Center, Code 662, Greenbelt, MD 20771, USA <proga@sobolev.gsfc.nasa.gov>

Quirrenbach, Andreas,
Center for Astrophysics and Space Sciences, Mail Code 0424, University of California, San Diego, La Jolla, CA 92093, USA <aquirrenbach@ucsd.edu>

Risaliti, Guido,
Smithsonian Astrophysical Observatory, 60 Garden Street, Cambridge, MA 02138, USA <grisaliti@cfa.harvard.edu>

Ruiz, José,
Department of Physics, The Catholic University of America, 200 Hannan Hall, Washington, DC 20064, USA <ruiz@yancey.gsfc.nasa.gov>

Sabra, Bassem,
Department of Astronomy, University of Florida, Gainesville, FL 32611, USA <sabra@astro.ufl.edu>

Schmitt, Henrique,
National Radio Astronomy Observatory, PO Box 0, Socorro, NM 87801, USA <hschmitt@aoc.nrao.edu>

Shields, Joesph,
Department of Physics and Astronomy, Ohio University, Athens, OH 45701, USA <shields@phy.ohiou.edu>

Sivron, Ran,
Bucknell University, 1 Derr Drive, Lewisburg, PA 17837, USA <rsivron@bucknell.edu>

Sol, Helene,
Observatoire de Paris, Place J. Janssen, Meudon, France 92295, France <helene.sol@obspm.fr>

Starling, Rhaana,
Mullard Space Science Laboratory, University College London, Holmbury St. Mary, Dorking, Surrey, RH5 6NT, UK <rlcs@mssl.ucl.ac.uk>

Torricelli, Guidetta,
Osservatorio Astrofisico di Arcetri, L.go E. Fermi 5, 50125 Firenze, Italy <torricel@arcetri.astro.it>

Turner, T. Jane,
Joint Center for Astrophysics, Department of Physics, University of Maryland, Baltimore County, 1000 Hilltop Circle, Baltimore, MD 21250, USA *and* NASA/Goddard Space Flight Center, Mail Code 662, Greenbelt, MD 20771, USA <turner@lucretia.gsfc.nasa.gov>

van Bemmel, Ilse,
Kapteyn Astronomical Institute, PO Box 800, NL-9700 AB Groningen, The Netherlands <bemmel@astro.rug.nl>

Verner, Ekaterina,
Department of Physics, The Catholic University of America 200 Hannan Hall, Washington, DC 20064, USA
<kverner@fez.gsfc.nasa.gov>

Vestergaard, Marianne,
Ohio State University, 140 West 18th Avenue, Columbus, OH 43210, USA <marianne@astronomy.ohio-state.edu>

Wei, Jianyan,
Beijing Astronomical Observatory, A20 Datun Road, Chaoyang District, Beijing 100012, China <wjy@bao.ac.cn>

Weymann, Ray,
Carnegie Observatories, 813 Santa Barbara Street, Pasadena, CA 91101, USA <rjw@ociw.edu>

Woodgate, Bruce,
NASA/Goddard Space Flight Center, Code 681, Greenbelt, MD 20771, USA <woodgate@stars.gsfc.nasa.gov>

Conference Photo

Mass Outflow in Active Galactic Nuclei: New Perspectives
ASP Conference Series, Vol. 255, 2002
D.M. Crenshaw, S.B. Kraemer, and I.M. George

High-Resolution X-ray Spectroscopy of the Warm Gas in Seyfert 1 Galaxies

Jelle S. Kaastra

SRON, Sorbonnelaan 2, 3584 CA Utrecht, The Netherlands

Abstract. The important role of high-resolution X-ray spectroscopy in the study of warm gas in the centers of active galactic nuclei is demonstrated. This is illustrated with the examples of IRAS 13349+2439 and NGC 5548. These data reveal a complex structure of the warm absorber, both in terms of ionization structure and dynamical structure.

1. Introduction

It is known for a long time that the dominant spectral component in Active Galactic Nuclei (AGN) has a power law shape. More recently, in several cases a soft X-ray excess has been detected attributed to emission from the accretion disk. In particular observations with the *ROSAT* and *ASCA* satellites have shown the presence of another component in the X-ray spectrum of AGN: the so-called "warm absorber". The low- to medium-resolution X-ray spectra indicated a flux deficit as compared to a power law spectral fit in the 0.7−2 keV energy band. This flux deficit has been interpreted as absorption edges from O VII and/or O VIII.

However recent observations with the grating spectrometers onboard the *Chandra X-ray Observatory* (*CXO*) and *XMM-Newton* do not always confirm this scenario. In some cases there is evidence for spectral features from the accretion disk (relativistically broadened lines) that mimic a warm absorber when observed with low spectral resolution (Branduardi-Raymont et al. 2001, 2002). In this contribution we focus upon the genuine warm absorber that is present in AGN, albeit not always as strong as previously thought.

2. The Importance of Absorption-line Spectroscopy

Figure 1 shows the continuum transmission of a homogeneous slab consisting of pure O VIII. It is evident from this figure that in order to detect absorption the column density must be at least 10^{22} O VIII atoms per m^2, and a significant absorption edge is present if the column density is 10^{23} m^{-2}. Whenever such a large column density is present, O VIII is visible through its continuum absorption, with an overall flux decrease of at least 10% in the 6–15 Å band.

A slab of material does not only produce continuum absorption due to the photoelectric effect, however, but it also produces resonance line absorption. Figure 2 shows the transmission due to line absorption in the O VIII Ly-α resonance line for the same set of column densities as depicted in Figure 1. It

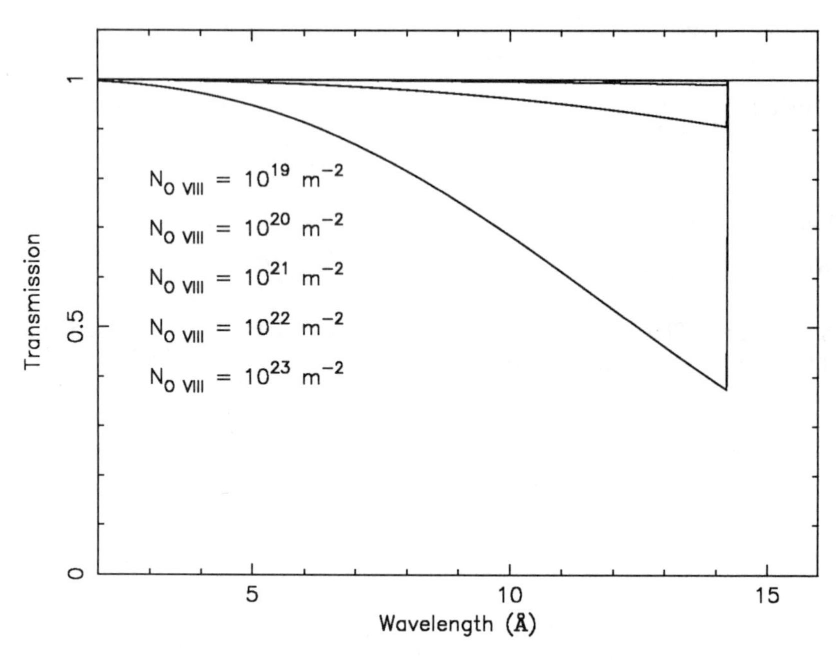

Figure 1. Continuum transmission of a slab of O VIII for different column densities as indicated from top to bottom. The first two curves almost coincide.

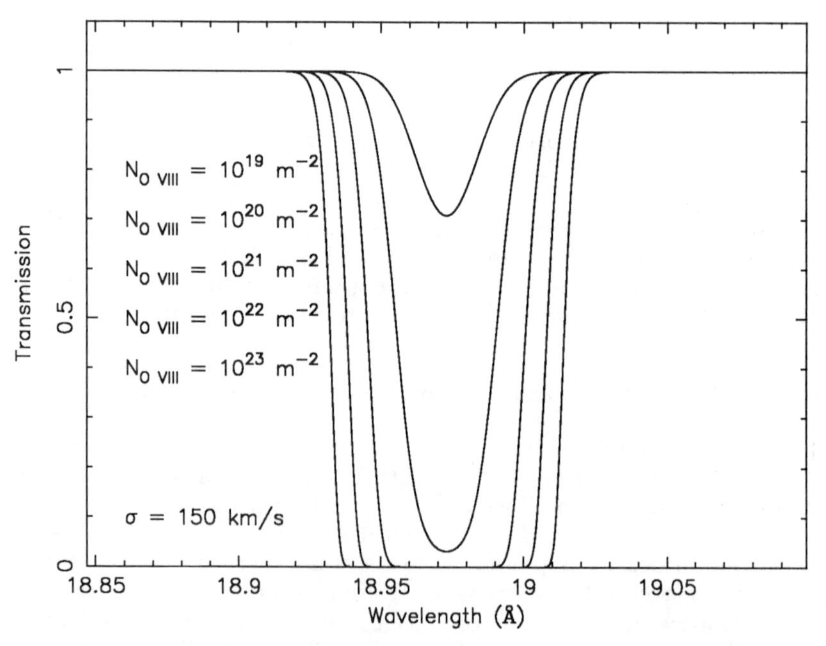

Figure 2. Line transmission of a slab of O VIII for different column densities as indicated from top to bottom.

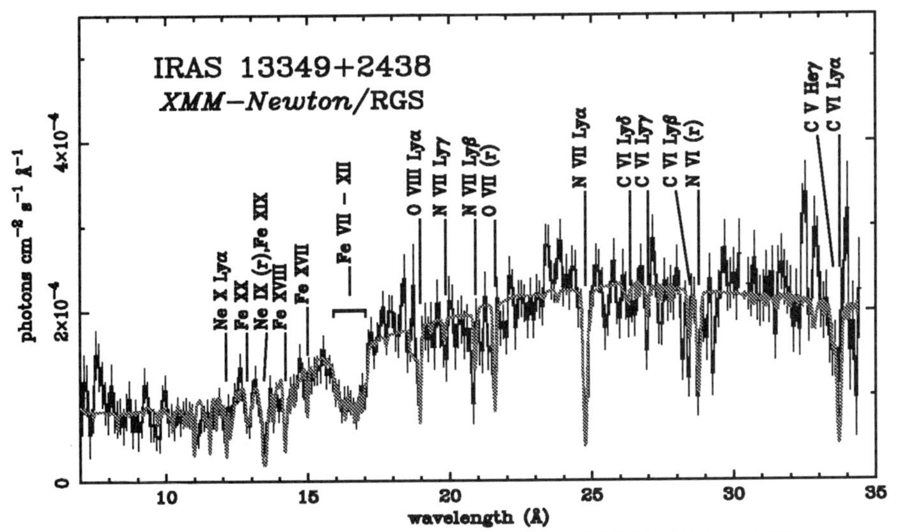

Figure 3. *XMM-Newton* RGS spectrum of IRAS 13349+2438, corrected for the cosmological redshift (from Sako et al. 2001).

is evident that for small column densities such as 10^{19} m^{-2} noticeable line absorption is visible; for the same column density, the continuum absorption is almost undetectable. At the column densities where the continuum starts becoming visible (10^{22} m^{-2}) the line is already completely saturated and has a "black" core. The point to make here is that these lines are only visible with high spectral resolution, as is possible with the *CXO* and *XMM-Newton* grating spectrometers. When measured with low or medium spectral resolution (such as CCD cameras) the lines cannot be detected at all. However by comparing Figures 1 and 2 it is immediately evident that line spectroscopy gives orders of magnitude more sensitivity for detecting small columns of warm gas.

3. New Components of the Warm Absorber

After the discovery of narrow absorption lines in Seyfert 1 galaxies (the *CXO* Low Energy Transmission Grating Spectrometer [LETGS] measurement of NGC 5548; Kaastra et al. 2000) several other AGN have been observed both by *CXO* and *XMM-Newton*. A very interesting case is the quasar IRAS 13349+2438, studied with the Refelection Grating Spectrometers (RGS) of *XMM-Newton* by Sako et al. (2001). This radio quiet object at a redshift of 0.108 was known to have a complex warm absorber. The *XMM-Newton* data are shown in Figure 3. The spectrum shows broad absorption lines with a Gaussian σ_v of 600 km $s^{-1}$and an average blueshift (outflow) velocity of 200 km s^{-1}. This is qualitatively similar to what has been seen in NGC 5548. The higher ionization lines tend to have smaller outflow velocities (consistent with zero velocity) while the lower ionization lines have the largest outflow velocities (up to 400 km s^{-1}). This can be understood in the framework of a radiatively driven, accelerating wind model.

Interestingly, the spectrum shows a deep and narrow absorption feature near 17 Å. If this would have been observed with low spectral resolution, it would have been identified as the O VII absorption edge at 16.77 Å. However, with the high spectral resolution of the RGS such an interpretation can be excluded (it is at a significantly higher wavelength than the O VII edge). In addition, the observed feature is rather deep while the O VII edge should give a long blueward tail in the transmission curve.

The structure can be identified instead as an "unresolved transition array" (UTA) from inner shell 2p–3d resonance absorption lines in relatively cool iron (Fe VII–Fe XII). It is strikingly similar to the laboratory absorption measurements of a heated iron foil (Chenais-Popovics et al. 2000).

The discovery of this UTA has important implications. It offers the opportunity to study material at low ionization parameters. Note that the other resonance absorption lines from the same ions have much longer wavelengths, above 60 Å where interstellar absorption becomes important and the effective area of the only high-resolution spectrometer available in that band (the *CXO* LETGS) is not very high.

The presence of these ions in IRAS 13349+2438 yields additional support that the warm absorber consists of different components, in this case a low and a high ionization component. The low ionization component could originate in the dusty torus, since its column density is close to the column derived from the observed optical reddening of the nucleus. However, at this moment such an association is hard to prove due to the fact that either a density or distance estimate of the cool component is lacking.

4. NGC 5548

NGC 5548 was the first Seyfert galaxy to be observed at high spectral resolution in the X-ray band. The *CXO* LETGS spectrum has been published by Kaastra et al. (2000). The spectrum shows many strong absorption lines from highly ionized ions such as C V, C VI, N VI, N VII, O VII, O VIII, Ne IX, Ne X, as well as several iron ions and others. The ions that are detected indicate the absorbing medium is photoionized. The observed blueshift of the lines indicates that the absorber has the form of an outflowing, warm, photoionized wind. After a thorough calibration of both the wavelength scale and the effective area of the LETGS, performed at SRON, it is now possible to study the dynamics of the warm absorber in NGC 5548 in more detail. Figure 4 shows the measured line centroids derived from unblended lines of several interesting ions. A comparison with the centroids of the six UV components identified by Crenshaw & Kraemer (1999) shows that most of the lines from the C, N, O and Ne ions are consistent with the UV components 2−−5. It is hard to tell how much column density each component contributes (the spectral resolution of the LETGS limits the velocity resolution to a few hundred km s^{-1}). Also, the relative contribution of each component may vary from ion to ion. An indication is the tendency of larger blueshifts for the lowly ionized iron ions, similar to what has been found in IRAS 13349+2438. A different view to this problem is presented in Figure 5. This shows the C VI Lyα line, one of the strongest lines at longer wavelengths and thereby relatively well resolved. The spectral resolution of the LETGS at this

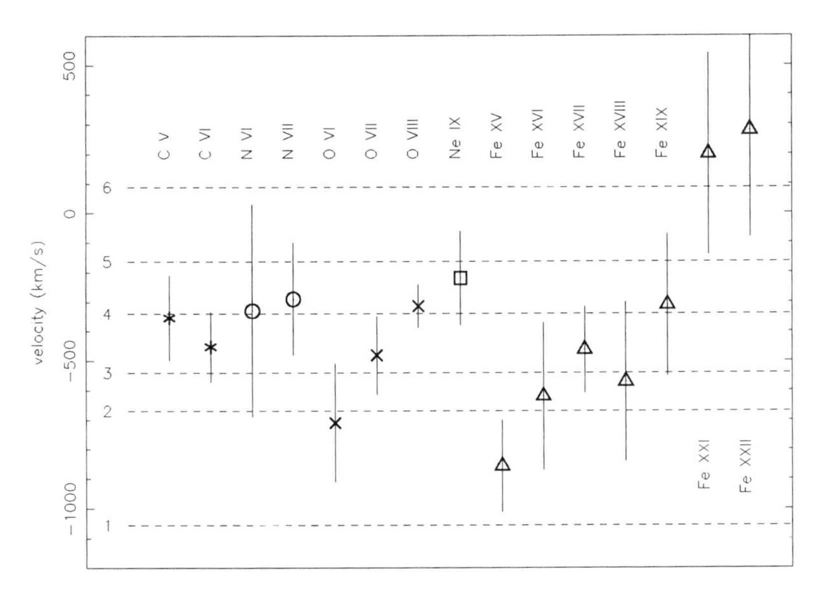

Figure 4. Velocity centroids for ions in the warm absorber of NGC 5548, as obtained with the *CXO* LETGS. The 6 dashed lines indicate the velocity components of the UV lines as identified by Crenshaw & Kraemer (1999). From Kaastra et al., in preparation.

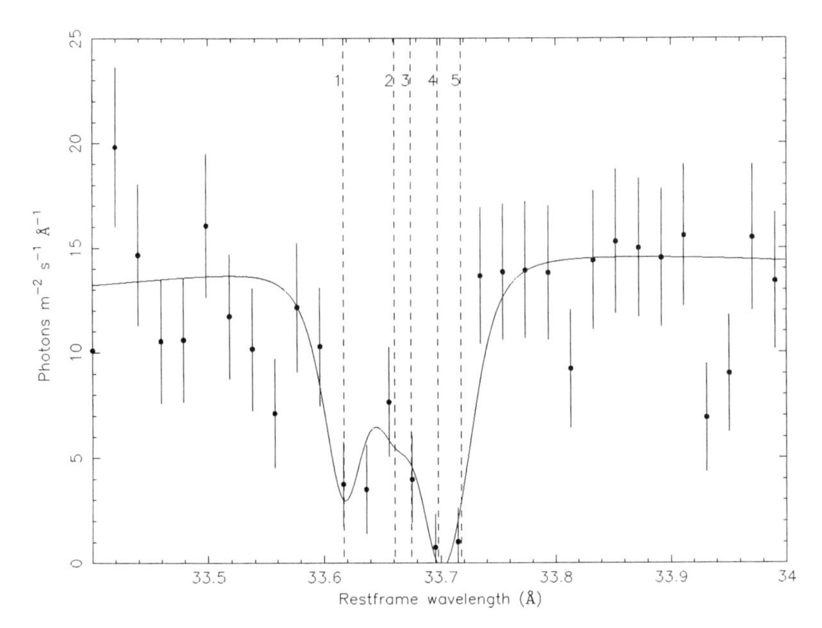

Figure 5. Line profile of the C VI Lyman-α line in NGC 5548. The velocities of five of the UV components identified by Crenshaw & Kraemer (1999) are indicated. From Kaastra et al., in preparation.

wavelength is still not good enough to clearly separate the different components, but it is clear from Figure 5 that all five dynamic components contribute to the C VI profile. This complexity of the line profile makes it difficult to make a quantitative model where both the X-ray data as well as the higher resolution UV data are taken into account properly.

A more quantitative analysis of the LETGS data of NGC 5548 is underway (Kaastra et al., in preparation). This analysis shows that apart from the dominant warm absorber component (yielding most of the O VII and O VII), there is also a low ionization component (visible by a weak UTA similar to IRAS 13349+2438 mentioned in the previous section), as well as a higher ionized component, mainly yielding absorption in ions such as Fe XX.

A key issue in understanding the complex warm absorber is a study of its time variability. The long *CXO* High Energy Transmission Grating Spectrometer (HETGS) observation of NGC 5548 showed a two times smaller flux as compared to the earlier LETGS observation, but the properties of the warm absorber do not appear to have changed that much. A similar conclusion can be drawn from a short 20 ks exposure by *XMM-Newton* a year later; again the source had a lower continuum flux but similar warm absorber properties as during the LETGS observation.

5. Conclusion

The new high-resolution X-ray spectra of Seyfert galaxies show that the warm photoionized wind has rich absorption line spectra. These spectra reveal the dynamical and ionization structure of this wind, thereby opening unique possibilities to a better understanding of the innermost regions of AGN.

Acknowledgments. I thank my collaborators in the RGS and LETGS teams for providing me with some of the material presented here, in particular Masao Sako from Columbia University. This work is based on observations obtained with *XMM-Newton*, an ESA science mission with instruments and contributions directly funded by ESA Member States and the USA (NASA). The Laboratory for Space Research Utrecht is supported financially by NWO, the Netherlands Organization for Scientific Research.

References

Branduardi-Raymont, G., Sako, M., Kahn, S.M., Brinkman, A.C., Kaastra, J.S., Page, M.J. 2001, A&A 365, L140

Branduardi-Raymont, G., et al. 2002, in Mass Outflow in Active Galactic Nuclei: New Perspectives, eds. D.M. Crenshaw, S.B. Kraemer, & I.M. George (San Francisco: ASP), p. 31

Chenais-Popovics, C., et al., 2000, ApJS 127, 275

Crenshaw, D.M., Kraemer, S.B. 1999, ApJ 521, 572

Kaastra, J.S., Mewe, R., Liedahl, D.A., Komossa, S., Brinkman, A.C. 2000, A&A 354, L83

Sako, M., et al. 2001, A&A 365, L168

Mass Outflow in Active Galactic Nuclei: New Perspectives
ASP Conference Series, Vol. 255, 2002
D.M. Crenshaw, S.B. Kraemer, and I.M. George

High-Resolution X-ray Spectroscopy of the Outflow in NGC 3783

Shai Kaspi

Department of Astronomy and Astrophysics, The Pennsylvania State University, University Park, PA 16802, USA

Abstract. The high-resolution X-ray spectrum of NGC 3783 shows several dozen absorption lines and a few emission lines from the H-like and He-like ions of O, Ne, Mg, Si, and S as well as from Fe XVII–Fe XXIII L-shell transitions. Combining several lines from each element, we clearly demonstrate the existence of the absorption lines and determine they are blueshifted relative to the systemic velocity by -610 ± 130 km s^{-1}. We model the absorption and emission lines as arising from two highly ionized gas components which have an order of magnitude difference in their ionization parameters. The two components are radially outflowing away from the nucleus and thus contribute to both the absorption and the emission. We discuss the relations between the UV and X-ray absorbers and conclude that although the currently available observations had increased our knowledge they cannot, yet, firmly resolve this issue.

1. Introduction

NGC 3783 is a well studied bright Seyfert 1 galaxy ($V \approx 13.5$ mag). It has one of the most strongest X-ray absorption features around 0.7–1.5 keV which have been typically attributed to O VII (739 eV) and O VIII (871 eV) edges — the so called warm absorber. NGC 3783 was observed in the X-rays with *ROSAT* (Turner et al. 1993) and *ASCA* (e.g., George et al. 1998). The 2–10 keV spectrum is fitted by a power law with photon index $\Gamma \approx 1.7$–1.8. The flux in this energy band varies in the range $\sim (4\text{–}9) \times 10^{-11}$ ergs cm^{-2} s^{-1}, and the mean X-ray luminosity Active Galactic Nucleus (AGN) is $\sim 3 \times 10^{43}$ ergs s^{-1} (for $H_0 = 50$ km s^{-1} Mpc^{-1}). Modeling the apparent O VII and O VIII edges indicates a column density of ionized gas of $\sim 2 \times 10^{22}$ cm^{-2}.

The UV spectrum of NGC 3783 shows intrinsic absorption features due to C IV, N V, and H I (Kraemer, Crenshaw, & Gabel 2001, and references therein). The absorption strength of these systems is known to vary. Currently there are three known absorption systems in the UV at radial velocities of approximately -560, -720, and -1400 km s^{-1} (blueshifted) relative to the optical redshift [throughout this study we use a redshift of 0.009760 ± 0.000093 (2926 ± 28 km s^{-1})]. The strength of the absorption is found to be variable over time scales of months to years. The relations between the X-ray and the UV absorbers are not clear. It was suggested by several studies that both absorptions might arise from the same region in the AGN (e.g., Mathur, Elvis, & Wilkes 1995; Shields & Hamann 1997) but no conclusive results could be reached.

In this contribution I report on the 56 ks observation of NGC 3783 through the High Energy Transmission Grating Spectrometer (HETGS) on the *Chandra X-ray Observatory* (*CXO*) with the Advanced CCD Imaging Spectrometer (ACIS-S) as the detector. The observation, the high-resolution spectrum, and the analysis are described in detail by Kaspi et al. (2000, 2001).

2. High Resolution X-ray Spectrum

NGC 3783 was observed continuously with the *CXO*/HETGS on 2000 January 21 for 56 ks as part of the GTO program of the ACIS team. The data were reduced with the *Chandra X-ray Center* software `CIAO` (version 1.1.4). The HETGS produces high order spectra from two grating assemblies, the medium-energy grating (MEG) and the high-energy grating (HEG). Both positive and negative orders are imaged by the ACIS-S array. We combined the +1st and −1st orders in each of the MEG and HEG spectra by averaging them using a $1/\sigma^2$ weighted mean to produce the mean MEG and HEG spectra for each observation. To increase the signal-to-noise ratio we also binned the MEG and HEG spectra to 0.01 Å bins and took a weighted average of them (Figure 1).

The most interesting features discovered in this observation are the dozens of narrow absorption lines from He-like and H-like ions of all the abundant elements from oxygen to silicon (O, Ne, Mg, S, Si, and a tentative identification of Al). The blueshift of the absorption lines relative to their expected theoretical wavelengths is clearly seen in Figure 2. We also identified several emission lines, mainly from the triplets of oxygen, neon, and magnesium. The emission lines are consistent with being at the systemic velocity. The combination of the blueshifted absorption lines with the emission lines causes the lines to have P-Cygni profiles that are characteristic of an ionized gas sphere which is radially outflowing from the center of the AGN.

To better measure and study the absorption lines we have co-added, in velocity space, several absorption lines from the same element. These velocity spectra were built up on a photon-by-photon basis by computing the velocity shift of each photon relative to the line in question and then binning the photons into 200 km s^{-1} bins. The velocity spectra are shown in Figure 2. The blueshifted absorption is clearly detected from oxygen, neon, magnesium, and silicon. There is also a hint of absorption by sulfur. The mean blueshift is -610 ± 130 km s^{-1}. The top panel of Figure 2b shows a Gaussian absorption line which represents the line response function of the HEG at the neon resolution. Given this instrumental profile, the neon line is clearly resolved with FWHM of 840^{+490}_{-360} km s^{-1}. All other lines are not resolved and have FWHM upper limits which are consistent with FWHM of a few hundred km s^{-1}.

3. Modeling the High Resolution X-ray Spectrum

In order to model the high-resolution X-ray spectrum we first determined the continuum. We used the wavelength bands in the spectrum where no lines are present (or expected to be present) to define a line free zone (LFZ). It is important to note that this LFZ provides a measure of the *observed* contin-

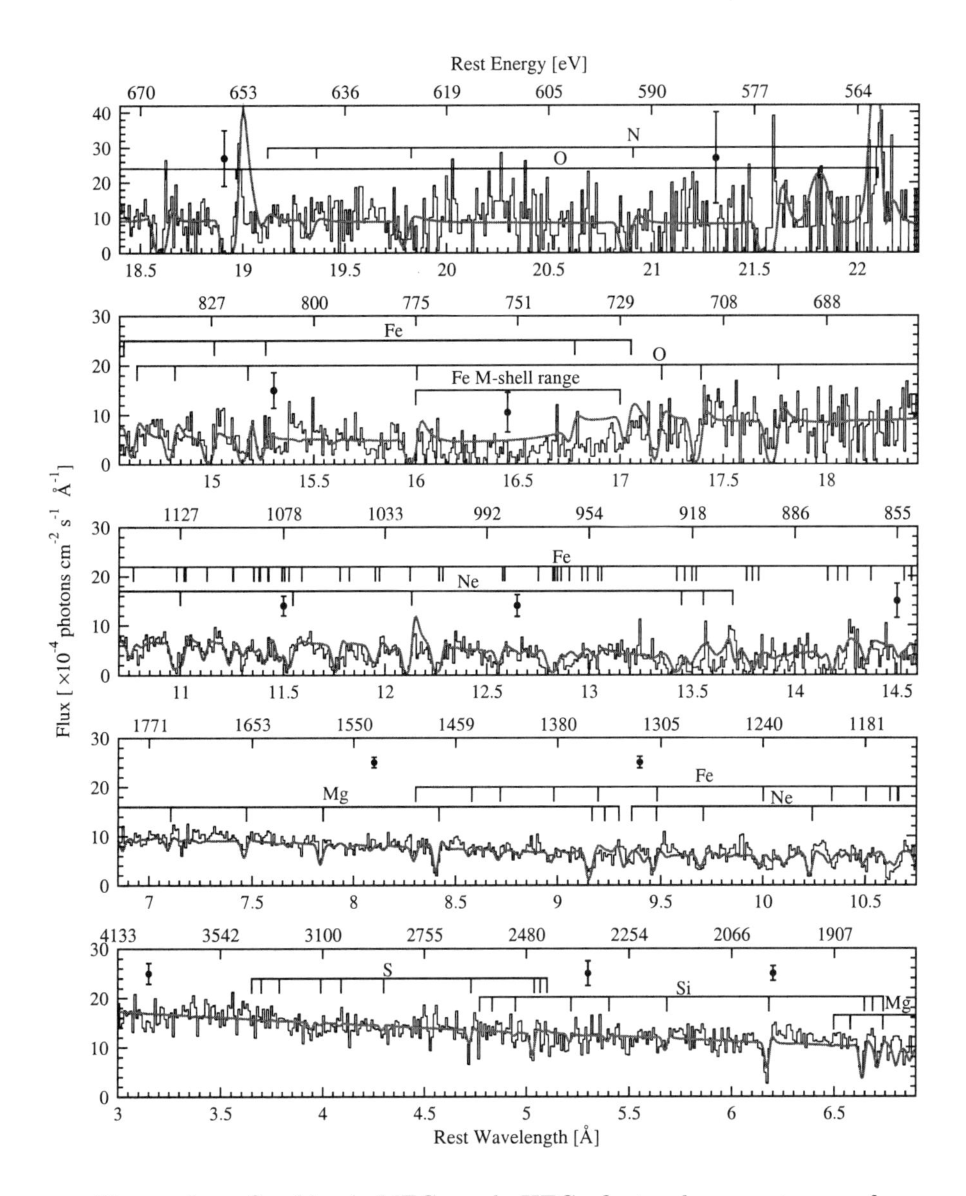

Figure 1. Combined MEG and HEG first-order spectrum of NGC 3783 (black histogram). Data are binned into 0.01 Å bins and are not smoothed. The spectrum has been corrected for Galactic absorption. Dots with error bars show the typical $\pm 1\sigma$ statistical errors at various wavelengths. The two-component photoionization model discussed in § 3 is overplotted (gray line) to emphasize its good agreement with the data. The H-like and He-like lines of N, O, Ne, Mg, Si, and S, as well as the strongest Fe lines contributing to the model are marked.

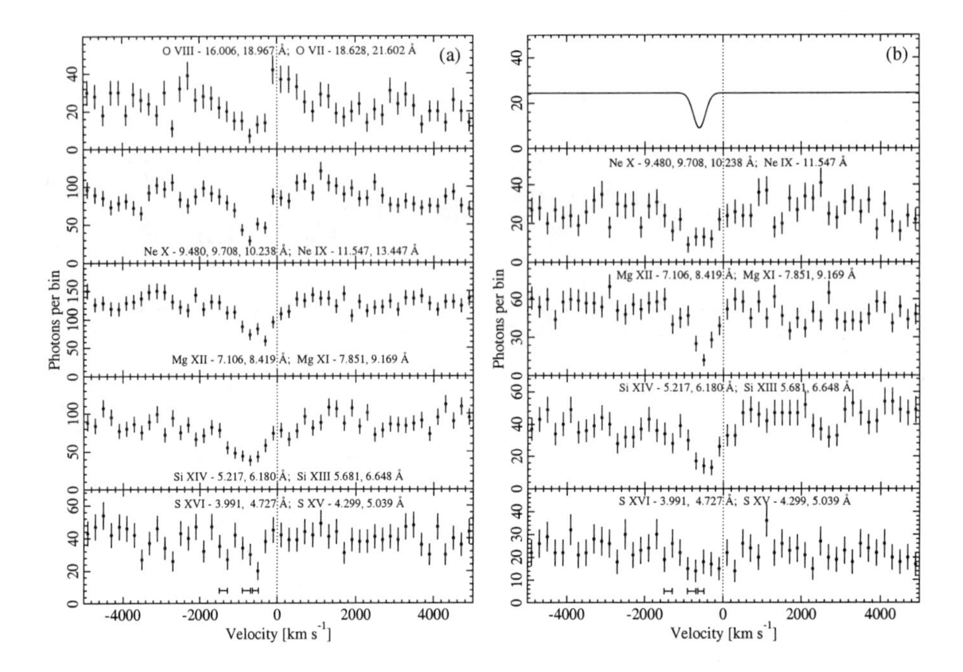

Figure 2. (a) MEG and (b) HEG velocity spectra showing co-added lines. The bin size is 200 $\mathrm{km\,s^{-1}}$. Note that absorption is clearly detected from oxygen, neon, magnesium and silicon. There is also a hint of absorption by sulfur, although this is not significant. Note also the P Cygni profile for the oxygen lines. The top panel on the right demonstrate a Gaussian absorption line representing the line response function of the HEG at 9.480 Å (the Gaussian FWHM is $364\,\mathrm{km\,s^{-1}}$); this is the poorest line response function applicable to the co-added velocity spectrum for neon, shown in the panel immediately below. Note that the neon absorption line appears significantly broader than the line response function. The velocity shifts and FWHMs of the UV absorption systems are marked as three horizontal lines in the lower panels.

uum, which will include spectral curvature resulting from bound-free absorption edges caused by the warm absorber and any "Compton-reflected" continuum at high energies. The spectral model adopted for the LFZ therefore consisted of a power-law continuum (of photon index Γ), an ionized absorber, and a reflection continuum from neutral material. The best-fitting parameters are $\Gamma = 1.77^{+0.10}_{-0.11}$, $\log N_{\mathrm{H}} = 21.94^{+0.15}_{-0.20}$ and $\log U_{\mathrm{oxygen}} = -2.01^{+0.25}_{-0.28}$. This model is also consistent with the models fitted to the simultaneous observations performed with *ASCA* and *RXTE* (to within the calibration limitation of the different instruments they were consistent at around 10%), as well as with past X-ray observations of NGC 3783.

The X-ray spectrum spans more than an order of magnitude in energy and the ions identified in the spectrum are from a large range of ionization.

Hence, we modeled the absorption and emission lines with a multi-component photoionization model. For the photoionization calculations we used `ION2000` (the 2000 version of `ION`; see Netzer 1996; Netzer, Turner, & George 1998). The model consist of two components of radially outflowing shells which both have an outflow velocity of 610 km s^{-1}. The outer shell (low-ionization component) has ionization parameter, defined over the 0.538–10 keV band (as described in George et al. 2000) of $U_{\rm oxygen} = 0.018$ and a global covering factor of 50%. The inner shell (high-ionization component) has an order of magnitude higher ionization parameter, $U_{\rm oxygen} = 0.18$, and a global covering factor of 30%. Both components have hydrogen column density of $10^{22.2}$ cm^{-2}, line-of-sight covering factor of 95%, gas density of 10^8 cm^{-3}, and turbulence velocity of 300 km s^{-1} (corresponding to FWHM of 500 km s^{-1}).

The spectrum produced by the model was convolved with the line spread function of the MEG to account for the broadening of the lines by the instrument. The resulting model is shown as a gray line in Figure 1. The very good agreement of the model with the data is qualitatively evident from the plot. We also tested the model quantitatively by comparing the equivalent widths of the lines from the model to the equivalent widths that are measured from the observed spectrum; the agreement between the model and the data is good with a reduced χ^2 of 1.14.

4. Relations between the UV and X-ray Absorbers

An intriguing question discussed extensively in the past few years is regarding the relations between the UV and X-ray absorbers. There is a one-to-one correspondence between objects that show intrinsic UV absorption and those that show X-ray warm absorbers (e.g., Crenshaw et al. 1999). However, little is known about any relation between them, or if the absorption in the two wavelength bands arise from the same zone of the AGN, or from two distinct zones.

In an *HST*/STIS UV spectrum of NGC3783 obtained 37 days after the *CXO*/HETGS observations, Kraemer et al. (2001) find three intrinsic UV absorption systems at −548, −724, and −1365 km s^{-1}. These systems have FWHMs of 170, 280, and 193 km s^{-1}, respectively. The UV systems at −548 and −724 km s^{-1} are consistent with the X-ray absorption blueshift but the UV line widths seem to be much narrower than at least the resolved neon X-ray line. The low-ionization component of our X-ray model predicts UV line EWs which are consistent with the UV observations, hence suggesting that the UV and X-ray lines may originate from the same region. However, we note that this result is *extremely* sensitive to the unobservable (because of Galactic hydrogen) continuum between the extreme-UV and the soft X-ray.

It is interesting to compare the relations between the UV and X-ray absorbers in NGC 3783 to the same relation in NGC 4051 (a narrow line Seyfert 1 galaxy that its high resolution UV [*HST*/STIS] and X-ray [*CXO*/HETGS] observations are reported by Collinge et al. 2001). In NGC 3783 there is a much stronger X-ray absorption than UV absorption and the low velocity UV absorption systems have a corresponding X-ray velocity absorption system, while there is no X-ray absorber corresponding to the high velocity UV absorption system.

In NGC 4051 there are two X-ray absorption systems one at -600 km s^{-1} and one at -2340 km s^{-1}. The UV absorption is resolved to as many as nine different intrinsic absorption systems with velocities between -650 and $+30$ km s^{-1}. The UV absorption is much stronger than the X-ray absorption in this object. Although the low-velocity X-ray absorption is consistent in velocity with many of the UV absorption systems, the high-velocity X-ray absorption seems to have no UV counterpart. In summary, although we are gaining much more information about the absorption systems in the X-ray and UV with the most advanced high-resolution spectrographs, the relation between the two absorbers is yet to be determined.

As part of a large collaboration we are currently (summer 2001) monitoring NGC 3783 for several months with *RXTE*, *CXO*, *FUSE*, *HST*, and ground based observations. With *CXO*/HETGS we will have five observations each of 170 ks (a total of 850 ks). We will look for variations in the continua and lines both in the UV and the X-ray in order to better understand the relations between the two absorbers. With the help of these new observations, we will be able to study further and resolve the X-ray spectrum of NGC 3783. Together with the simultaneous *HST*/STIS observations, this campaign should enable to determine the nature, origin, and evolution of the X-ray and UV absorbers by a direct comparison of their dynamical and physical properties.

Acknowledgments. I am grateful to my collaborators in this study W. N. Brandt, Hagai Netzer, Ian M. George, George Chartas, Ehud Behar, Rita M. Sambruna, Gordon P. Garmire, and John A. Nousek. I gratefully acknowledge the financial support of NASA grant NAS 8-38252 (G.P. Garmire, PI), NASA LTSA grant NAG 5-8107, and NASA grant NAG 5-7282.

References

Collinge, M. J., et al. 2001, ApJ, 557, ApJ, 557, 2

Crenshaw, D. M., Kraemer, S. B., Boggess, A., Maran, S. P., Mushotzky, R. F., & Wu, C.-C. 1999, ApJ, 516, 750

George, I. M., Turner, T. J., Mushotzky, R., Nandra, K., & Netzer, H. 1998, ApJ, 503, 174

George, I. M., Turner, T. J., Yaqoob, T., Netzer, H., Laor, A., Mushotzky, R. F., Nandra, K., & Takahashi, T. 2000, ApJ, 531, 52

Kaspi, S., Brandt, W. N., Netzer, H., Sambruna, R., Chartas, G., Garmire, G. P., & Nousek, J. A. 2000, ApJ, 535, L17

Kaspi, S., et al. 2001, ApJ, 554, 216

Kraemer, S. B., Crenshaw, D. M., & Gabel, J. R. 2001, ApJ, 557, 30

Mathur, S., Elvis, M., & Wilkes, B. 1995, ApJ, 452, 230

Netzer, H. 1996, ApJ, 473, 781

Netzer, H., Turner, T. J., & George, I. M. 1998, ApJ, 504, 680

Shields, J. C., & Hamann, F. 1997, ApJ, 481, 752

Turner, T. J., Nandra, K., George, I. M., Fabian, A. C., & Pounds, K. A. 1993, ApJ, 419, 127

Mass Outflow in Active Galactic Nuclei: New Perspectives
ASP Conference Series, Vol. 255, 2002
D.M. Crenshaw, S.B. Kraemer, and I.M. George

Chandra HETGS Observations of NGC 1068

Patrick M. Ogle

University of California Santa Barbara, Physics Department, Santa Barbara, CA 93106, USA

Claude Canizares, Dan Dewey, Julia Lee, & Herman Marshall

Massachusetts Institute of Technology, Center for Space Research, 77 Massachusetts Avenue, NE80, Cambridge, MA 02139, USA

Abstract. We present X-ray grating spectroscopy and imaging of the Seyfert 2 galaxy NGC 1068. A spectrum of the nucleus reveals the predominance of photoionization through strong forbidden emission lines and recombination continua. A spectrum of the off-nuclear regions is not dominated by recombination and shows strong Fe L emission. Strong resonance lines may be due to resonance scattering of the nuclear continuum or thermal emission from a hot plasma. The structure of the narrow line region varies with energy. Emission in the 1 keV band is more diffuse than at lower energies. In addition, Fe Kα and Fe XXV emission are strongly concentrated to the nuclear hot spot.

1. Introduction

NGC 1068 is perhaps the most studied galaxy of its class. It is a bright, nearby (z=0.0038) Seyfert 2 galaxy with a central starburst and a strong photoionization cone illuminated by a hidden Seyfert 1 nucleus (Antonucci & Miller, 1985). The direct X-ray continuum from the nucleus is completely obscured by Compton-thick material. The optical narrow-line region (NLR) consists of a biconical outflow, as revealed by imaging and spectroscopy with *HST* (e.g. Crenshaw & Kraemer, 2000). The NLR clouds reach velocities in excess of 2000 km s^{-1}. However, optical emission lines probe only the low-ionization component of the mass outflow in NGC 1068.

ROSAT observations (Wilson et al., 1992) revealed that the X-ray emission from NGC 1068 is extended and corresponds to the starburst disk on arcminute scales. Spectroscopy with *BBXRT* and *ASCA* showed strong Fe Kα emission and a soft X-ray component containing unresolved and blended emission lines (Marshall et al. 1993; Ueno et al. 1994). It was difficult to distinguish between thermal emission from the starburst and emission from the smaller scale NLR with these observations.

The *Chandra X-ray Observatory* (*CXO*), with its High Energy Transmission Grating Spectrometer (HETGS), enables spatially and spectrally resolved emission line spectroscopy of NGC 1068. This provides crucial new information on the spatial distribution, temperature, ionization state, and density of the outflow region. Our grating spectroscopy is complementary to the *CXO*/ACIS imaging spectroscopy recently published by Wilson et al. (2001).

2. Nuclear and Off-nuclear Spectra

We observed NGC 1068 for 46 ks with *CXO*/HETGS, with the dispersion axis roughly perpendicular to the axis of the extended NLR. We extracted spectra (Figures 1 & 2) from two 3 arcsec regions. The first region is centered on the nuclear hot spot, the second on the bright cloud 3 arcsec NE of the nucleus (Figure 3).

The nuclear spectrum of NGC 1068 (Figure 1) is dominated by recombination from a photoionized plasma. This is made evident by the strong narrow forbidden lines (f) and narrow radiative recombination continua (RRCs) of H-like and He-like ions from N VII to Fe XXV. In addition, there are strong fluorescent Kα lines from nearly neutral Fe, S, and Si. Fe Kβ is also detected.

However, the nuclear spectrum is not completely described by recombination and fluorescence. The H-like and He-like resonance lines are stronger than what is expected from pure recombination. The strength of the extra resonance line emission varies from ion to ion. This effect is also seen in NGC 4151 and Mkn 3 (Ogle et al. 2000; Sako et al. 2000). It is likely that the enhanced resonance lines are primarily from emission following photoexcitation (resonance scattering). This is supported by the relative strength of the high order Lyman lines.

The off-nuclear spectrum of NGC 1068 (Figure 2) is dramatically different from the nuclear spectrum. Fe Kα emission is very weak compared to the nucleus. This is consistent with an extended NLR component which is optically thin and highly ionized. The recombination emission in forbidden lines and RRCs is also quite weak, suggesting a relatively small column of photoionized plasma. The resonance lines are enhanced off-nucleus, which could be from photoexcitation, as described above. Alternatively, the resonance emission could come from collisionally excited ions in a 1 keV plasma. This second explanation is attractive because a diffuse, hot plasma could provide pressure confinement for the cooler X-ray and optical NLR clouds. It is also supported by the relatively strong Fe L emission in the off-nuclear spectrum. It is difficult to disentangle the effects of photoexcitation and collisional excitation in the off-nuclear spectrum because the high order Lyman lines are too weak to measure.

3. Broad-band X-ray Imaging

We created images in 4 broad bands from the HETGS 0th-order image of NGC 1068 (Figure 3), showing the central 10 arcsec of the NLR. An adaptive smoothing algorithm was used to enhance the extended emission. We clearly see that the structure of the X-ray NLR changes with energy (Figure 3). The soft and medium X-ray bands (0.4-0.6, 0.6-0.8 keV) contain strong line emission (Figures 1 & 2) from NLR clouds which correspond to those seen in optical and UV images (Macchetto et al. 1994). Detailed differences between these bands may be due to differences in ionization or obscuring column density.

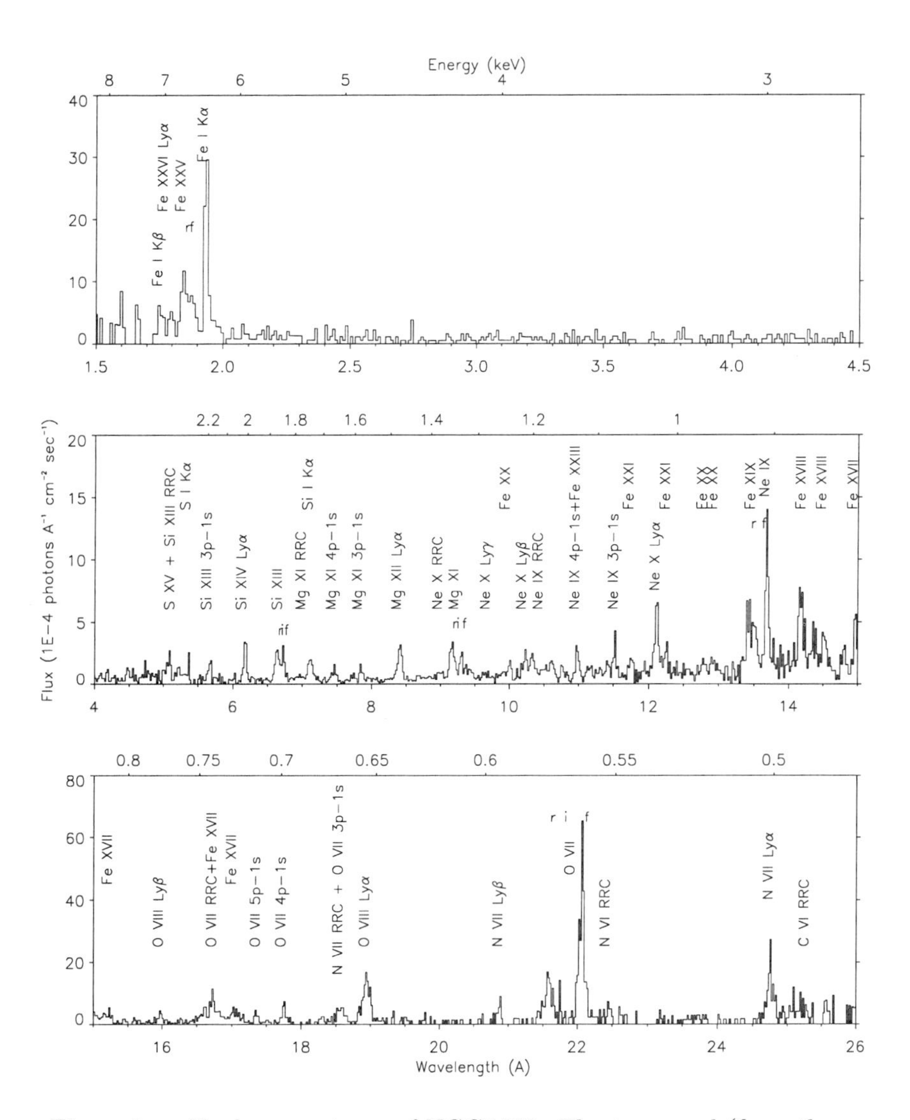

Figure 1. Nuclear spectrum of NGC 1068. The top panel (from the High Energy Grating, HEG) shows strong neutral and highly ionized Fe emission lines. The middle and bottom panels (from the Medium Energy Grating, MEG) are dominated by recombination emission from H-like and He-like ions. Note the strong forbidden lines (f) and recombination continua (RRCs).

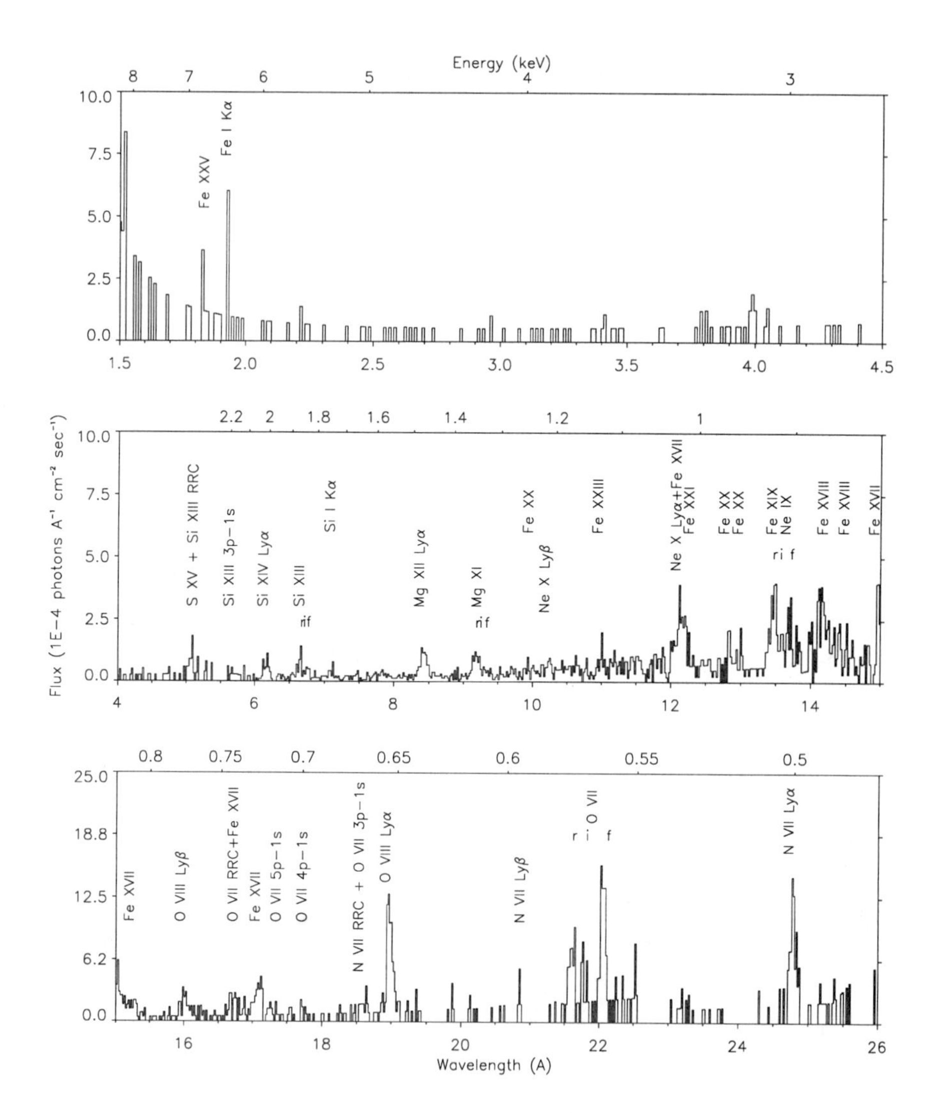

Figure 2. Off-nuclear spectrum of NGC 1068. Note how weak the Fe I Kα line (top panel) is relative to the nuclear spectrum (Fig. 1). Also note the relative weakness of the forbidden lines (f) and recombination continua (RRCs). The resonance lines (e.g. O VII r) are relatively strong. Strong Fe L lines (0.7-1.3 keV) are suggestive of collisionally excited emission from a hot plasma.

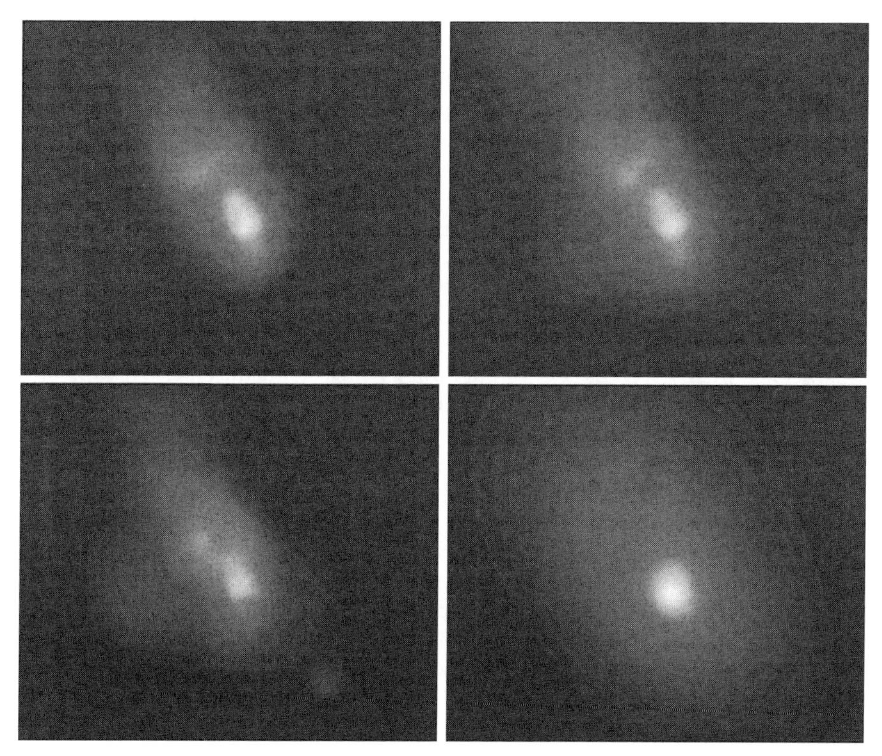

Figure 3. Broad-band images of NGC 1068. Clockwise from top left: E< 0.6keV, 0.6-0.8 keV, 0.8-1.3 keV, E>6 keV. The field of view is 20×17 arcsec (1.6×1.3 kpc), with N at top, E at left.

The 0.8-1.3 keV band has a markedly different structure from the other bands. This is partly due to emission from the nuclear starburst. One of the spiral arms can be seen as the extension trailing to the east of the nucleus. The NLR cone is still prominent but is also less clumpy at 1 keV than in the lower energy bands. Strong Fe L emission dominates this band in the off-nuclear spectrum. Its diffuse spatial distribution is compatible with an origin in a hot intercloud medium. It may be difficult to explain the structure at 1 keV by photoexcitation.

The high energy band ($E > 6$ keV) is dominated by emission from Fe I Kα and Fe XXV (Figure 1). This emission is concentrated in the nucleus and peaks at the optical hot spot. It is possible that the hot spot corresponds to the inner wall of a circumnuclear molecular torus, which is illuminated by the hidden nucleus. If this is the case, then we expect a large range of ionization in this region for a Compton-thick torus. The skin would be highly ionized and emit strongly in Fe XXV, while the deeper layers would be neutral and a strong source of fluorescent emission from Fe I Kα. There is also weak extended emission in the $E > 6$ keV image, aligned with the NLR, which may be due Fe I Kα fluorescence. Its smooth appearance is likely due to the adaptive smoothing algorithm. Note that the high-energy photons penetrate the obscuration from the host galactic disk, giving us a view of the far-side (SW) cone which is blocked at lower energies.

4. Conclusions

CXO grating spectroscopy gives us a high-resolution view of the nuclear and extended X-ray emission in NGC 1068. We see that photoionization and fluorescence are primarily responsible for X-ray emission from the nuclear NLR. However, the roles of collisional excitation and photoexcitation in the off-nuclear and intercloud regions need to be clarified. This may be accomplished by longer observations with *CXO* HETGS and Low Energy Transmission Grating Spectrometer (LETGS) which have been proposed. Such observations would allow us to measure the high order Lyman line and Fe L line strengths in several spatially resolved regions in the NLR. This will be important for determining the nature of the intercloud medium, and whether or not it can provide pressure confinement of the NLR clouds.

Further studies of the ionization state, density, kinematics, and abundances of the X-ray NLR in Seyfert 2 galaxies will have important applications for the warm absorbers seen in the spectra of Seyfert 1 galaxies. Both systems appear to have similar column densities, ionization, and other parameters. We now see that both are manifestations of the outflow phenomenon in Seyfert galaxies. Our *CXO* observations of NGC 1068 show that these outflows are very inhomogeneous and have a large range in ionization level. Once a complete survey of the outflow has been made at wavelengths from optical to X-ray, it will become possible to characterize the complete range of conditions and total outflow rate without prejudice to ionization state.

References

Antonucci, R. R. J., & Miller, J. S., 1985, ApJ, 297, 621

Crenshaw, D.M. & Kraemer, S. 2000, ApJ, 532, L101

Macchetto, F., Capetti, A., Sparks, W. B., Axon, D. J., & Boksenberg, A. 1994, ApJ, 435, L15

Marshall et al. 1993, AJ, 405, 168

Ogle, P. M., Marshall, H. L., Lee, J. C., & Canizares, C. R. 2000, ApJ, 545, L81

Sako, M., Kahn, S. M., Paerels, F., & Liedahl, D. A. 2000, ApJ, 543, L115

Ueno, S., Mushotzky, R. F., Koyama, K., Iwasawa, K., Awaki, H., & Hayashi, I. 1994, PASJ, 46, L71

Wilson, A. S., Elvis, M., Lawrence, A., & Bland-Hawthorn, J. 1992, ApJ, 391, L75

Wilson et al. 2001 ApJ, in press, astro-ph/0104027

Mass Outflow in Active Galactic Nuclei: New Perspectives
ASP Conference Series, Vol. 255, 2002
D.M. Crenshaw, S.B. Kraemer, and I.M. George

A *Chandra* Snapshot Survey of Broad Absorption Line Quasars

Paul J. Green, Thomas L. Aldcroft

Harvard-Smithsonian Center for Astrophysics, 60 Garden St., Cambridge, MA 02138, USA

Smita Mathur

The Ohio State University, 140 West 18th Avenue, Columbus, OH 43210-1173, USA

Belinda J. Wilkes and Martin Elvis

Harvard-Smithsonian Center for Astrophysics, 60 Garden St., Cambridge, MA 02138, USA

Abstract. The profound soft X-ray silence of quasars that show broad absorption lines (BALs) in their optical/UV spectra has been reasonably attributed to large column densities of absorbing material intrinsic to the quasar. However, estimates of the absorbing column have been founded on many assumptions, the most important being that the strength and spectrum of the underlying X-ray emission is identical to that of the more typical non-BAL QSOs. Solid proof of both the existence of an absorber, and the presumed similarity of the underlying spectral energy distribution have been lacking, except by analogy to quasars with weaker optical/UV absorbers. From a *Chandra X-ray Observatory* (*CXO*) snapshot of 10 optically bright BAL QSOs, we show that the X-ray spectral energy distributions of typical high ionization BAL QSOs, both the underlying power-law slope and the X-ray to optical flux ratio, are affected by strong intrinsic absorption, and that their underlying X-ray emission is consistent with non-BAL QSOs. By contrast, removal of the best-fit absorption column detected in the high ionization BAL QSOs still leaves the 4 low-ionization BAL QSOs in our sample as unusually X-ray faint for their optical luminosities.

1. Introduction

Quasars are immensely powerful and complicated structures, yet so distant as to appear faint and pointlike. So their study to date is mostly informed by spectroscopy. But quasar optical/UV emission line spectra are almost certainly subject to a profound averaging process, or maybe a Darwinian selection whereby clouds in the best position to radiate efficiently dominate the emission spectrum (e.g., Baldwin et al. 1995). Absorption lines caused by material intrinsic to quasars probably hold greater promise for revealing the conditions near the supermassive black holes. The velocity and ionization structure of the absorbers can be studied with spectra of adequate signal-to-noise ratio and resolution, if the absorption is not too complex or saturated. Regardless, some of us prefer to study the messiest, most spectacular cases, the broad absorption lines (BALs). Since BAL QSOs host strong absorbing clouds flowing outward from the nucleus along our line of sight with velocities of 5,000 up to $\sim 50,000$ km s^{-1}, their phenomenology is both rich and extreme.

While the intrinsic absorbing columns inferred from the UV BALs ($N_H^{intr} \sim 10^{20-21}$cm^{-2}; Korista et al. 1992) appear low enough that we would have expected very little X-ray absorption, BAL QSOs instead proved virtually undetectable in soft X-rays (Green et al. 1995; Green & Mathur 1996). We could nevertheless derive the implied X-ray absorbing columns by assuming that BAL QSOs are "normal underneath", i.e. have intrinsic luminosities and spectral energy distributions (SEDs) similar to non-BAL QSOs, particularly the optical-to-X-ray ratio parameterized by $\alpha_{\rm ox}$ [1] and X-ray continuum photon index Γ. The large implied X-ray columns mean that the UV BALs sample an ionization state that holds for just a tiny fraction of the absorbing material - increasing the estimates of the mass outflow rate by 2 to 3 orders of magnitude. Outflows must then represent a significant component of the QSO energy budget. While evidence exists to confirm the absorption scenario (e.g. Mathur et al. 2000), the crucial assumption that BAL QSOs are normal underneath has eluded verification.

We therefore proposed for *CXO* Cycle 1 time to image a sample of optically bright BAL QSOs with the Advanced CCD Imaging Spectrometer (ACIS). Our goal was to survey a representative sample of BAL QSOs with short exposure times so that we might test the absorption scenario, and identify some most X-ray bright BAL QSOs for further followup. *CXO* is well-suited to these goals due to its $\sim 1''$ point-response profile and low background, which provide excellent sensitivity that extends into the more penetrating hard X-ray bandpass.

2. Observations

We compiled a list of QSOs with unmistakably strong *bona fide* BALs and magnitudes (usually B or m_{pg}) brighter than 17th. We calculated our proposed *CXO* exposure times to result in a detection for each source, making the usual assumptions outlined above. The resulting sample spans redshifts from 0.1 to 2.4, and a wide range of BAL QSO phenomena (Green et al. 2001). Four QSOs with low-ionization BALs (loBALs) are included, which also show broad absorption in lower ionization lines of Mg II or Fe II. LoBAL QSOs are reddened and tend to have particularly high polarization (Schmidt & Hines 1999). All 10 sources were observed between 1999 December 30 and 2000 May 15 using the back-illuminated S3 chip of ACIS on board *CXO*.

---

[1] $\alpha_{\rm ox}$ is the slope of a hypothetical power-law from 2500 Å to 2 keV; $\alpha_{\rm ox} = 0.384 \log(\frac{L_{2500\,Å}}{L_{2\rm keV}})$.

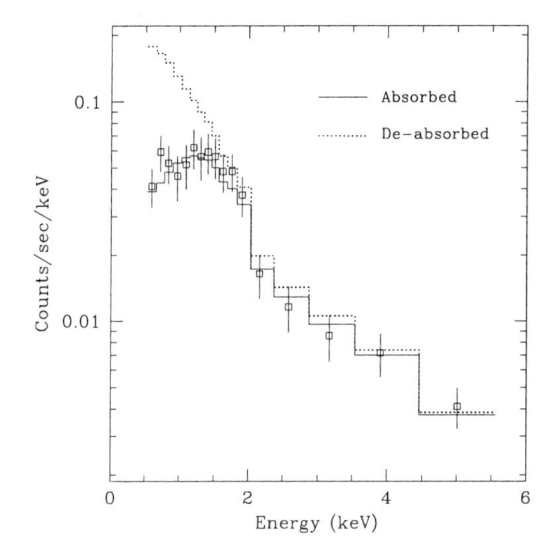

Figure 1. Summed *CXO* X-ray spectrum for the BAL QSOs with more than 20 counts. The sum of the all the individual source models is plotted over the merged event lists of all 6 objects. The solid line shows the global best-fit model (Model C). The dashed line shows the "de-absorbed" model spectrum, where the intrinsic absorption component is removed from Model C *after* fitting.

3. Analysis

We performed spectral modeling for the six sources with more than 20 counts. Since each source spectrum taken individually has insufficient counts to usefully constrain the intrinsic absorption or power-law spectral index, instead we *simultaneously* fitted all 6 spectra. Note that these brightest sources are all hiBAL QSOs, so the spectral parameters we derive may not apply to loBAL QSOs. We tested several source models. A single global power-law with an individual flux normalization for each QSO and ($z = 0$) absorption fixed to the Galactic value for each QSO (Model A) yields a hard photon index $\Gamma = 1.1 \pm 0.1$. At 99.7% (3σ) confidence, a better fit is achieved by allowing for a global intrinsic absorption column fixed at each quasar's redshift. The best-fit slope of Model (B) is $\Gamma = 1.44 \pm 0.23$, with intrinsic (rest-frame) absorption $N_H^{intr} = 6.5^{+4.5}_{-3.8} \times 10^{22}\,\mathrm{cm}^{-2}$. Inclusion of a *global* partial covering parameter C_f for the redshifted absorbers (Model C) further improves the fit, again at 99.5% confidence (F-test). The "composite" BAL QSO has intrinsic (rest-frame) absorption $N_H^{intr} = 6.5^{+4.5}_{-3.8} \times 10^{22}\,\mathrm{cm}^{-2}$ covering 80^{+9}_{-17}% of the source, whose intrinsic power-law energy index $\Gamma = 1.80 \pm 0.35$. This final best fit model is plotted in Figure 1 over the summed spectrum.

We note that one QSO represents 30% of the counts fitted, and that the next most dominant 20%. We therefore explored the suitability of the global fit for each QSO by thawing the power-law slope and amplitude in Model C and refitting individual QSOs. All the QSOs included in the simultaneous fit have power-law slopes and intrinsic columns consistent with the final global fits Model C except Q0254-334 with $\Gamma = 2.6 \pm 0.4$. Most likely, this QSO is not strongly absorbed in X-rays, but warrants further investigation. It represents about 15% of the counts in the global fit, and our conclusions are unchanged if we exclude it entirely.

Our most important result, which has never been demonstrated before for a representative sample of BAL QSOs, is that the best-fit power-law index Γ for our BAL QSO sample is entirely consistent with the (*ASCA*) mean for non-BAL radio quiet quasars of $\sim 1.89 \pm 0.05$ with dispersion $\sigma = 0.27 \pm 0.04$ (Reeves & Turner 2000; Vignali et al. 1999). Gallagher et al. (2001, 2002) find similar results for PG2112+059, which appears to have weak BALs and perhaps the brightest flux of any BAL QSO. Their current *CXO* sample of BAL QSOs should considerably further these studies.

4. Broadband Energy Distributions

Now that we have a *measured* mean spectral shape for hiBAL QSOs, for the first time we can calculate fluxes consistently using the best-fit model with the redshifted absorption component removed. This tells us what $\alpha_{\rm ox}$ values BAL QSOs would have without their intrinsic absorption, since their (de-absorbed) intrinsic SEDs are well-characterized by the above slope. With the modeled intrinsic absorption removed, the high ionization BAL QSOs in our sample fit reasonably well along the empirical trend of increasing α_{ox} (weakening X-ray emission) with increasing L_{opt}. On the other hand, the four low-ionization BAL QSOs in our sample are extremely X-ray weak. Two are not detected at all (for which we assign 5 counts as an upper limit). Of the two loBAL QSOs that are detected, one is the most nearby object (at $z = 0.148$), and the other is a radio-intermediate BAL QSO.

The *apparent* L_X of loBAL QSOs are so low that they are comparable to bright starburst galaxies. Since their broad emission lines betray the presence of an AGN, the minimum difference between our hiBAL and loBAL QSOs is $\Delta\alpha_{\rm ox} \sim 0.3$, could be due to additional absorption approaching $10^{24}{\rm cm}^{-2}$ in loBALs that is unaccounted for in our hiBAL spectral model. The decrease in polarization toward longer wavelengths in some loBALs suggests edge-on dust-scattering models where the scattered line of sight is less reddened, so that loBAL QSOs have been proposed as the *most* edge-on QSOs (Brotherton et al. 1997). On the other hand, loBAL QSOs may be nascent QSOs embedded in a dense, dusty star formation region (e.g., Voit, Weymann, & Korista 1993). The expected strong extinction has been seen (Sprayberry & Foltz 1992; Boroson & Meyers 1992), and could explain their low (1-2%) incidence in optically-selected samples. Deep, higher S/N X-ray spectroscopy of these objects should solidify their tie-in to hiBAL QSOs, and begin to clarify the relationship between merger/starburst activity and outflows in AGN.

4.1. Orientation, Outburst, and Evolution

Most interpretations of the optical and UV observations imply that all QSOs host BAL clouds, at least for the more common high ionization BALs (hiBALs

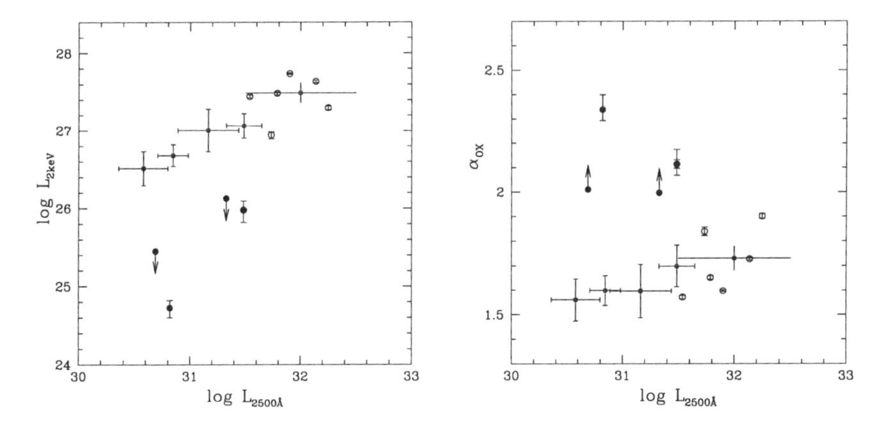

Figure 2. LEFT: The log of the monochromatic (2keV) X-ray luminosity plotted against the log of monochromatic 2500Å optical luminosity for quasars (both in units of ergs sec^{-1} Hz^{-1}). RIGHT: The optical to X-ray spectral slope α_{ox} (from 2500 Å to 2 keV), also plotted against log$L_{2500Å}$. In both panels, circles depict the 10 BAL QSOs in our *CXO* sample. Filled circles are those objects known to have low-ionization broad absorption lines (loBALs). Errorbars represent the poisson error in the counts. Arrows mark limits to X-ray luminosity in our *CXO* exposures. The open boxes with error bars are means from co-added subsamples of radio-quiet LBQS QSOs observed in the *ROSAT* All Sky Survey (Green et al. 1995). The errorbars are the RMS dispersion of the QSOs in each bin. We also add one mean point at higher luminosity (log$L_{2500Å}$ > 31.5) for *ROSAT*-observed radio-quiet QSOs from Yuan et al. (1998).

henceforth). This 'unification' picture is supported by (1) the covering factor of the BAL region (< 20% Hamann et al. 1993), (2) by the optical/UV emission lines and continuum slopes of hiBAL QSOs, remarkably similar to those of non-BAL QSOs (Weymann et al. 1991), and (3) by the higher average polarization of BAL QSOs (Schmidt & Hines 1999) and increase of polarization in the BAL troughs (Ogle et al. 1999).

An alternative or perhaps complementary approach to BAL QSOs is that, rather than representing an orientation effect, they are in a phase of high accretion rate (outburst). If the BAL phase represents a high accretion rate period in a quasar's lifetime, than an intrinsic power-law steeper than that for non-BAL QSOs might be expected, by analogy to narrow line Seyferts and Galactic black hole candidate binary systems in outburst (Leighly 1999; Mathur 2000). Our composite sample analysis does not support this, with the possible exception of Q 0254-334. Mathur discusses one other such case PHL 5200 in this volume. We also note that a slope as steep as $\Gamma = 2.5$ is allowed at 3σ if the intrinsic column reaches $N_H^{intr} \sim 10^{23}\mathrm{cm}^{-2}$ (Green et al. 2001). We thus consider the possibility open that both orientation *and* high accretion rate may contribute to the BAL phenomenon. The latter may affect the evolution of the observed space density of quasars, and is very sensitive to selection effects.

Higher accretion rates would tend to be associated with mergers, which in turn should be found more frequently at early cosmic times in a scenario of hierarchical structure formation. Thus a larger fraction of BAL QSOs should exist at higher redshifts. Hints abound that this may be the case (Maiolino et al. 2001; Chartas 2000). The detections presented here in relatively short *CXO* exposures indicate that deep, large-area X-ray surveys with good sensitivity above 2keV should help find an increasing number of high-z BAL QSOs.

It's important to note that naïve individual spectral fits to the few count X-ray spectra of BAL QSOs yield hard spectral slopes such that especially at high redshifts, BAL QSOs might represent a significant fraction of absorbed AGN postulated to contribute to the cosmic X-ray background.

Acknowledgments. We thank the entire *Chandra* team, and acknowledge support by *CXO* grant GO 0-1030X and NASA contract NAS8-39073.

References

Baldwin, J. A., Ferland, G., Korista, K., & Verner, D. 1995, ApJ, 455, 119

Boroson, T.A. & Meyers, K.A. 1992, ApJ, 397, 442

Brotherton, M.S., et al. 1997, ApJL 487, L113

Chartas, G. 2000, ApJ, 531, 81

Gallagher, S. C., et al. 2001, ApJ, 546, 795

Gallagher, S. C., et al. 2002, in Mass Outflow in Active Galactic Nuclei: New Perspectives, eds. D.M. Crenshaw, S.B. Kraemer, & I.M. George (San Francisco: ASP), p. 25

Green, P. J., et al. 1995, ApJ, 450, 51

Green, P. J. & Mathur, S. 1996, ApJ, 462, 637

Green, P. J., et al. 2001, ApJ, in press

Hamann, F., Korista, K. T., & Morris, S. L. 1993, ApJ, 415, 541

Korista, K. T. et al. 1992, ApJ, 401, 529

Leighly, K. M. 1999, ApJS, 125, 317

Maiolino, R., et al. 2001, A&A, 372, L5

Mathur, S. 2000, MNRAS, 314, L17

Mathur S. et al. 2000, ApJ, 533, 79

Ogle, P. M., et al. 1999, ApJS, 125, 1

Reeves, J. N. & Turner, M. J. L. 2000, MNRAS, 316, 234

Schmidt, G. & Hines, D. 1999, ApJ, 512, 125

Sprayberry, D. & Foltz, C.B. 1992, ApJ, 390, 39

Vignali, C., et al. 1999, ApJ, 516, 582

Voit, G. M., Weymann, R. J., & Korista, K. T. 1993, ApJ, 413, 95

Weymann, R. J. et al. 1991, ApJ, 373, 23

Yuan, W., Brinkmann, W., Siebert, J., & Voges, W. 1998, A&A, 330, 108

Mass Outflow in Active Galactic Nuclei: New Perspectives
ASP Conference Series, Vol. 255, 2002
D.M. Crenshaw, S.B. Kraemer, and I.M. George

X-ray Spectroscopy of BAL and Mini-BAL QSOs

S. C. Gallagher, W. N. Brandt, G. Chartas, & G. P. Garmire

The Pennsylvania State University, Department of Astronomy & Astrophysics, 525 Davey Laboratory, University Park, PA 16802, USA

Abstract. Broad-absorption line (BAL) QSOs are notoriously faint X-ray sources, presumably due to extreme intrinsic absorption. However, several objects have begun to appear through the obscuration with recent X-ray observations by the *Chandra X-ray Observatory* (*CXO*) and *ASCA*. Where enough counts are present for X-ray spectroscopy, the signatures of absorption are clear. The evidence is also mounting that the absorbers are more complicated than previous simple models assumed; current absorber models need to be extended to the high-luminosity, high-velocity, and high-ionization regime appropriate for BAL QSOs.

1. Introduction

Since the first surveys with *ROSAT*, BAL QSOs have been known to be faint soft X-ray sources compared to their optical fluxes (Kopko, Turnshek, & Espey 1994; Green & Mathur 1996). Given the extreme absorption evident in the ultraviolet, this soft X-ray faintness was assumed to result from intrinsic absorption. Based on this model, the intrinsic column densities required to suppress the X-ray flux, assuming a normal QSO spectral energy distribution, were found to be $\gtrsim 5 \times 10^{22}$ cm^{-2} (Green & Mathur 1996). Due to the 2–10 keV response of its detectors, a subsequent *ASCA* survey was able to raise this lower limit by an order of magnitude for some objects, to $\gtrsim 5 \times 10^{23}$ cm^{-2} (Gallagher et al. 1999). In all of these studies, the premise of an underlying typical QSO spectral energy distribution and X-ray continuum was maintained. The strong correlation found by Brandt, Laor, & Wills (2000) between C IV absorption equivalent width (EW) and faintness in soft X-rays further supported this assumption.

ASCA observations of individual BAL QSOs such as PHL 5200 (Mathur, Elvis, & Singh 1995) and Mrk 231 (Iwasawa 1999; Turner 1999) provided suggestive evidence that intrinsic absorption was in fact to blame for X-ray faintness. However, limited photon statistics precluded a definitive diagnosis. The observation of PG 2112+059 with *ASCA* on 1999 October 30 provided the first solid evidence for intrinsic X-ray absorption in a BAL QSO.

2. X-ray Spectroscopy of a BAL QSO: PG 2112+059

PG 2112+059 is one of the most luminous low-redshift Palomar-Green QSOs with $M_V = -27.3$. Ultraviolet spectroscopy with *HST* clearly revealed broad,

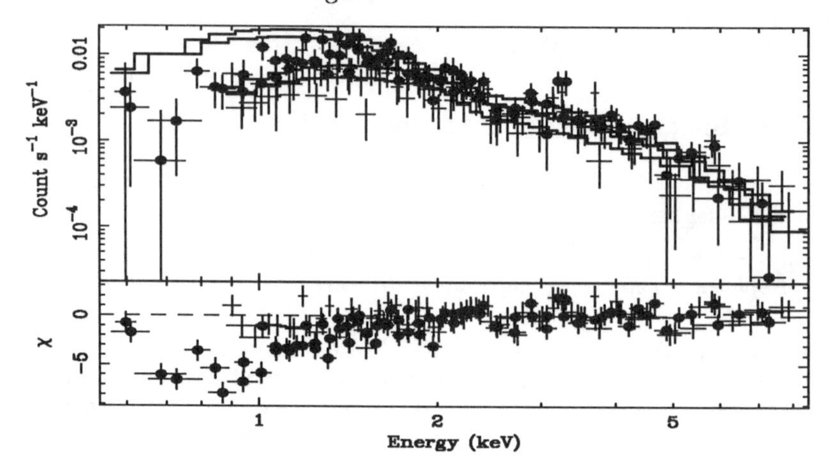

Figure 1. *ASCA* SIS and GIS observed-frame spectra of PG 2112+059 fitted with a power-law model above 2 keV, which has then been extrapolated back to lower energies. Note the significant negative residuals (lower panel) below ~ 2 keV suggestive of a complex absorber.

shallow C IV absorption (Jannuzi et al. 1998) with EW of 19 Å (Brandt et al. 2000). A 21.1 ks *ROSAT* PSPC observation on 1991 Nov 15 detected PG 2112+059, unusual for a BAL QSO. A 31.9 ks *ASCA* observation provided ~ 2000 counts in all four detectors, enough for spectroscopic analysis (described in detail in Gallagher et al. 2001).

To model the continuum, the data were fit above 3 keV (2 keV in the observed frame) with a power law. The resulting photon index, $\Gamma = 1.94^{+0.23}_{-0.21}$, was consistent with those of typical radio-quiet QSOs (e.g., Reeves & Turner 2000). The power law was extrapolated to the lowest energies in the *ASCA* bandpass to investigate potential intrinsic absorption (see Figure 1). The significant negative residuals are indicative of strong absorption, and the spectrum was subsequently fit with an intrinsic, neutral absorber with a column density, $N_{\rm H} \approx 10^{22}\,{\rm cm}^{-2}$. The structure in the residuals is suggestive of complexity in the absorption, but the signal-to-noise ratio was insufficient to investigate this fully.

This analysis of PG 2112+059 provided the first direct evidence of a normal X-ray continuum suffering from intrinsic absorption in a BAL QSO. In addition, correcting the X-ray flux for this absorption also demonstrated that this BAL QSO had an underlying spectral energy distribution typical of radio-quiet QSOs. Though PG 2112+059 has a high optical flux with $B = 15.4$, many of the optically brightest BAL QSOs have been undetected in *ASCA* observations of similar or greater exposure time. Notably, PG 0946+301 ($B = 16.0$) was barely detected by *ASCA* with ~ 100 ks (Mathur et al. 2000).

The launch of the *Chandra X-ray Observatory* (*CXO*) has opened a new era in X-ray observations of BAL QSOs. The low background and excellent spatial resolution allow *CXO* to probe 2–10 keV fluxes approximately twice as faint as *ASCA* in one quarter of the time.

Table 1. Basic Properties of BAL and Mini-BAL QSOs[a]

Name	R	z	Intrinsic N_H (10^{22} cm^{-2})	f_{cov}[b]	F_{2-10}[c]
APM 08279+5255[d]	15.2	3.87	$7.0^{+2.6}_{-2.2}$		41
RX J0911.4+0551[d,e]	18.0	2.80	19^{+28}_{-18}	$0.71^{+0.20}_{-0.39}$	5.8
PG 1115+080[d,e]	15.8	1.72	$3.8^{+2.5}_{-2.2}$	$0.64^{+0.11}_{-0.16}$	30
PG 2112+059	15.4[f]	0.457	$1.1^{+0.5}_{-0.4}$		75

NOTES – [a]X-ray errors given are for 90% confidence taking all parameters except normalization to be of interest. [b]Covering fraction is only provided when partial-covering absorption models provided a better fit to the data than simple neutral absorption. [c]Flux measured in the 2–10 keV band from the best-fitting X-ray spectral model in units of 10^{-14} erg cm^{-2} s^{-1}. [d]Gravitational lens system. The listed optical magnitude is for the brightest image, and the X-ray spectral information is for all images combined. [e]Mini-BAL QSO, see § 3.2. [f]B magnitude.

3. Gravitationally Lensed QSOs with Broad Absorption

As part of a *CXO* GTO program, gravitationally lensed QSOs were observed with ACIS-S3 to take advantage of the power of the High Resolution Mirror Assembly to resolve the individual lensed images. As an added benefit, several of these targets contain broad absorption lines. The magnifying effect of the lensing allows us to probe fainter intrinsic X-ray luminosities than would otherwise be possible, and these targets provided some of the best prospects for spectroscopic analysis. The X-ray spectral analysis for each object was done on all images combined to increase the signal-to-noise ratio, and the results are summarized in Table 1.

3.1. APM 08279+5255

Since its recent discovery in 1998, the BAL QSO APM 08279+5255 has inspired more than 20 publications. Its apparently incredible bolometric luminosity, which seemed to exceed $10^{15}\ L_{\odot}$, was found to be magnified by a factor of $\gtrsim 40$ by gravitational lensing (Irwin et al. 1998). On 2000 Oct 11, APM 08279+5255 was observed for 9.3 ks with the ACIS-S3 intrument. *CXO* revealed this luminous QSO to be sufficiently bright for spectral analysis. A result similar to that obtained for PG 2112+059 was found: APM 08279+5255 showed a typical QSO continuum with the signature of strong absorption. The best-fitting model is comprised of a power law with 7×10^{22} cm^{-2} of neutral absorbing gas (see Figure 2). Increasing the complexity of the spectral model to include a partially covering or ionized absorber did not improve the fits. At such high redshift, $z = 3.87$, the diagnostics for such models have passed below the *CXO* bandpass.

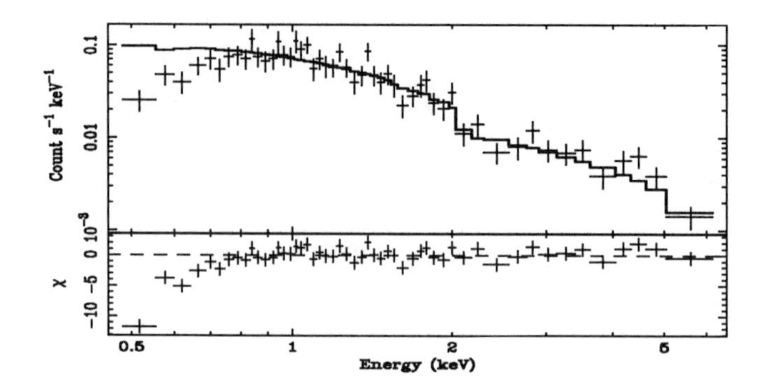

Figure 2. *CXO* ACIS-S3 spectrum of APM 08279+5255 fit with a power law above 5 keV (1 keV in the observed-frame) which has then been extrapolated back to lower energies. The high redshift of APM 08279+5255 has shifted the signatures of absorption almost completely out of the ACIS bandpass.

3.2. RX J0911.4+0551 and PG 1115+080

In contrast to APM 08279+5255, RX J0911.4+0551 and PG 1115+080 have luminosities more comparable to those of Seyfert 1 galaxies, and they are both properly classed as mini-BAL QSOs. Although the C IV absorption troughs are obviously broad ($\Delta v \gtrsim 3000\,\mathrm{km\,s^{-1}}$), these objects do not formally meet the BAL QSO criteria of Weymann et al. (1991). With less extreme ultraviolet absorption, mini-BAL QSOs might be expected to be stronger X-ray sources than bona-fide BAL QSOs. In fact, the mini-BAL QSO PG 1411+442 was successfully observed with *ASCA*; spectral analysis indicated a substantial intrinsic absorber with a column density, $N_{\mathrm{H}} \approx 10^{23}\,\mathrm{cm^{-2}}$, and an absorption covering fraction, $f_{\mathrm{cov}} \approx 97\%$ (Brinkmann et al. 1999; Gallagher et al. 2001).

RX J0911.4+0551 was discovered as part of a program to identify bright *ROSAT* sources (Bade et al. 1997), though the 29.2 ks *CXO* observation of 1999 November 2 showed it be a factor of ≈ 8 fainter than during the *ROSAT* All Sky Survey (Chartas et al. 2001). Spectral analysis revealed an X-ray continuum with a typical photon index overlaid with absorption. However, the absorption was not adequately modeled with neutral gas. The low-energy residuals from a power-law fit above rest-frame 5 keV suggested some complexity in the absorption such as would result from either a partially covering or an ionized absorber. Both models were significant improvements over the neutral absorber with the first model being slightly preferred (Chartas et al. 2001). The best-fitting intrinsic column density, $N_{\mathrm{H}} = 2 \times 10^{23}\,\mathrm{cm^{-2}}$, is the largest of the four QSOs presented in this paper.

PG 1115+080 is notable for significant variability of the ultraviolet O VI emission and absorption lines (Michalitsianos, Oliversen, & Nichols 1996) as well as X-ray flux changes (Chartas 2000). This target has been observed with *CXO* on two occasions, for 26.2 ks on 2000 June 2 and 9.7 ks on 2000 November 3. The data were analyzed following the procedure outlined above with a similar result as for RX J0911.4+0551; a partially covering absorber model with $N_{\mathrm{H}} =$

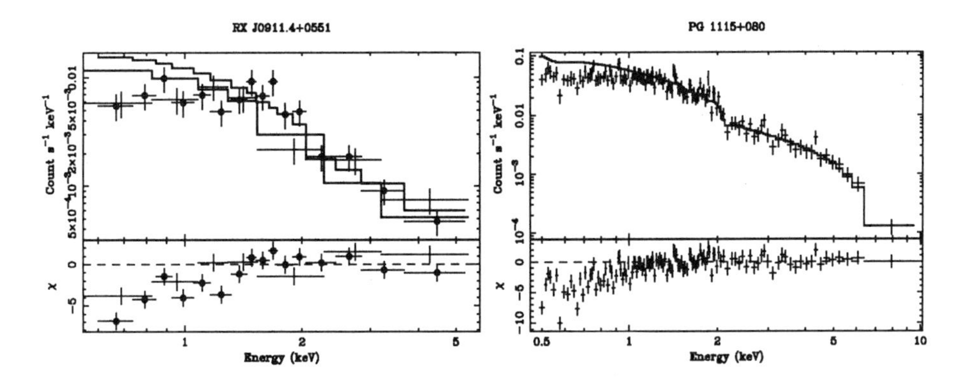

Figure 3. *CXO* ACIS-S3 spectra of the two mini-BAL QSOs, RX J0911.4+0551 and PG 1115+080. Both spectra have been fit above rest-frame 5 keV with a power-law model, which has then been extrapolated back to the lowest energies. A partial-covering absorption model provides a better fit than neutral absorption for both objects.

$4 \times 10^{22}\,\mathrm{cm}^{-2}$ was preferred over neutral absorption. PG 1115+080 has fairly good photon statistics (see Figure 3), thus making it a good target for additional observations to investigate X-ray spectral variability.

4. General Picture and Conclusions

As the number of BAL QSOs detected with enough X-ray photons for spectral analysis grows, a consistent picture is beginning to emerge. The X-ray continua can be well modeled by power-law models with photon indices consistent with those of other radio-quiet QSOs, $\Gamma \approx 2$. In addition, correcting the X-ray spectra for absorption reveals normal ultraviolet-to-X-ray flux ratios, thus indicating that the underlying spectral energy distributions of BAL QSOs are not unusual. Both of these observations support the scenario whereby broad absorption line outflows are common components of the nuclear environments of radio-quiet QSOs.

Confirming these generalizations will require additional, long spectroscopic observations of the brightest BAL QSOs. To complement this endeavor, large, exploratory surveys of well-defined samples of BAL QSOs will offer enough information to examine which multi-wavelength properties are related to the X-ray characteristics. As of yet, predicting which BAL QSOs will be productive targets for spectroscopic X-ray observations remains a black art. Connecting the X-ray properties, such as flux and coarse spectral shape, to properties in other spectral regimes will ultimately help us to understand the nature of the intrinsic absorption in the X-ray and ultraviolet. We have begun such a program in the *CXO* Cycle 2 observing round with 18 BAL QSO targets from the Large Bright Quasar Survey, and the data that have arrived thus far are promising.

In terms of the column density of the absorbing gas, the best-fitting values range from $(1\text{–}20)\times10^{22}\,\mathrm{cm}^{-2}$. Though relatively simple models can adequately explain the observations, the physical absorber in each system is likely to be complex. Partial-covering absorption models suggest that multiple lines of sight are present; the direct view suffers from heavy obscuration while a second, scattered line of sight could be clearer. In addition, if the X-ray absorbing gas is close to the nucleus and associated with the ultraviolet-absorbing gas, it must also be highly ionized. In this case, the column density measurements can only be considered lower limits. Additionally, significant velocity dispersion of the absorber would increase the continuum opacity of bound-bound absorption lines, and thus further complicate determining an accurate value of the column density.

X-rays are powerful probes of the inner regions of BAL QSOs as they are sensitive to molecular, neutral, and partially ionized gas. Estimates of the absorption column density from X-ray observations (compared to ultraviolet spectral analysis) suggest that the bulk of the absorbing gas is more readily accessible in the high-energy regime. Thus, X-ray observations offer the greatest potential for determining the true mass outflow rate of QSOs. To constrain this value, a measure of the velocity structure of the X-ray absorbing gas is essential. To this end, a gratings observation with *XMM-Newton* of the X-ray brightest BAL QSO, PG 2112+059, offers the most promise.

Acknowledgments. This research was supported by NASA grant NAS8-38252, Principal Investigator, GPG. SCG also gratefully acknowledges NASA GSRP grant NGT5-50277.

References

Bade, N., Siebert, J., Lopez, S., Voges, W., & Reimers, D. 1997, A&A, 317, L13

Brandt, W. N., Laor, A., & Wills, B. J. 2000, ApJ, 528, 637

Brinkmann, W., Wang, T., Matsuoka, M., & Yuan, W. 1999, A&A, 345, 43

Chartas, G. 2000, ApJ, 531, 81

Chartas, G., et al. 2001, ApJ, 558, 119

Gallagher, S. C., et al. 1999, ApJ, 519, 549

Gallagher, S. C., et al. 2001, ApJ, 546, 795

Green, P. J. & Mathur, S. 1996, ApJ, 462, 637

Irwin, M. J., Ibata, R. A., Lewis, G. F., & Totten, E. J. 1998, ApJ, 505, 529

Iwasawa, K. 1999, MNRAS, 302, 96

Jannuzi, B. T., et al. 1998, ApJS, 118, 1

Kopko, M., Turnshek, D. A., & Espey, B. R. 1994, in IAU Symp. 159, Multi-Wavelength Continuum Emission of AGN, ed. T. Courvoisier & A. Blecha (Dordrecht: Kluwer), 450

Mathur, S., Elvis, M., & Singh, K. P. 1995, ApJ, 455, L9

Mathur, S. et al. 2000, ApJ, 533, L79

Michalitsianos, A. G., Oliversen, R. J., & Nichols, J. 1996, ApJ, 461, 593

Morgan, N. D., et al. & Schecter, P. L., 2001, ApJ, 555, 1

Reeves, J. N. & Turner, M. J. L. 2000, MNRAS, 316, 234

Turner, T. J. 1999, ApJ, 511, 142

Weymann, R. J., Morris, S. L., Foltz, C. B., & Hewett, P. C. 1991, ApJ, 373, 23

Mass Outflow in Active Galactic Nuclei: New Perspectives
ASP Conference Series, Vol. 255, 2002
D.M. Crenshaw, S.B. Kraemer, and I.M. George

XMM-Newton RGS observations of Soft X-ray Emission Lines from Relativistic Accretion Disks

G. Branduardi-Raymont, M. J. Page

Mullard Space Science Laboratory, University College London, Holmbury St. Mary, Dorking, Surrey, RH5 6NT, UK

M. Sako, S. M. Kahn

Department of Physics and Columbia Astrophysics Laboratory, Columbia University, 550 West 120th Street, New York, NY 10027, USA

J. S. Kaastra, A. C. Brinkman

SRON, Sorbonnelaan 2, 3584 CA Utrecht, The Netherlands

Abstract. *XMM-Newton* Reflection Grating Spectrometer (RGS) spectra of the Narrow Line Seyfert 1 galaxies MCG−6-30-15 and Mrk 766 are shown to be inconsistent with standard AGN models comprising a power-law continuum absorbed by either cold or ionized matter. Both objects display remarkably similar spectral features in the 5–35 Å band: these require to be interpreted as H-like oxygen, nitrogen, and carbon emission lines, gravitationally redshifted and broadened by relativistic effects in the vicinity of a Kerr black hole. The underlying continuum shape, and the presence of narrow absorption lines, can be explained in terms of a power-law absorbed by ionized material, but at a much lesser level than implied by pure warm absorption modeling previously carried out.

1. Introduction

High resolution X-ray spectroscopy coupled with large collecting area offers the potential to unravel the emission and absorption processes which underline the low-energy X-ray spectra of Active Galactic Nuclei (AGN). Such capability is provided for the first time by the *XMM-Newton* RGS. Until now, the soft X-ray spectra of AGN have generally been modeled by a power-law continuum absorbed by neutral and/or partially ionized material (e.g. Halpern 1984; Reynolds 1997). However, the details of this description are far from being clear. In particular, the origin and location of the "warm absorber", and its relation to the absorbing gas observed in the UV, are still very much matters of debate (e.g. Otani et al. 1996, Shields & Hamann 1997).

As we show below, RGS observations of MCG−6-30-15 and Mrk 766, carried out early in the *XMM-Newton* mission, have changed our perspective of the physical processes taking place in the inner regions of AGN. This new view of their soft X-ray spectra has led us to recognise the signatures of relativistic line emission, once believed to be confined to higher energies.

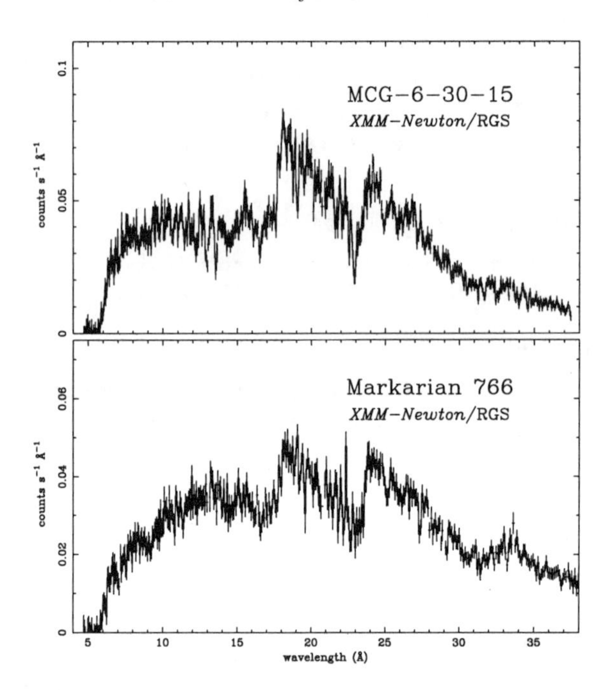

Figure 1. Raw first order RGS spectra of MCG−6-30-15 and Mrk 766, plotted in the observer's frame.

2. The Targets: MCG−6-30-15 and Mrk 766

MCG−6-30-15 and Mrk 766 are bright and nearby Narrow Line Seyfert 1 galaxies with relatively low Galactic absorption in the line of sight (see Table 1 for details). Both objects display large and rapid X-ray variability, in flux and in the slope of their power-law continua. Of the two, only Mrk 766 possesses a soft excess, which is less variable than the higher energy flux, and only MCG −6-30-15 shows Compton reflection in its spectrum. Broad features in the < 1 keV spectra of both sources have been interpreted in terms of absorption in ionized material at some distance from the nuclear black hole. The shape of the broad fluorescent Fe Kα line which is observed in both objects at 6 – 7 keV is attributed to relativistic effects in a disk surrounding the central black hole (Tanaka et al. 1995).

Table 1. Parameters of target sources

	MCG−6-30-15	Mrk 766
z	0.00775	0.01293
$N_{\mathrm{H_{Gal}}}$ (cm^{-2})	4.1×10^{20}	1.8×10^{20}
$L_{2-10\mathrm{keV}}$ (erg s^{-1})	$1 - 2 \times 10^{43}$	$0.5 - 1.3 \times 10^{43}$

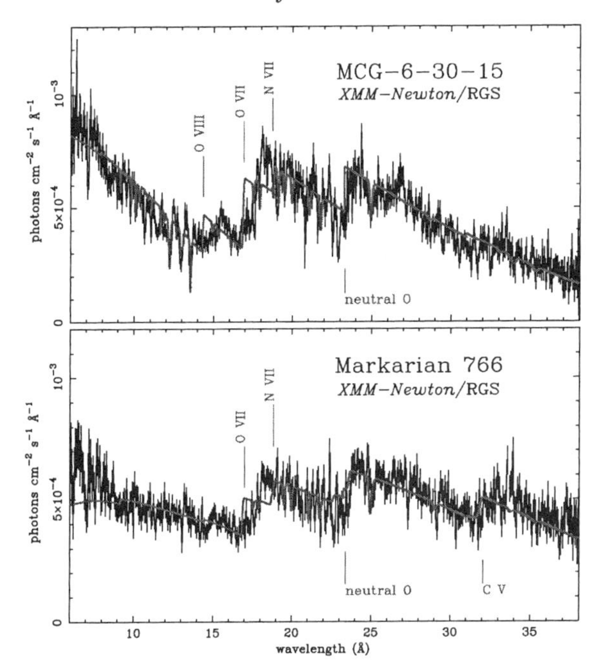

Figure 2. 'Fluxed' spectra of the two sources (corrected for effective area) with the best-fit pure warm absorber model, plotted in the observer's frame.

3. *XMM-Newton* RGS observations and Spectral Analysis

XMM-Newton observed MCG −6-30-15 in July 2000 for a total of 120 ks, and Mrk 766 in May 2000 for 55 ks. The RGS data have been processed and calibrated with the *XMM-Newton* Science Analysis Software. The current RGS wavelength scale is accurate to $\sim$ 8 mÅ (see den Herder et al. 2001, for a description of the instrumentation).

The raw first order spectra of the two galaxies for both RGS-1 and -2 combined are shown in Figure 1: they are remarkably similar, displaying prominent "saw-tooth" features, which peak at around 18, 24 and 33 Å. A single power-law fit with neutral absorption is clearly unacceptable; in particular, the neutral oxygen edge at 23 Å implies a column density in excess of what is required by the continuum fit, and there is no evidence of neutral absorption edges from other elements at their expected positions.

Next, we tried to fit the spectra with a pure warm absorber model including absorption edges and absorption lines associated with all ions of abundant elements (C, N, O, Ne, Mg, and Fe). The fits are shown in Figure 2 for both galaxies (for details see Branduardi-Raymont et al. 2001, BR01 hereafter): they are unacceptable for several reasons. Firstly, the observed, putative O VIII and O VII edges, implying column densities of the order of $3 - 4 \times 10^{18}$ cm^{-2}, are redshifted with respect to their expected positions by very large amonts (~ 1 Å), corresponding to velocities of the order of $\sim 16,000$ km s^{-1} towards the central

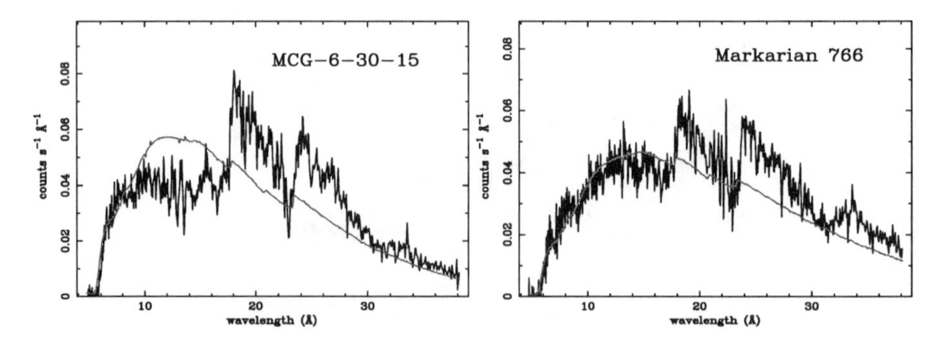

Figure 3. Extension of the EPIC PN high energy power-law continuum superposed on the observed RGS spectra.

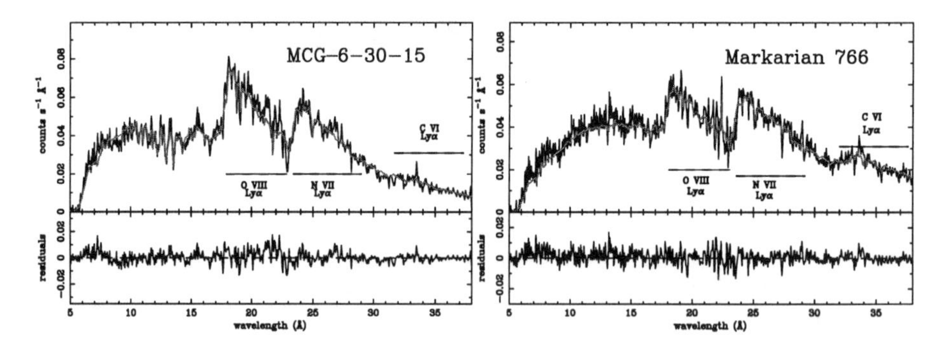

Figure 4. Best fit models, including relativistic lines and warm absorption, plotted with the observed RGS spectra.

source. However, at the redshifts of the putative edges, no associated absorption lines are seen in either object, with upper limits of $\sim$ 20 mÅ on the equivalent width; at the column densities derived from the 'edges' the absorption lines are saturated, thus the upper limit on the equivalent width translates to a limit of $\leq$ 60 km s^{-1} for the velocity width of the infalling material: such a small value is hard to reconcile with the apparent large redshifts of the edges, implying a radial inflow at one particular velocity. Excess neutral absorbing material is still required to fit the data at longer wavelengths. An excess of flux is also present between 18 and 19 Å in MCG−6-30-15.

Surprisingly, we have found that the RGS spectra of both MCG−6-30-15 and Mrk 766 can be well fitted with an alternative model consisting of an absorbed power-law, and emission lines gravitationally redshifted and broadened by relativistic effects, in a medium which is encircling a massive, rotating black hole. In this interpretation, the saw-toothed features in Figure 1 are attributed to (in ascending wavelength order) H-like Lyα lines of O VIII, N VII and C VI (BR01). We find that the line wavelengths are all consistent with the galaxies' systemic velocities and that all the lines have the same broad profiles: in practice we see the identical spectral structure in the two separate sources. No additional neutral column density is required in excess of the Galactic value. The fits residuals are also much less obvious and systematic than for the pure warm absorber interpretation.

The lack of Fe L and He-like K emission lines in the RGS spectra suggests that the observed emission lines are most likely due to radiative recombination onto fully stripped ions. A preliminary analysis of this model's physical self-consistency is presented in BR01, and is encouraging. However, this initial interpretation requires the underlying power-law continuum to flatten below $\sim$ 2.5 keV, which is unusual and difficult to explain. We have recently found a solution to this problem by including a degree of warm absorption in our model: this is indeed required by the presence of the narrow absorption lines from Ne, Fe, O, N, C detected in the RGS spectra of both objects. This absorption has the effect of 'eating' into the continuum at short wavelengths, an effect previously produced by the continuum flattening required in the fits when no warm absorption was included. This is clearly illustrated in Figure 3 where the extension of the high energy power-law continuum from the EPIC PN instrument (Strüder et al. 2001) is superposed on the observed RGS spectra, and is seen to 'cut through' the O, N and C relativistic lines.

Table 2. RGS best fit parameters for a model including relativistic lines and warm absorption

	MCG−6-30-15	Mrk 766
Γ	1.8	2.1
Inclination i	38°	34°
Emissivity index q	3.1	3.5
$R_{\rm in}$	1.24 $R_{\rm g}^{\rm a}$	1.24 $R_{\rm g}^{\rm a}$
$R_{\rm out}$	56 $R_{\rm g}$	45 $R_{\rm g}$

NOTES – [a] Minimum physical value allowed in the fit.

Table 2 lists the best fit parameters obtained when we add a warm absorber in our model, which already includes a power-law continuum, with cold absorption fixed at the Galactic value, and three emission lines originating near a maximally rotating Kerr black hole (Laor 1991). The line wavelengths are fixed at their expected values at the redshifts of the sources. The continuum power-law slope is fitted, as are the disk inclination angle i, the emissivity index q and the inner and outer limits $R_{\rm in}$ and $R_{\rm out}$ of the disk emission region. These parameters are tied for all the lines in the fit. It is worth noting that the disk line parameters are in very good agreement with those derived for the Fe Kα line ($i = 34^{\circ}{}^{+5}_{-6}$ and $36^{\circ}{}^{+8}_{-7}$, $q = 2.8 \pm 0.5$ and $3.0^{+0.8}_{-0.4}$ for MCG−6-30-15 and Mrk 766 respectively, Nandra et al. 1997). Plots of the best fit models and the observed RGS spectra are shown in Figure 4. Details of this refined interpretation of the RGS spectra of MCG−6-30-15 and Mrk 766 will be reported elsewhere (Sako et al., in preparation).

4. Conclusions

The Narrow Line Seyfert 1 galaxies MCG−6-30-15 and Mrk 766 are found to possess very similar RGS spectra, which cannot be explained in terms of a pure warm absorber model. However, broad line emission from a relativistic disk surrounding a maximally rotating Kerr black hole appears to explain the data remarkably well when a degree of warm absorption is also included. This novel result demonstrates how the combination of large effective area and high energy resolution afforded by the *XMM-Newton* RGS is truly revolutionising our understanding of the many processes responsible for the soft X-ray spectra of AGN.

References

Branduardi-Raymont, G., Sako, M., Kahn, S.M., Brinkman, A.C., Kaastra, J.S., Page, M.J. 2001, A&A 365, L140

den Herder, J. W., et al. 2001, A&A, 365, L7

Halpern, J. P. 1984, ApJ, 281, 90

Laor, A. 1991, ApJ, 376, 90

Nandra, K., George, I.M., Mushotzky, R.F., Turner, T.J., Yaqoob, T. 1997, ApJ, 477, 602

Otani, C., et al. 1996, PASJ, 48, 211

Reynolds, C. S. 1997, MNRAS, 286, 513

Shields, J. C. & Hamann, F. 1997, ApJ, 481, 752

Strüder, L., Briel, U., Dennerl, K. et al. 2001, A&A, 365, L18

Tanaka, Y., et al. 1995, Nature, 375, 659

Mass Outflow in Active Galactic Nuclei: New Perspectives
ASP Conference Series, Vol. 255, 2002
D.M. Crenshaw, S.B. Kraemer, and I.M. George

A *Chandra* Observation of the Red Quasar 3C 212

Thomas L. Aldcroft, Steve S. Murray, and Martin Elvis

Harvard-Smithsonian Center for Astrophysics, 60 Garden Street, Cambridge, MA 02138, USA

Abstract.
The red, absorbed quasar 3C 212 was observed with ACIS-S3 on board the *Chandra X-ray Observatory* (*CXO*) for 19.5 ks, and $\sim$ 4000 counts were detected. The X-ray absorption in this object has previously been interpreted as a highly ionized warm absorber. Spectral fitting of the *CXO* data confirms that excess absorption over the Galactic value is required at very high confidence. The spectrum is well fit by a power law with Galactic absorption and a neutral absorbing column of $4.2 \pm 0.4 \times 10^{21}$ cm^{-2} at the quasar redshift. Other non-powerlaw models give a poor fit to the data. The optical depth of an O VII absorbing edge at the quasar redshift is less than 0.9 at 90% confidence. This is marginally consistent with the value $3.9^{+3.9}_{-3.0}$ (1-σ) found with *ROSAT* data, but the warm absorber interpretation in this object may need re-examination.

1. Introduction

3C 212 (Q0855+143) is a radio-loud quasar at redshift $z = 1.049$. It is one of the prototype "red quasars" studied by Smith & Spinrad (1980). These objects have very faint optical counterparts, but are bright in the near infrared (1-2μm) (Rieke et al. 1979). Red quasars could be missed in typical surveys, and Webster et al. (1995) proposed that up to 80% of radio-loud quasars are hidden in the optical. Understanding the cause of this reddening is important for revealing the true nature of AGNs and their overall population statistics.

In addition to being red in the optical, 3C 212 harbors an associated Mg II absorption system (Aldcroft et al. 1994) and soft X-ray absorption detected with the *ROSAT* PSPC (Elvis et al. 1994). Mathur (1994) extended the analysis of the PSPC data and found evidence for intrinsic O VII absorption. A warm absorber solution was found which could reproduce both the the UV and X-ray absorbing columns. However, the signal-to-noise in the PSPC spectrum was insufficient to strongly constrain the absorbing column, nor even to rule out other source models such as a blackbody spectrum or a Raymond-Smith plasma.

In order to further understand this interesting object, 3C 212 was observed as a *CXO* GTO target using the Advanced CCD Imaging Spectrometer (ACIS). In this paper we describe the results of the observation and our spectral and imaging analysis.

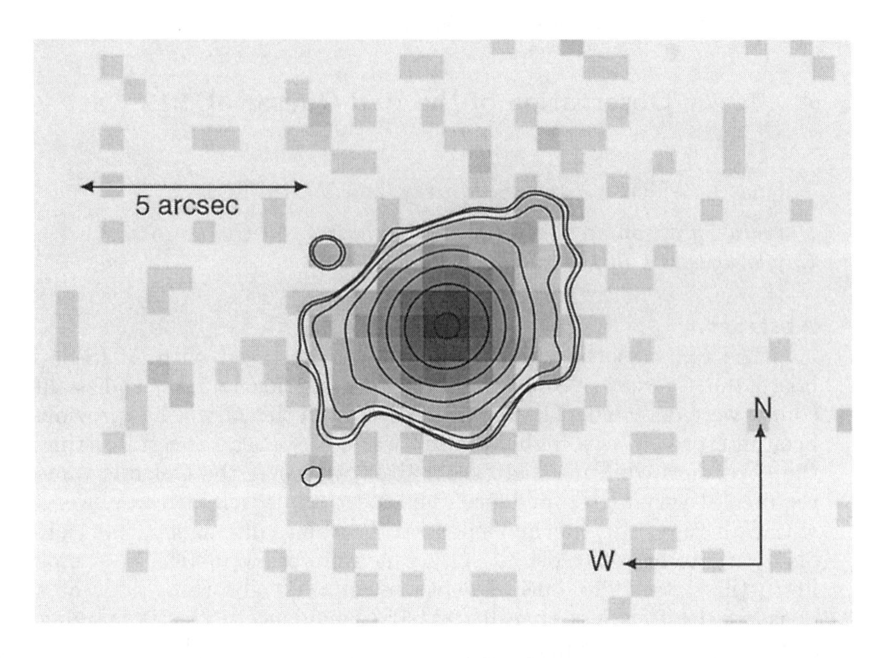

Figure 1. Image from the 20 ks ACIS-S exposure of 3C 212. The contours were derived after smoothing, and are logarithmically spaced.

2. Observations and Data Reduction

3C 212 was observed as a *CXO* GTO target, using ACIS-S for 19.5 ks on 2000 October 26 at 08:10:51 GMT. The source was placed on the ACIS-S3 chip, and a 1/8 sub-array readout mode with 0.5 s integrations was selected in order to mitigate pile-up. A total of 4049 counts in the source region were detected, with the X-ray celestial coordinates of the source matching the optical 3C 212 coordinates to within ~ 1 arcsec.

An image of the source is shown in Figure 1, with 0.5 arcsec spatial binning, corresponding to the ACIS CCD pixel size. Overlayed on the image are logarithmically spaced X-ray contours. These contours were derived using a smoothed version of raw event data.

The X-ray observations were reduced following the standard `CIAO` software thread to extract an ACIS spectrum: (1) Extract source events within a 4.0 arcsec radius; (2) Create the aspect histogram file; (3) Create the Redistribution Matrix and Ancillary Response Files (RMF and ARF) appropriate to the time-dependent source position on chip. Calibration data from the `CALDB` v2.0 release were used. In addition to the standard thread, the event data were filtered on energy to use the range 0.3–8.0 keV, and they were grouped to a minimum of 20 counts per bin.

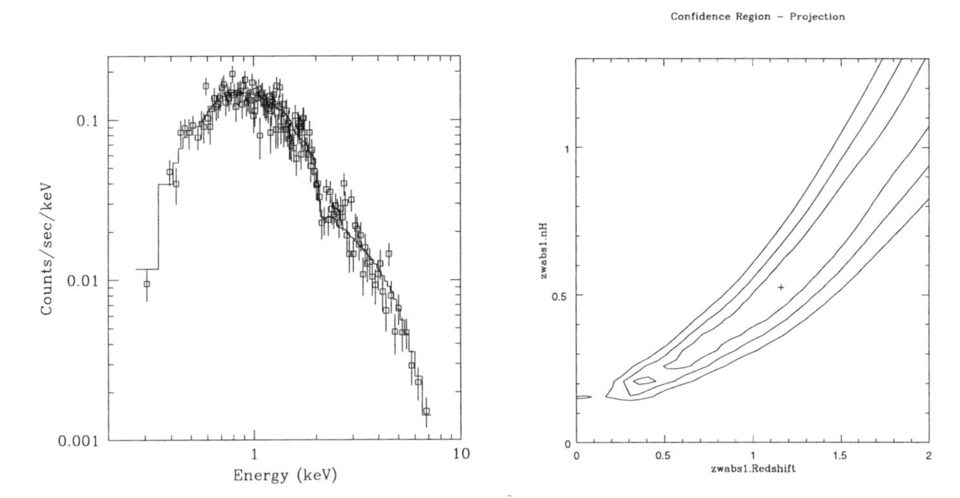

Figure 2. LEFT: Extracted spectrum of 3C 212 and best-fit powerlaw model including Galactic and redshifted neutral absorption. RIGHT: 1,2 & 3-σ contours for the absorbing column versus intrinsic absorber redshift. The absorber is constrained to have $z \gtrsim 0.25$.

3. Spectral Fitting

Figure 2 shows the extracted spectrum for 3C 212 along with the fitted model consisting of a powerlaw with Galactic absorption and neutral absorption at the quasar redshift. Spectral fitting and analysis were done with the `CIAO Sherpa` software, using Powell minimization and the χ^2 model variance statistic.

The results of fitting a power law with absorption are shown in Table 1. In these fits, the redshift of the intrinsic absorber is frozen at the quasar redshift. These fit values are consistent with the *ROSAT* PSPC results given in Elvis et

Table 1. Spectral fit parameters

Model	Γ (1)	$N_{H,Gal}$ (10^{21} cm^{-2}) (2)	$N_{H,z}$ (10^{21} cm^{-2}) (3)	χ^2 (d.o.f) (4)
$N_{H,Gal}$ (fixed)	1.24 ± 0.04	(0.36)	...	396.0(136)
$N_{H,Gal}$ (free)	1.78 ± 0.07	1.64 ± 0.15	...	181.9(135)
$N_{H,Gal} + N_{H,z}$	1.66 ± 0.06	(0.36)	4.2 ± 0.4	163.1(135)

NOTES– Uncertainties are 90% confidence limits. Fit values in parentheses are frozen. (1) Power law photon index. (2) Absorbing column at $z = 0$. (3) Absorbing column at quasar redshift. (4) χ^2-statistic (degrees of freedom)

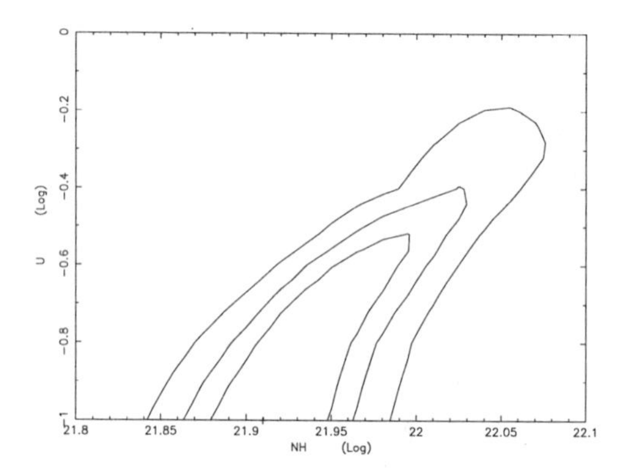

Figure 3. Plot of 1,2,3-σ contours for a warm absorber model fit of 3C 212 data. Mathur (1994) found $\log N_H = 21.94$ and $\log U = -0.5$.

al. (1994) and Mathur (1994): $\Gamma = 1.4^{+0.8}_{-0.6}$ and $N_{H,z} = 8.7^{+7.5}_{-6.3} \times 10^{21}$ cm^{-2}. We also fit the third model (Galactic + Intrinsic N_H) with the redshift of the intrinsic absorber free to vary. Although not well constrained, the best fit value of absorber redshift is 1.2, which matches the quasar redshift ($z = 1.049$) quite well. The absorber redshift is at $z_{abs} > 0.25$ at 95% confidence. Other spectral models, such as a Raymond-Smith plasma or blackbody spectrum, give poor fits to the data and can be ruled out.

It should be be noted that the region used to extract the spectrum includes the extended emission discussed in Section 4. However, this component is extremely faint compared to the core and does not significantly affect the spectral fitting.

3.1. A Warm Absorber?

Mathur (1994) found that the fit to *ROSAT* PSPC data could be improved by adding an absorption edge corresponding to O VII (0.74 keV) at the quasar redshift. The best-fit edge optical depth was $\tau = 3.9^{+3.9}_{-3.0}$ (1-σ). We added an edge component (at 0.36 keV observed frame) to the best-fit model and found a best-fit value of $\tau = 0.0$ with 90% upper limit of 0.9.

We also fit the data with a warm absorber model (kindly provided by F. Nicastro), which is based on `CLOUDY` calculations. A contour plot showing absorber N_H versus ionization parameter U is given in Figure 3. This again points to a neutral absorber, although the ionization parameter of $\log U = -0.5$ found by Mathur (1994) is not ruled out.

4. Extended Emission

Figure 4 shows that the 3C 212 X-ray emission is elongated in the soft band (0.3 - 1.0 keV), but not in the hard band (2.5 - 8.0 keV). This is in contrast with the point-spread function (PSF) of the High Resolution Mirrow assembly (HRMA) on *CXO*, which is essentially circular at the off-axis angle of the source.

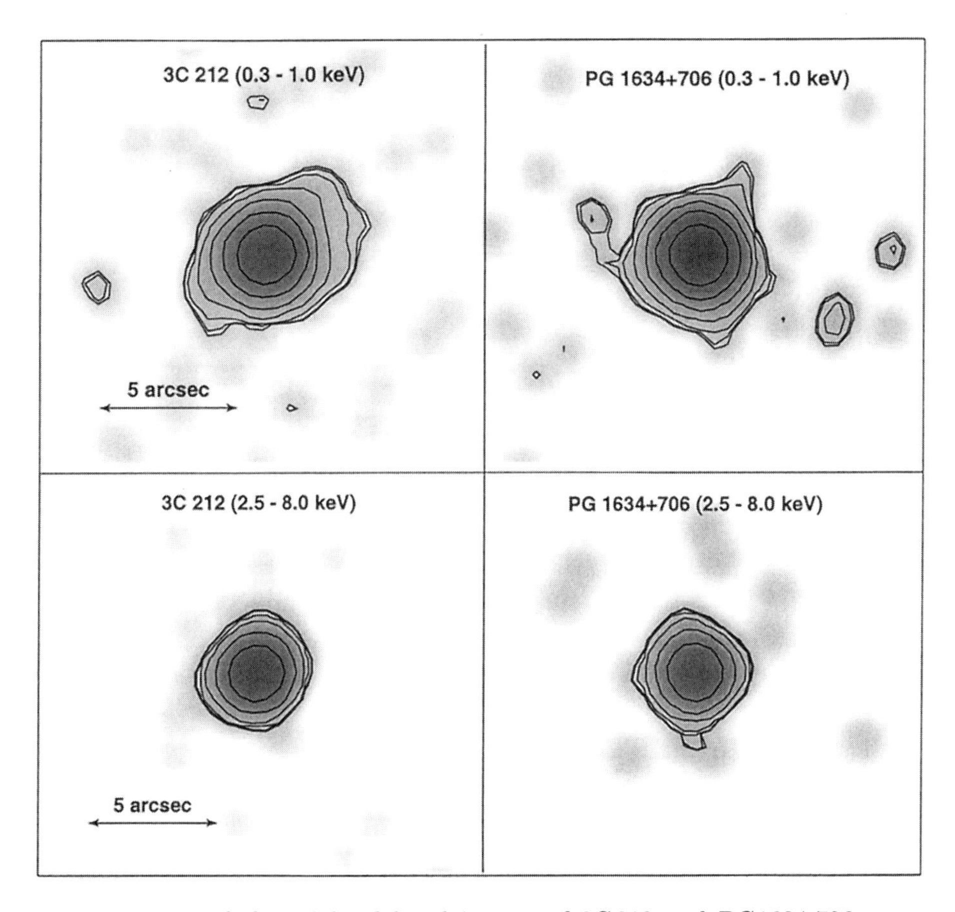

Figure 4. Soft and hard band images of 3C 212 and PG1634-706. Contour levels are logarithmically spaced and are the same for 3C 212 and PG 1634-706.

For visual comparison, we have taken a calibration observation of PG 1634+704, processed it in the same way, scaled it to the same total counts, and plotted with the same contour levels. The hard emission and core of the soft emission are identical. Note that in these images the data have been smoothed to bring out low surface brightness features. The soft nature of the extended emission rules out instrumental effects such as bad aspect. Note also that the HRMA PSF is broader at *hard* energies.

The extended emission could arise in the jet of 3C 212. Figure 1 in Stockton & Ridgway (1998) shows a deep optical image of 3C 212 overlayed with contours of VLA imaging. There are several prominent radio knots within 2 arcsec, as well as the radio lobes at about 4-5 arcsec from the nucleus. The position of the lobes are consistent with the extended X-ray emission.

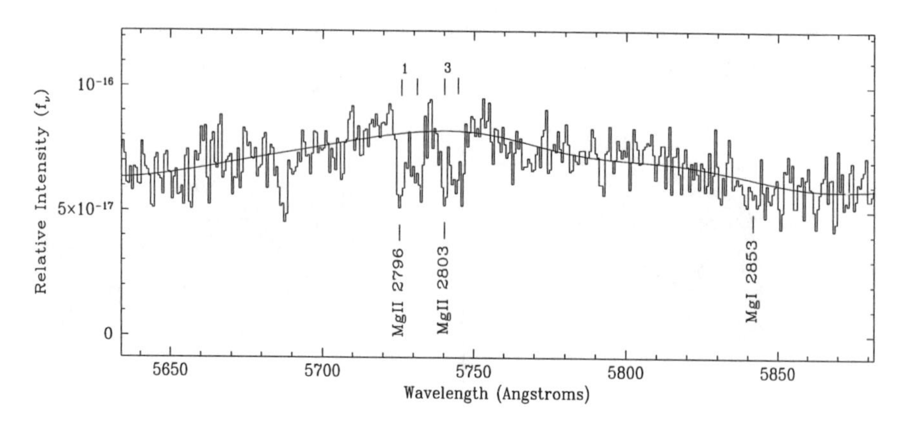

Figure 5. Mg II absorption complex in the optical spectrum of 3C 212, with components at $z = 1.0477$ and $z = 1.0491$

5. Associated Mg II Absorber

3C 212 is known to harbor an associated Mg II absorption complex, with components at $z = 1.0491$ and $z = 1.0477$, as shown in Figure 5. Stockton & Ridgway (1998) used the *Keck* LRIS to carry out a comprehensive redshift survey of faint galaxies in the 3C 212 field. They found only one galaxy within 100 arcsec with a redshift very near that of 3C 212. However, the galaxy redshift of $z = 1.053$ is inconsistent with the Mg II absorption redshifts. We conclude that the Mg II absorption is intrinsic to the quasar or its host galaxy. The question of whether the UV (Mg II) and X-ray absorption are produced in the same material (Mathur 1994) is still open. Their analysis which assumed a substantial O VII absorbing column should be revisited in light of the *CXO* data. High-resolution X-ray spectroscopy is needed, but for 3C 212 we will have to wait for the next generation of large area X-ray telescopes

Acknowledgments. The authors would like thank Aneta Siemiginowska for help with `Sherpa`, as well as the entire *Chandra* team for making possible the observations and data analysis. This work was supported by NASA grant NAS8-39073.

References

Aldcroft, T. L., Bechtold, J., & Elvis, M. 1994, ApJS, 93, 1

Elvis, M., Fiore, F., Mathur, S., & Wilkes, B. 1994, ApJ, 425, 103

Mathur, S. 1994, ApJ, 431, L75

Rieke, G. H., Lebofsky, M. J., & Kinman, T. D. 1979, ApJ, 232, L151

Smith, H. E., & Spinrad, H. 1980, ApJ, 236, 419

Stockton, A., & Ridgway, S. E. 1998, ApJ, 115, 1340

Webster, R. L., Francis, P. J., Peterson, B. A., Drinkwater, M. J., & Masci, F. J. 1995, Nature, 375, 469

Mass Outflow in Active Galactic Nuclei: New Perspectives
ASP Conference Series, Vol. 255, 2002
D.M. Crenshaw, S.B. Kraemer, and I.M. George

A Physically-Consistent Model for X-ray Emission by Seyfert 2 Galaxies Demonstrated using NGC 1068

E. Behar, A. Kinkhabwala, M. Sako, F. Paerels, S.M. Kahn

Columbia University, 550 W 120th St., New York, NY 10027, USA

A.C. Brinkman, J. Kaastra, R. van der Meer

SRON, Sorbonnelaan 2, 3548 CA, Utrecht, The Netherlands

Abstract. Preliminary analysis of the X-ray spectrum of NGC 1068 obtained by the Refelection Grating Spectrometer (RGS) on board *XMM-Newton* is presented. A physically consistent model is developed in order to quantitatively describe the reprocessing of the central AGN continuum source into the discrete X-ray emission observed in Seyfert 2 galaxies. All the important atomic processes are taken into account, including photoexcitation, which has been neglected in some previous models. The model fits the high resolution NGC 1068 data very well, which implies that the contribution of hot, collisional gas[1] to the X-ray spectrum of NGC 1068 is negligible.

1. Introduction

Recent high-resolution X-ray observations of Seyfert 2 galaxies with the grating spectrometers on board *XMM-Newton* and the *Chandra X-ray Observatory* (*CXO*) reveal an emission spectrum rich in discrete features. This finding supports the unified AGN model, according to which the continuum nuclear source in Seyfert 2 galaxies is completely blocked from our line of sight and we can see only the reprocessed light. To date, discrete emission has been detected in Markarian 3 (Sako et al. 2000), NGC 4151 (Ogle et al. 2000), and the Circinus Galaxy (Sambruna et al. 2001). NGC 1068, which is the brightest Seyfert 2 galaxy and therefore best suited for detailed spectroscopic studies, has been observed by all X-ray grating spectrometers: the Low- and High-energy Grating Spectrometers (LETGS & HETGS) on board *CXO* (Brinkman et al., in preparation; Ogle et al. 2001, respectively), and the RGS on board *XMM-Newton* (Kinkhabwala et al., in preparation and the present work). The conspicuous radiative recombination continuum (RRC) features observed in all Seyfert 2 galaxies verify the photoionized nature and low-temperature ($kT_e \sim$ few eV) of the X-ray emitting plasma.

[1]Throughout this paper we use the term "collisional gas" to denote a a hot plasma in which ionization occurs by means of collisions between atoms/ions and free electrons, and in which the spectral lines are formed by electron–ion collisional excitation.

Spectra emitted by photoionized plasmas have been occasionally modeled assuming that the spectral lines and RRC's are solely produced by recombination and subsequent radiative cascades (e.g., Liedahl et al. 1990; Sako et al. 1999; Porquet & Dubau 2001). When comparing these pure-recombination models to the Seyfert 2 X-ray spectra, significant residuals are found, which are mostly in lines with high oscillator strengths that are also among the strongest lines in hot, collisional plasmas. Sako et al. (2000) were the first to suggest that these lines arise from photoexcitation (resonant scattering), while Ogle et al. (2000, 2001) interpret them as evidence for hot (~keV) plasma confining the much colder photoionized gas. Neither explanation has been tested with a detailed model yet. It is only natural that in a strong radiation field environment, the same flux that ionizes the atoms (bound-free transitions) will also excite them (bound-bound). In fact, emission by recombination and subsequent cascades *without* photoexcitation would be possible only if the plasma is in a transient phase of cooling and recombining. However, as a general rule in ionization balance, both of these processes contribute appreciably to the emitted spectrum. The relative importance of photoexcitation, recombination, and collisional excitation can be assessed only through careful modeling of the various line (and RRC) intensities, while correctly accounting for the atomic state kinetics. The model suggested in this paper is similar in spirit to the models that have been employed to interpret the spectra emitted by laboratory laser-produced plasma (Doron et al. 1998) and stellar coronae (Behar, Cottam, & Kahn 2001). Only here, we add the transitions induced by the radiation field, namely photoionization and photoexcitation. We focus on the unique spectroscopic signatures of photoexcitation and point out their importance in the NGC 1068 spectrum obtained with the RGS. The full analysis of the NGC 1068 RGS observation will be presented elsewhere (Kinkhabwala et al., in preparation).

2. The Atomic-State Kinetic Model

For each ion, we use a steady-state model, which includes all of the excited levels of the emitting ion in addition to the ground level of the next charge state. For the simple case of H-like ions, that would be the nl configurations in addition to the bare nuclei state. The atomic processes included in the model are radiative decays, radiative recombination (RR), photoexcitation (PE), and photoionization (PI). Additional processes, such as collisional excitations, de-excitations, or dielectronic recombination processes can be easily incorporated, but their effect in plasmas typical of Seyfert 2 galaxies is negligibly small. Self absorption of the reprocessed light can be included, but has been found not to be important for NGC 1068. The electron energy distribution is assumed to be Maxwellian, corresponding to an electron temperature T_e. We denote the electron density as n_e, the Einstein coefficient for spontaneous radiative emission as A, and the rate coefficient for RR as $\alpha^{RR}(T_e)$. The set of rate equations for the density n_i^{+q} (in cm^{-3}) of an ion with positive charge q in a level i can then be written as given in Eqn. (1):

$$\frac{d}{dt}n_i^{+q} = \sum_{j>i} n_j^{+q}A_{ji} + n_e\sum_{k>i} n_k^{+(q+1)}\alpha_{ki}^{RR}(T_e) + \sum_{j<i} n_j^{+q}R_{ji}^{PE} - n_i^{+q}\left(\sum_{j<i} A_{ij} + \sum_{j>i} R_{ij}^{PE} + \sum_{k>i} R_{ik}^{PI}\right) = 0 \quad (1)$$

The rates R^{PE} for PE are calculated from the illuminating photon flux $F(E)$, which is a function of the photon energy E and is expressed in photons sec^{-1} cm^{-2} keV^{-1},

$$R_{ji}^{PE} = \int F(E)\sigma_{ji}^{PE}(E)dE \quad (2)$$

The cross section for PE from level j to level i is,

$$\sigma_{ji}^{PE}(E) = \frac{\pi e^2}{m_e c} f_{ij}\phi(E) \quad (3)$$

where e and m_e are the electron charge and mass, c is the speed of light, f_{ji} is the oscillator strength of the line, and $\phi(E)$ is the line profile. Analogously, the rates R^{PI} for PI are calculated using the PI cross section and the same flux $F(E)$. $F(E)$ is taken to be a power law $[F_0(E)]$ absorbed along the ionization cone. The optical depth $\tau(E)$ is calculated taking both line (PE) and edge (PI) absorption into account. Since with the RGS, we measure the spectrum integrated over the entire ionization cone of NGC 1068, we use for Eqn. (1) the flux averaged over the optical path (perpendicular to our line of sight), i.e.,

$$F(E) = \frac{F_0(E)}{L}\int_0^L exp[-\tau(E)]dl = F_0(E)\frac{1 - exp[-\tau(E)]}{\tau(E)}. \quad (4)$$

For spatially resolved spectra obtained with the *CXO* gratings, a different flux needs to be used for each region.

Level-by-level calculations are carried out for each ion that appears in the spectrum. High lying levels with n-values up to 7 are included explicitly in the model. The contribution of higher levels, both via RR and via PE towards these levels, is included by extrapolation. The set of Eqns. (1) is solved by normalizing the sum of all level populations to unity. The power law normalization (or, alternatively, the electron density) is tuned to give the appropriate ionization balance between the two adjacent ions, which reproduces the observed RRC to line ratios. The line intensities in number of photons emitted per second and per cm^3 are finally given by $n_i^{+q}A_{ij}$. The elementary atomic quantities used in this work (namely the level energies, radiative decay rates, and photoionization cross sections), were obtained by means of the multi-configuration, relativistic HULLAC (Hebrew University Lawrence Livermore Atomic Code) computer package developed by Bar-Shalom et al. (2001).

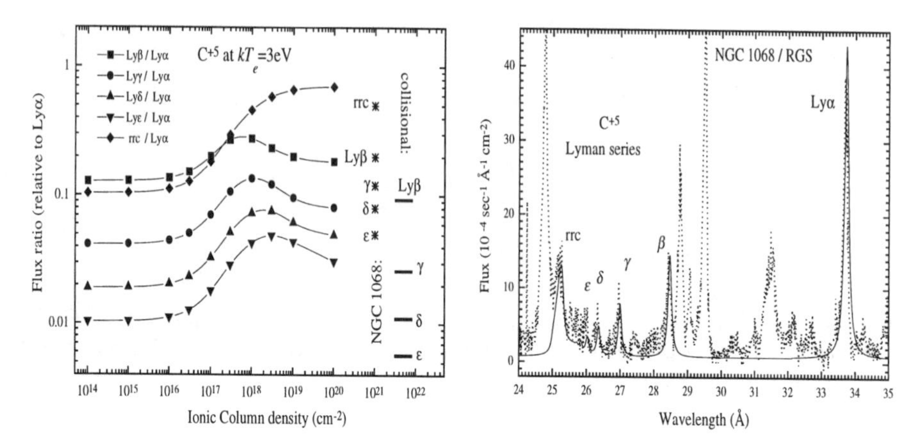

Figure 1. LEFT: Calculated intensity ratios of the Lyman series lines and RRC (circles) of C^{+5} to the Lyα intensity. The NGC 1068 values for these ratios, as obtained from the RGS spectrum, are also plotted (asterisks), as well as the values for collisional C^{+5} gas at 100 eV (horizontal bars). RIGHT: The NGC 1068 spectrum (dotted line) obtained by the RGS in the region of the C^{+5} lines and RRC compared with the present model (solid line) at 10^{18} cm^{-2}.

3. Signature of Photoexcitation in the High Rydberg Series Lines

Looking at a Rydberg series of lines, for instance the Lyman series, the ratios of high-n lines to that of (say) Lyα provide a probe for PE. At high ionic column density, beyond that in which the absorption of Lyα is saturated, the high-order lines can still be enhanced by PE resulting in the increase of their intensity ratio to Lyα. This effect is illustrated in the left panel of Figure 1 for C^{+5}. At the point where Lyα saturates ($\sim 10^{16}$ cm^{-2}), the ratios rise, eventually reaching values of up to a few times higher than the low column-density values. As the column density increases further, the other members of the series progressively saturate as well. The actual column densities for which all of these saturations occur depend on $\phi(E)$ (essentially the turbulent velocity, taken here to be 800 km s^{-1}), but the intensity ratio trend seen in Figure 1 is generic to the scenario for Seyfert 2 galaxies, in which photons from a power law source are reprocessed by means of PE and PI. Next to the theoretical ratio curves in the left panel of Figure 1, the ratios measured for NGC 1068 are shown (asterisks). These ratios are clearly in the high ionic column density regime at about 10^{18} cm^{-2}. On the right hand side of the left panel in Figure 1, the same ratios are indicated for collisional conditions where C^{+5} forms ($kT_e = 100$ eV). Clearly, these are lower than the ratios for photoionized conditions, even at their low column-density limit. Hence, an additional collisional component will only worsen the agreement of the model with the observed ratios.

Indeed, the current, rather simple model for 10^{18} cm^{-2} fits the measured spectrum of NGC 1068 fairly well, as can be readily seen in the right panel of

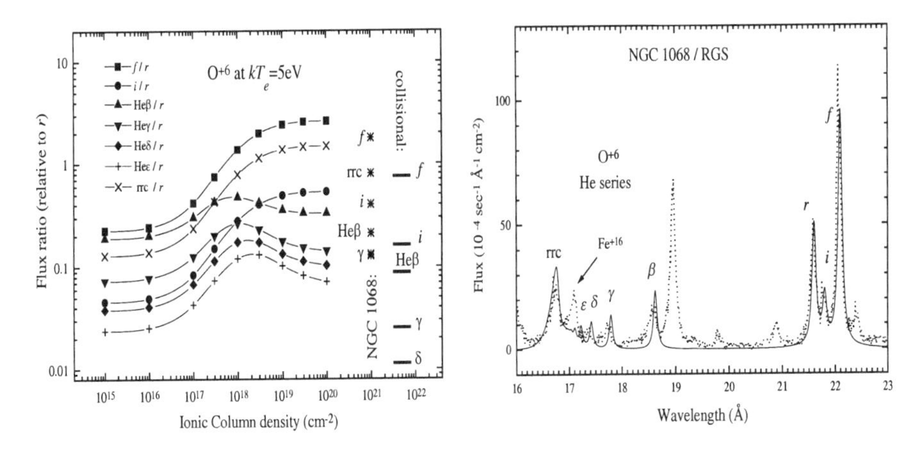

Figure 2. LEFT: Calculated intensity ratios of the Lyman series lines and RRC (circles) of O^{+6} to the Lyα intensity. The NGC 1068 values for these ratios, as obtained from the RGS spectrum, are also plotted (asterisks), as well as the values for collisional O^{+6} gas at 100 eV (horizontal bars). RIGHT: The NGC 1068 spectrum (dotted line) obtained by the RGS in the region of the O^{+6} lines and RRC compared with the present model (solid line) at 3 x 10^{18} cm^{-2}.

Figure 1. The small remaining discrepancies could be due to the approximate line profiles used here. The complete analysis of the NGC 1068 spectrum, which includes all of the observed ions as well as the corrections for the line profiles and outflow velocity shifts will be presented in Kinkhabwala et al. (in preparation). Here, the slight velocity shifts actually come in handy and assist the comparison between the data and the model.

4. He-like species

The He-like species feature, in addition to the leading 1s - np Rydberg series, the so-called forbidden (f) and intercombination (i) lines that together with the resonance (r) 1s - 2p line form the (n= 2 - 1; J = 1 - 0) He-like triplet. For O^{+6}, these lines fall at 22.097, 21.804, and 21.602 Å, respectively. Among these three, r has the highest oscillator-strength by far and, thus, is most affected by PE. The intensity ratios of the various O^{+6} lines to the r-line as a function of column density are plotted in the left panel of Figure 2. An effect similar to that for the H-like species is seen for the high order lines (β, γ, δ, etc.), where their intensity with respect to the r-line increases when the latter saturates, and reaching a plateau when they saturate as well. The line ratios for O^{+6} measured from the NGC 1068 spectrum are also plotted in the left panel of Figure 2 (asterisks). As for C^{+5}, the measured O^{+6} ratios are found to be higher than the collisional ratios, reducing the likelihood of hot collisional gas. The measured ratios of O^{+6} imply a column density of $\sim 3 \times 10^{18}$ cm^{-2}. The O^{+6} model at this column density is plotted versus the data in the right panel of Figure 2 and, evidently, the agreement is fairly good.

5. Conclusions

We have shown that a physically consistent model for reprocessing of the nuclear continuum and subsequent emission from Seyfert 2 galaxies, which includes all of the important atomic processes and particularly PE, reproduces the high-resolution RGS X-ray spectrum of NGC1068 very well. A more complete analysis of this spectrum is soon to appear in Kinkhabwala et al. (in preparation). Moreover, the traditional models, which did not include PE, clearly produce erroneous spectra and line ratios. This, in turn, could possibly lead to wrong astrophysical interpretation, such as the invoking of hot collisional gas, which to the best of our understanding is not evident from the X-rays emitted by Seyfert 2 galaxies.

References

Bar-Shalom, A., Klapisch, M., & Oreg, J. 2001, J. Quant. Spectr. Radiat. Transfer, 71, 169

Behar, E., Cottam. J., & Kahn, S.M. 2001, ApJ, 548, 966

Doron, R., Behar, E., Fraenkel, M., Mandelbaum, P., Zigler, A., Schwob, J.L., Faenov, A. Ya., Pikuz, T.A. 1998, Phys.Rev.A58, 1859

Liedahl, D.A., Kahn, S.M., Osterheld, A.L., Goldstein, W.H. 1990, ApJ, 350, L37

Ogle, P.M., Marshall, H.L., Lee, J.C., Canizares, C.R. 2000, ApJ, 545, L81

Ogle, P.M., Canizares, C.R., Duey, D., Lee, J.C., Marshall, H.L. 2002, in Mass Outflow in Active Galactic Nuclei: New Perspectives, eds. D.M. Crenshaw, S.B. Kraemer, & I.M. George (San Francisco: ASP), p. 13

Porquet, D., & Dubau, J. 2000, A&AS, 143, 495

Sako, M., Liedahl, D.A., Kahn, S.M., Paerels, F. 1999, ApJ, 525, 921

Sako et al. 2000, ApJ, 543, L115

Sambruna, R.M., Netzer, H., Kaspi, S., Brandt, W.N., Chartas, G., Garmire, G.P., Nousek, J.A., & Weaver, K.A. 2001, ApJ, 546, L13

Mass Outflow in Active Galactic Nuclei: New Perspectives
ASP Conference Series, Vol. 255, 2002
D.M. Crenshaw, S.B. Kraemer, and I.M. George

CXO HETGS Observations of Seyfert 1 Galaxies

Ian M George[1] & T. Jane Turner[1]

Joint Center for Astrophysics, Department of Physics, University of Maryland, Baltimore County, 1000 Hilltop Circle, Baltimore, MD 21250, USA

Tahir Yaqoob[1]

Department of Physics & Astronomy, Johns Hopkins University, 3400 North Charles Street, Baltimore, MD 21218, USA

Abstract. We present spectra from NGC 5548 and NGC 3227 obtained using the High-Energy Transmission Grating Spectrometer (HETGS) on board the *Chandra X-ray Observatory* (*CXO*). Spectra from two "control" sources, 3C 273 and NGC 3783, are also shown for direct comparison. Here, we concentrate on the strongest, narrow absorption features expected from He– & H–like ions of the abundant elements in the 0.7–3 keV band. In all cases the data from the Medium- and High-Energy Gratings (MEG & HEG) are shown separately to illustrate their relative quality, and to serve as an independent check on the reality of any features.

Our results confirm and illustrate the ionized outflow reported in NGC 5548 by Kaastra et al., and that in NGC 3783 by Kaspi et al. We also confirm a lack of any such material in the case of 3C 273. NGC 3227 was observed during a low, heavily-absorbed state, preventing any definitive statements being made at this time.

1. Introduction

With the recent launch of the *CXO* and *XMM-Newton*, X-ray astronomy has finally entered an era where detailed spectroscopy can be performed. Publications of results obtained using grating observations of Active Galactic Nuclei (AGN) are already appearing in the literature (e.g. Kaastra et al. 2000; Kaspi et al. 2000; Branduardi-Raymont et al 2001). The main purpose of this poster is to show a selection of *CXO* HETGS data from AGN. We concentrate on the spectra close to the $1s \leftrightarrow 2s$ resonances from abundant elements, and show the data "warts and all" to illustrate the quality (and limitations) of the data currently available, primarily for readers not yet familiar with HETGS data.

---

[1]also Laboratory for High Energy Astrophysics, Code 662, NASA/Goddard Space Flight Center, Greenbelt, MD 20771, USA.

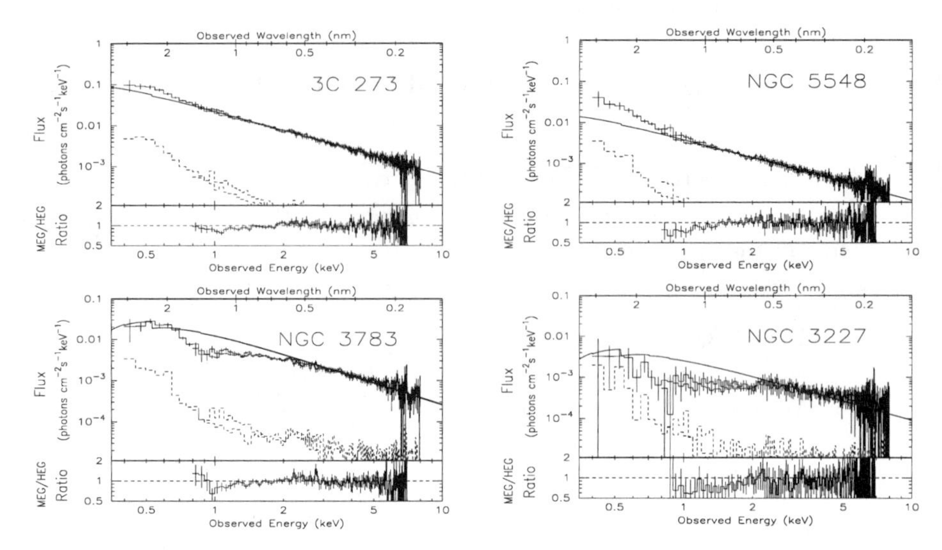

Figure 1. The upper panels show (separately) the MEG and HEG 1st-order spectra from the source extraction cell, and the (normalized) background spectra (dashed). In both cases 50 eV bins are used to illustrate the overall form (but which obvious smooth out any narrow features). The spectra have been corrected for the response of the detector, but not Galactic absorption. The lower panels show the ratio of the MEG/HEG spectra. This ratio is consistently less than unity in the ~0.9–2 keV band, indicating some remaining calibration issues.

2. The Observations & Data Processing

The data were obtained either through our own observations (NGC 3227 & NGC 5548) or from the *Chandra* Public Archive (3C 273 & NGC 3783). These data were then corrected for all known errors and deficiencies in the pipeline reprocessing at the time of writing using the `CIAO` (v2.1.1) package and the *Chandra* `CALDB` (v2.3). Specifically we corrected for the incorrect order-sorting tables used in the pipeline reprocessing of data prior to 2000 Dec 07, and the bug in the `asp_apply_sim` task. Only events with "*ASCA* Grades" 0, 2, 3, 4, and 6 are considered here. In all the analysis presented here, the source data were extracted from within $\theta_{cd} \pm 5 \times 10^{-4}$ degrees of the centroids in the cross dispersion direction of the Medium- and High-Energy Gratings (MEG and HEG). Background data were extracted from parallel strips with $|\ \theta_{cd}\ | = 5\text{–}25 \times 10^{-4}$ degrees.

3. The Form of the Overall X-ray Continua

Flux-calibrated spectra (but not corrected for Galactic absorption, $N_{\rm Gal}$(HI), along the line of sight to each source) are shown in the upper panels of Figure 1. For clarity and easy comparison with non-dispersed CCD data sets, the spectra are plotted using 50 eV bins. For illustration, in each case a power law (attenuated by $N_{\rm Gal}$(HI)) is overlaid. Throughout this work we adopt $H_0 = 75$ km s^{-1} Mpc^{-1}.

The power law shown for **3C 273** ($z = 0.15834$; $N_{\rm Gal}$(HI) $= 1.8\times10^{20}$ cm^{-2}) was derived from simultaneously fitting the MEG & HEG 1$^{\rm st}$-order spectra from the "line-free-zones" in the 2–5 keV band, hereafter LFZ(2–5 keV). These LFZs were chosen so as to be free of absorption and emission lines from cosmically abundant elements. The boundaries of these energy-ranges, or were further restricted to allow for intrinsic velocity shifts of $\pm$500km s^{-1} in any features. A bin size of 1 pm was used for both the MEG & HEG within each LFZ, but the data were then rebinned such that there were at least 30 photons in each new bin. From Figure 1 it is evident that the power law derived from the LFZ(2–5 keV), $\Gamma = 1.59^{+0.08}_{-0.08}$, extrapolates well down to ~1 keV, below which there is an excess of emission. Thus, here we use 3C 273 a bright "pure continuum" control source. The observed flux in the 2–10 keV band is F(2–10 keV) $= 1.1^{+0.1}_{-0.1}\times10^{-10}$ erg cm^{-2} s^{-1}, and the ($N_{\rm Gal}$–corrected) luminosity in the rest-frame is L(2–10 keV) $= 5.4^{+0.5}_{-0.5}\times10^{45}$ erg s^{-1}. We find the ratio of the normalizations derived for the MEG & HEG from the analysis of LFZ(2–5 keV) to be $R_{\rm M/H} = 1.03^{+0.04}_{-0.03}$. However in the lower panels of Figure 1 we show the ratio of the MEG/HEG spectra as a function of energy. This ratio is consistently less than unity in the ~0.9–2 keV band, indicating some cross-calibration issues remain unsolved.

NGC 3783 ($z = 0.00976$, $N_{\rm Gal}$(HI) $= 8.7\times10^{20}$ cm^{-2}) is well known to contain a large column of highly-ionized, circumnuclear material (e.g. see Kaspi 2002). The power law shown in Figure 1 is that ($\Gamma = 1.77$, L(2–10 keV) $= 9.9\times10^{42}$ erg s^{-1}) derived for the underlying continuum from Kaspi et al (2001). It is worth noting that the strength of the attenuation by the ionized material in NGC 3783 is sufficiently large to cause the application of a simple power-law model to the LFZ(2–5 keV) to give misleading results for the underlying continuum (even though the statistically acceptable fits can be obtained). It should be stressed also that the bulk of the deficit of photons in the 0.6–2 keV band is due to bound-free absorption in the ionized material, rather than a the blending of the large number of (relatively narrow) resonant absorption lines in this object. The normalizations derived from the MEG & HEG data within LFZ(2–5 keV) is $R_{\rm M/H} = 1.10^{+0.03}_{-0.03}$, primarily due to slight descrepancies in the 2–3 keV band. Here a number of features in the spectrum of NGC 3783 are shown for comparison with the other sources[1].

Attenuation by ionized material is known to be lower (but present – see below) in the case of **NGC 5548** ($z = 0.01676$, $N_{\rm Gal}$(HI) $= 1.6\times10^{20}$ cm^{-2}). It can be seen that a power law determined from LFZ(2–5 keV), with $\Gamma = 1.50^{+0.10}_{-0.10}$ (F(2–10 keV) $= 2.1^{+0.3}_{-0.3}\times10^{-11}$ erg cm^{-2} s^{-1}; L(2–10 keV) $= 1.2^{+0.1}_{-0.1}\times10^{43}$ erg s^{-1}) extrapolates well down to ~1 keV, below which there is an excess emission. For these data we find $R_{\rm M/H} = 1.05^{+0.06}_{-0.06}$ within LFZ(2–5 keV).

---

[1] It should be stressed that the analysis reported here for NGC 3783 is much less rigorous than reported by Kaspi et al. (2000,01).

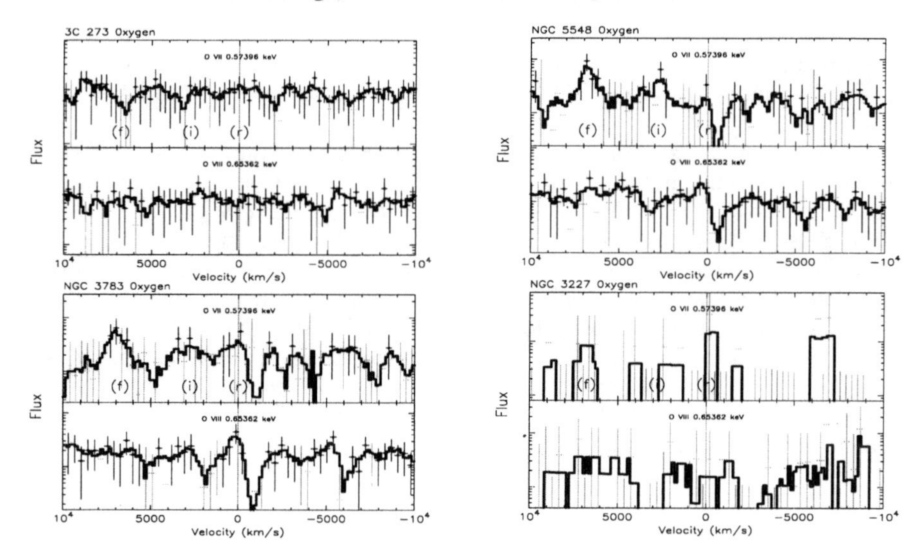

Figure 2. The upper panels are centered on the O VIII $1s \leftrightarrow 2s$ resonance. The locations of the forbidden (f) and intercombination (i) are also shown. The lower panels are centered on the O VIII $1s \leftrightarrow 2s$ resonance. The MEG data (only) are shown, using 3 pm bins (i.e. 3× the "1σ" resolution of the MEG). The histograms use 1 pm bins, but smoothed with a running mean over 5 pm. NOTE: blue/outflow is to the right, red/infall to the left; the y-axis is logarithmic.

NGC 3227 ($z = 0.00386$, $N_{\rm Gal}$(HI) $= 2.1 \times 10^{20}$ cm^{-2}) can be well represented by a heavily absorbed power law, similar to that seen by *ASCA* during 1995 (George et al., in preparation). The power law shown is that of the underlying continuum ($\Gamma = 1.53$, L(2–10 keV) $= 4.7 \times 10^{42}$ erg s^{-1}) as derived by George et al. (1998). We find the ratio of the normalizations derived from the MEG & HEG within LFZ(2–5 keV) to be $R = 0.95^{+0.12}_{-0.13}$. As for NGC 3783, the strength of the attenuation in NGC 3227 is sufficiently large to cause the application of a simple power-law model to the LFZ(2–5 keV) to give misleading results for the underlying continuum.

4. The He- and H-like absorption lines

Due to limitations on space, in Figures 2 & 3 we only[2] show spectra in velocity-space for the $n = 1 \leftrightarrow 2$ resonances of He- and H-like Oxygen and Neon (upper and lower panels respectively). Specifically, in the upper panels zero velocity is centered on the $1s2p\ ^1P_1 \leftrightarrow 1s^2$ in the rest-frame of the source. The locations of the "forbidden" ($1s2s\ ^3S_1 \leftrightarrow 1s^2$) and "intercombination" ($1s2p\ ^3P_{1,2} \leftrightarrow 1s^2$) lines of the He-like species are also shown. In the lower panels, zero velocity is centered on the $2p \leftrightarrow 1s$ transition in the rest-frame of the source.

[2]Similar plots for Magnesium, Silicon & Sulphur were shown at the workshop.

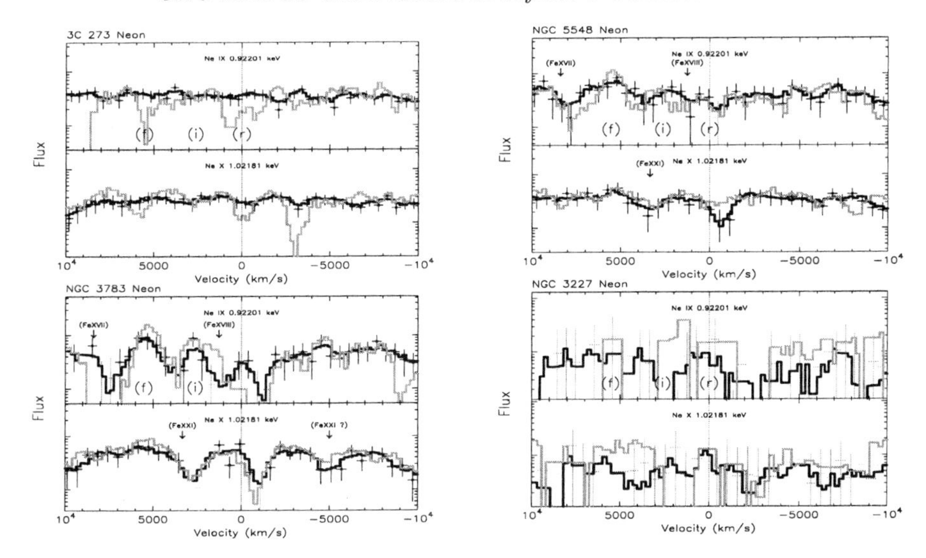

Figure 3. As for Fig.2, but for Ne IX $1s2p\ ^1P_1 \leftrightarrow 1s^2$ (upper panels), and Ne X $2p \leftrightarrow 1s$ (lower panels). In this case the HEG data are also shown (fainter lines).

Both figures show no features are reliably detected in 3C 273. The causes of the absorption-like features apparent in the HEG spectra (only) in the Ne IX and Ne X regions are under investigation. This contrasts with NGC 3783, for which the ionized material reported by Kaspi et al. (2001) is clearly evident in both the MEG and HEG data. All four resonant transitions are seen in absorption, and are offset from the systemic velocity by $\sim$few$\times 10^2$ km s^{-1}. The O VII and Ne IX forbidden and intercombination emission lines are also detected, and approximately centered on the systemic velocity. Close inspection also reveals emission from the resonant transitions in NGC 3783 centered at the systemic velocity (e.g. O VIII– see Figure 4). This is confirmed by detailed analysis, with the emission partly "filling in" the absorption lines. A number of other absorption features (e.g. those due to Fe XVII–XXII) are also evident in the regions of the NGC 3783 spectra plotted.

The data for NGC 5548 are somewhat similar to NGC 3783, except that all the features are weaker. The most noticeable features are the resonant transitions seen in absorption (e.g. O VII, O VIII & Ne X), again with a clear offset from the systemic velocity ($\sim$few$\times 10^2$ km s^{-1}). The O VII and Ne IX forbidden and intercombination emission lines are also detected, again approximately centered on the systemic velocity. Our results are in agreement with the results from the grating data presented by Kaastra et al. (2000). It should be noted that even though the NGC 5548 has a smaller column density of ionized material, the Ne X absorption line is most likely saturated (Figure 4). It does not look this way in Figure 3 simply due to the smearing by the line-response function of the instrument.

Finally, in the case of NGC 3227, the data are somewhat ambiguous (due to the lack of signal to noise). A detailed analysis of NGC 3227 is on-going.

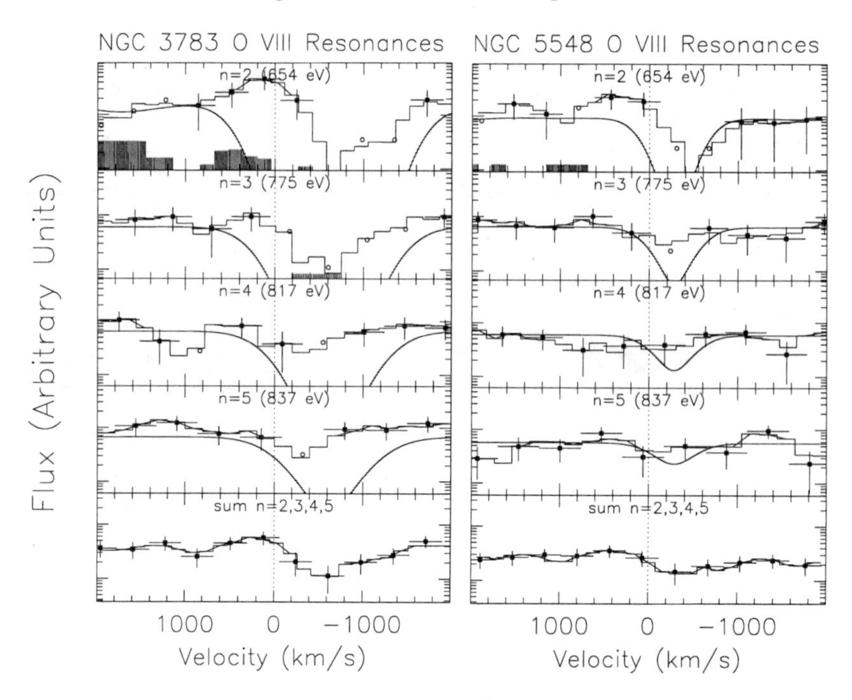

Figure 4. The O VIII resonant series for NGC 3783 & NGC 5548. The top four panels show the data centered on the $1s \leftrightarrow ns$ line for $n = 2, 3, 4, 5$. The lower panel shows the sum (in velocity-space). For NGC 3783, the model assumes N(O VIII $= 6.75 \times 10^{18}$ cm^{-2}, $\sigma_v = 358$ km s^{-1}, and $v = -610$ km s^{-1}. This is seen to be a slight overestimate of N(O VIII) and/or σ_v. For NGC 5548, the model assumes a column density N(O VIII) $= 1.26 \times 10^{18}$ cm^{-2}, a velocity dispersion of $\sigma_v = 140$ km s^{-1}, and an offset velocity $v = -280$ km s^{-1} (from Kaastra et al 2000). This is seen to be a fair representation of the data. Emission, especially in the $n = 2$ panels, is clearly evident centered on the systemic velocity.

References

Branduardi-Raymont, G., et al. 2001, A&A, 365, L140

George, I.M., et al. 1998, ApJ, 509, 146

Kaastra, J. S., Mewe, R., Liedahl, D. A., Komossa, S., Brinkman, A. C. 2000, A&A, 354, L83

Kaspi, S. 2002, in Mass Outflow in Active Galactic Nuclei: New Perspectives, eds. D.M. Crenshaw, S.B. Kraemer, & I.M. George (San Francisco: ASP), p. 7.

Kaspi, S., et al. 2000, ApJ, 535, L17

Kaspi, S., 2001, ApJ, in press

Mass Outflow in Active Galactic Nuclei: New Perspectives
ASP Conference Series, Vol. 255, 2002
D.M. Crenshaw, S.B. Kraemer, and I.M. George

Ubiquitous Column-Density Variability in Seyfert 2 Galaxies

G. Risaliti, M. Elvis, F. Nicastro

Harvard-Smithsonian Center for Astrophysics, 60 Garden Street, Cambridge, MA 02138, USA

Abstract. We present a study of X-ray column density variability in Seyfert 2 galaxies. We show that variations in N_H are observed in almost all the objects with multiple hard X-ray observations. Variation timescales (as short as a few months in several cases) are not in agreement with the standard scenario of a parsec-scale toroidal absorber. We propose that the X-ray absorber in Seyfert galaxies is located much nearer to the center than previously assumed, on the Broad Line Region Scale. An extension of the model by Elvis (2000) can explain the observed variability. We also show preliminary results of N_H variability search inside single X-ray observations, which suggest that variations can occour on timescales of a few $\times 10^4$ s.

1. Introduction

Strong obscuration is observed in the hard (2-10 keV) X-ray spectra of Type 2 Active Galactic Nuclei (AGN), where a photoelectric cut-off at energies E > 1-2 keV indicates the presence of a column density of absorbing gas $N_H > 10^{22}$ cm^{-2}.

The simplest geometry for this gas surrounding the nucleus is that of a torus covering $\sim$ 80% of the solid angle (in order to reproduce the 4:1 observed ratio between unobscured and obscured AGN, Maiolino & Rieke 1995). One of the unsolved questions about this putative torus is its typical dimensions. Detailed models have been proposed for both a 100 pc-scale torus (Granato et al. 1997) and for a parsec-scale one (Pier & Krolik 1992). Both models are supported by several pieces of observational evidence, so it is likely that both the components could be present in AGN. Here we investigate the variability of the X-ray absorbing column density in X-ray defined Seyfert 2 galaxies having column densities higher than $\sim 10^{22}$ cm^{-2}, but less than 10^{24} cm^{-2} (in order to have a measurement of the photoelectric cut-off in the 2-10 keV band). We collected all the data available in the literature for Seyfert 2 galaxies and we complemented them with the analysis of unpublished data in the *ASCA* and *BeppoSAX* public archives. In the following Sections we show the results and we show that they can be explained within a consistent physical picture only assuming that the absorber is located at a distance from the center typical of the Broad Emission Line Region (BELR).

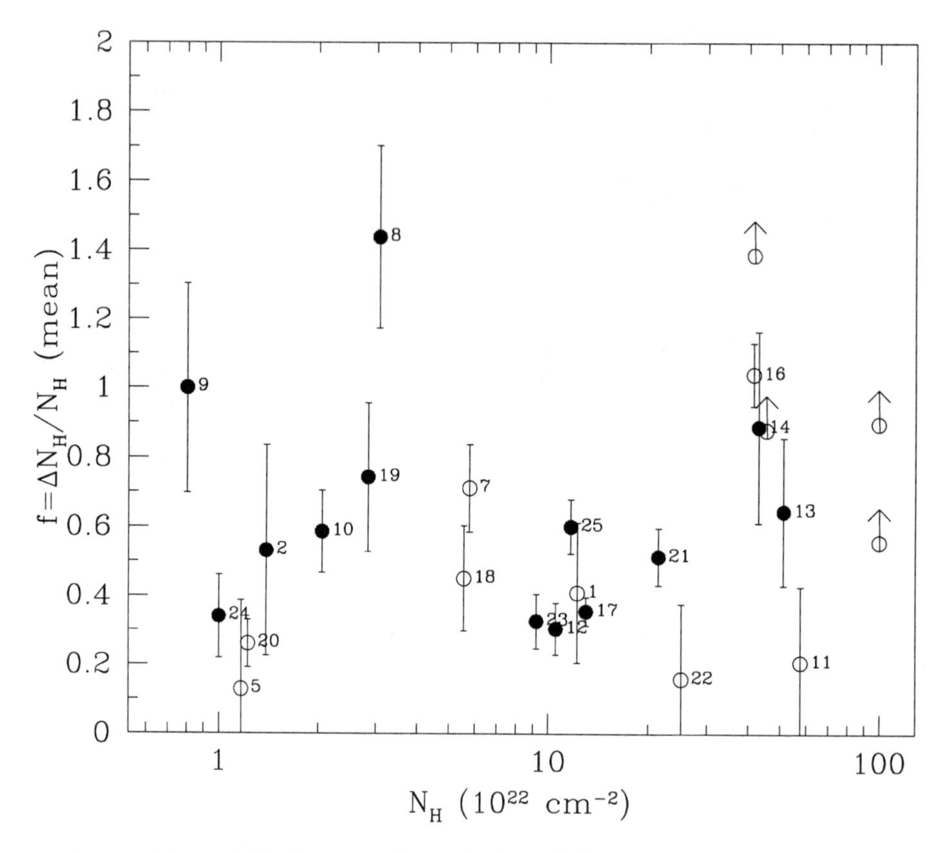

Figure 1. Ratio between the variation of N_H and the mean N_H for all the Seyfert 2s with multiple hard X-ray observations.

2. Data

We found that a sample of 25 sources were observed at least twice in the hard X-rays. Out of these 25 sources, 22 show N_H variability on timescales from a few months to years (Figure 1). In Table 1 we summarize the results: for each source we report the time interval between the two closest varations of N_H, the measured variation relative to the average, and the number of hard X-ray observations available for each source.

The full detailed analysis of the 139 observations of these 25 sources is described in Risaliti et al. (2001).

3. Analysis of the Results

There are two physical reasons that can explain the variability of the absorbing column density: variation in the ionization state of the absorber, due to variations in the ionizing continuum, and variations of the amount of gas along the line of sight. We ruled out the first possibility, since the N_H variations are not

Table 1. Data on N_H variability of our sample.

Name	Time interval	variation (%)	# Obs.
CENTAURUS A	2 months	46%	18
NGC 4258	5 months	10%	6
NGC 2110	6 months	15%	5
NGC 4941	6 months	> 81%	2
NGC 7582	6 months	122%	10
NGC 2992	6.5 months	40%	12
ESO 103-G35	6.5 months	19%	13
NGC 4507	11 months	25%	5
NGC 4388	1 year	50%	4
MCG-5-23-16	1.1 years	94%	11
NGC 1386	2 years	>112%	2
NGC 1365	2.5 years	> 85%	2
IRAS 13197-164	3 years	44%	2
NGC 1808	3.3 years	200%	2
NGC 5506	3.5 years	28%	11
IRAS 18325-5926	3.5 years	24%	3
NGC 7172	3.5 years	31%	6
NGC 5252	4 years	36%	2
NGC 7314	4.5 years	31%	4
IRAS 05189-2524	4.6 years	64%	2
NGC 256a	5 years	94%	8
MKN 348	8 years	20%	2
IRAS 04575-7537	–	–	2
NGC 3081	–	–	2
IC 5063	–	–	3

correlated with the flux variations (see Risaliti et al. 2001 for details). The second scenario — motions in a clumpy medium — is the only that can account for the observations. However, the results on the ubiquity of N_H variability in Seyfert 2 galaxies, together with the short (from 2 months to a few years) variability timescales, pose severe problems to the standard torus model. We can idealize the situation by assuming the typical timescale of variation, t, to be the crossing time of a discrete cloud across the line of sight. Assuming that the absorption is due to spherical clouds moving with Keplerian velocities, the distance from the central black hole of mass $M_\bullet$ is given by:

$$R \sim 3 \times 10^{16} \frac{M_\bullet}{10^9 M_\odot} \left(\frac{\rho}{10^6 \mathrm{cm}^{-3}}\right)^2 \left(\frac{t}{5 \mathrm{\ Ms}}\right)^2 \left(\frac{N_H}{10^{22} \mathrm{cm}^{-2}}\right)^{-2} \mathrm{cm} \qquad (1)$$

where ρ is the density of the cloud. The black hole mass and the cloud density have been normalized to extreme values for a putative torus in order to obtain the greatest distance.

The distance we obtained is typical for the BELR. Even if many parameters are poorly constrained, it is not possible to obtain a value of $R > 10^{18}$ cm within

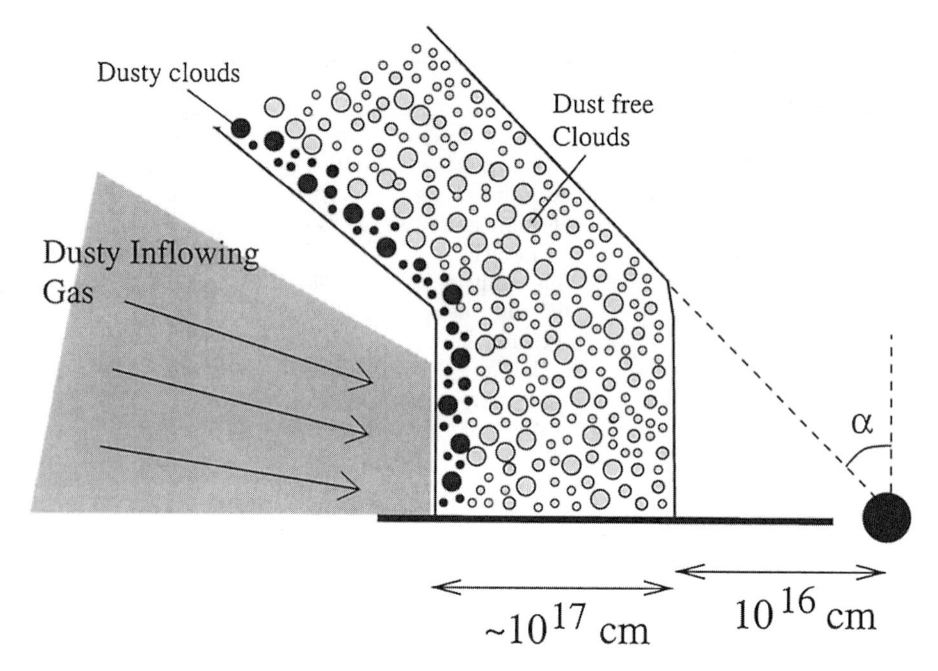

Figure 2. A simple model, derived from Elvis (2000), which explains the X-ray absorption properties of Seyfert 2 galaxies. The X-ray central emission is absorbed by the Broad Emission Line Clouds. The column density variability timescale is the average crossing time of a cloud along the line of sight.

a consistent physical scenario. Therefore, the parsec-scale torus model is not able to expain our data.

An alternative scenario, within the standard AGN model (Antonucci 1993), is that the X-ray absorber is located in the BELR, much nearer to the central black hole than the standard torus. If we assume that the broad line clouds are responsible for the absorption in the X-rays, we can find a consistent combination of the parameters in the previous equations, with higher cloud densities ($\rho \approx 10^9$ cm^{-2}) and shorter variability timescales ($t \approx 3$ days). We note that such timescales are not ruled out by our data, since we cannot investigate variations shorter than the time interval between two observations of the same source.

An absorber which is very compact (as requested by our data) and axisymmetric (as required by the arguments supporting the unified schemes) can be easily obtained extending the wind model by Elvis (2000, 02). In this model most of the phenomenology of type 1 AGNs is explained through a two-phase wind arising from the accretion disc. The cold phase of the wind is formed by the Broad Emission Line Clouds (BELC). A simple extension of this model could be that in type 2 AGNs the wind is thicker, and the BELC cover all lines of sight through the wind, as illustrated in Figure 2. The external part of the wind can well be cold enough for dust to survive, therefore this absorber can also explain the optical properties of Seyfert 2 galaxies. Interestingly, the average dust-to-gas

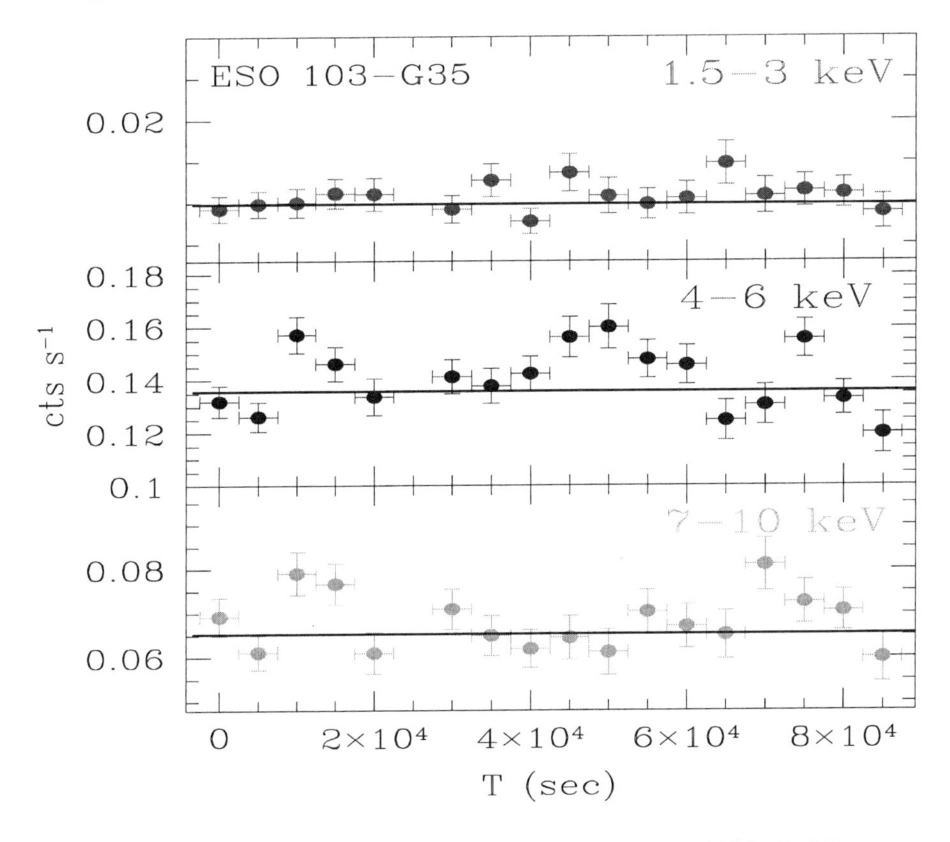

Figure 3. X-ray light curves from a *BeppoSAX* observation of ESO 103-G35.

ratio predicted by this model, assuming that the external part of the wind has a standard ISM composition, is lower than Galactic, in agreement with recent findings (Maiolino et al. 2001, Risaliti et al. 2001)

4. Conclusions and Future Work

Variability of X-ray absorbing column density appears to be an ubiquitous property in Seyfert 2 galaxies. Variation timescales can be as short as a few months. We have shown that these data rule out an X-ray absorber on a parsec scale. Instead, we propose that the Broad Emission Line Clouds, in an axisymmetric distribution, like in the model of Elvis (2000), are responsible for X-ray absorption. If this is the case, variations in N_H are expected on timescales of days. Our variability study is limited by the shortest time between two observations of the same source. However, our work can be significantly improved looking for N_H variations inside single, long observations of the brightest sources in our sample. This work is still in progress, however we can show some preliminary, very promising results. In Figure 3 we plot three light curves for three different energy intervals from a *BeppoSAX* observation of the Seyfert 2 galaxy ESO 103-

G35. The intervals are chosen in order to have the photoelectric cut-off at the separation energy between the first two lightcurves. Therefore, in the case of variations of N_H, we expect that one of the first two light curves to vary, while the other two remain constant (the third energy interval is little affected by small N_H variations). Instead, if we have a flux variation, we expect both the second and third light curves to vary, while the first remains constant, since the emission at energies lower that the cutoff is mainly due to an extended/reprocessed component. It is clear from Figure 3 that in the interval between 40 and 60 ks the light curves vary in a way suggesting N_H variability. We extracted two spectra, one in the 40—60 ks interval, and the other in the remaining time intervals. We performed a careful fit of these spectra, and concluded that the N_H measurements differ by $\sim 3 \times 10^{22}$ cm^{-2}, at a 5σ level of confidence. This N_H variation on a timescale of ~20 ks strongly suggests that the X-ray absorber is very close to the central black hole, in agreement with the model we have proposed.

Acknowledgments. This work was supported in part by NASA grant NAG5-4808.

References

Antonucci R. R. 1993, ARA&A, 31, 473

Elvis M. 2000, ApJ, 545, 63

Elvis M. 2002, in Mass Outflow in Active Galactic Nuclei: New Perspectives, eds. D.M. Crenshaw, S.B. Kraemer, & I.M. George (San Francisco: ASP), p. 303

Granato, G., et al., 1997, ApJ, 486, 147

Maiolino R., & Rieke, G.H. 1995, ApJ, 454, 95

Maiolino R., et al. 2001, A&A, 365, 28

Pier, E.A., & Krolik, J.H. 1992, ApJ, 401, 99

Risaliti G., et al. 2001, A&A, 371, 37

Risaliti G., Elvis M. & Nicastro F. 2001, ApJ, submitted

Mass Outflow in Active Galactic Nuclei: New Perspectives
ASP Conference Series, Vol. 255, 2002
D.M. Crenshaw, S.B. Kraemer, and I.M. George

X-Ray Absorption Associated with High-Velocity UV Absorbers

Bassem M. Sabra & Fred Hamann

Dept. of Astronomy, University of Florida, Gainesville, FL 32611, USA

Joseph C. Shields

Dept. of Physics & Astronomy, Ohio University, Athens, OH 45701, USA

Ian M. George[1]

Joint Center for Astrophysics, Department of Physics, University of Maryland, Baltimore County, 1000 Hilltop Circle, Baltimore, MD 21250, USA

Buell Jannuzi

NOAO, 950 North Cherry Ave., Tuscon, AZ 85719, USA

Abstract. We present *Chandra X-ray Observatory* (*CXO*) observations of two radio-quiet QSOs, PG 2302+029 and PG 1254+047. PG 2302+029 has an ultra high-velocity UV absorption system (~ -56000 km s^{-1}), while PG 1254+047 is a Broad Absorption Line (BAL) QSO with detached troughs. Both objects are X-ray weak, consistent with the known correlation between $\alpha_{\rm ox}$ and the strength of the UV absorption lines. The data suggest that there is evidence that both objects are intrinsically weak X-ray sources, in addition to being heavily X-ray absorbed. The X-ray absorption column densities are $N_{\rm H} > 10^{22}$ cm^{-2} for neutral gas and the intrinsic emission spectra have $\alpha_{ox} > 2$. The data are fit best by including ionized (rather than neutral) absorbers, with column densities $N_{\rm H}^{PG2302} > 2.98 \times 10^{22}$ cm^{-2} and $N_{\rm H}^{PG1254} > 17.3 \times 10^{22}$ cm^{-2}. The degrees of ionization are consistent with the UV lines, as are the total column densities if the strongest lines are saturated.

1. Introduction

Almost all quasi-stellar objects (QSOs) are X-ray sources. The reprocessing of X-rays by matter along the line of sight imprints informative features on the observed spectrum. We are examining the X-ray properties of QSOs with intrinsic UV absorption lines. Here we discuss *CXO* observations of the QSOs

---

[1] also Laboratory for High Energy Astrophysics, Code 662, NASA/Goddard Space Flight Center, Greenbelt, MD 20771, USA

PG 2302+029 and PG 1254+047. The BAL QSO PG 1254+047, displays detached BALs (FWHM $\sim 10^4$ km s^{-1} centered at -2×10^4 km s^{-1}) in the UV (Hamann 1998). PG 2302+029 is peculiar for its ultra-high velocity ($\sim -56 \times 10^3$ km s^{-1}) UV absorption lines with FWHM $\sim 4 \times 10^3$ km s^{-1} (Jannuzi et al. 1996). The intrinsic nature of the absorber in PG 2302+029 was confirmed recently by line variability (Jannuzi et al., in preparation). Both objects are characterized by being faint X-ray sources. Our aim is to determine the properties of the X-ray spectrum, search for any signs of absorption, and define the relationship between the UV/X-ray absorbers. If the X-ray and UV absorbers are the same, then the -56×10^3 km s^{-1} velocity shift of the UV lines in PG 2302+029 **potentially is** resolvable with the Advanced CCD Imaging Spectrometer (ACIS) on *CXO*, given sharp features **and** adequate signal-to-noise ratio.

The absorption from this type of objects originates in an outflow from the central engine. Determining the relation between the gases producing the X-ray and UV absorption has profound implications on the physics of wind formation and acceleration (e.g., Mathur et al. 1995; Murray et al. 1995). Standard analysis, using the absorption line troughs to derive the optical depths and column densities, typically implies the total column densities are ~ 100 times lower than the X-ray absorbing columns. However, the discrepancy could be alleviated if the UV lines are more optically thick than they appear, e.g. partial covering fills the troughs thus hidding larger column densities (Hamann 1998).

2. Observations

PG 2302+029 and PG 1254+047 were observed by *CXO* using ACIS on 2000 January 7 and 2000 May 29, respectively. The most recent (2000 November 2 and 2001 February 29, respectively) re-processed data were used. No filtering for high background or bad aspect times was neccessary. Data extraction and calibration was performed using the `CIAO` (v1.4) software package. The `XSPEC` package was used for rebinning and spectral analysis. We created the response matrix and ancillary response files using appropriate calibration data.

We extracted the source counts from circular regions with radii of 5″. Background regions were annuli with radii of 10″ to 20″. We obtained a total of 391 ± 21 counts for PG 2302+029 and 47 ± 8 counts for PG 1254+047. Table 1 lists pertinent information about the two objects. The spectra were binned to have at least 30 counts/bin (10 counts/bin for PG 1254+047). The spectral analysis discussed in this contribution includes energy bins below 0.5 keV. It has since come to our attention that these bins may suffer from calibration problems;

Table 1. PG 2302+029 and PG 1254+047

Object	Obs. Date	Exp. (ksec)	$^\dagger N_H^{Gal.}$ (cm^{-2})	$^\ddagger z_{em}$	z_{abs}	B (mag)
PG 2302	2000-01-07	48	5×10^{20}	1.044	0.695	16.30
PG 1254	2000-05-29	36	2×10^{20}	1.024	0.870	16.15

NOTES – † Lockman & Savage (1995). ‡ NASA Extragalactic Database.

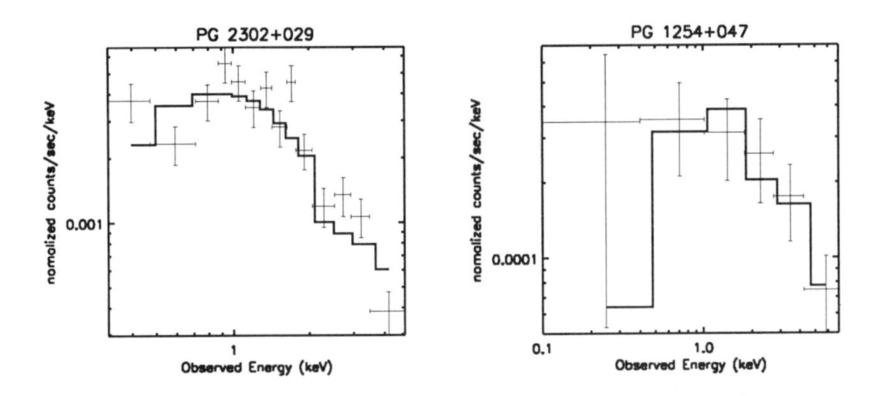

Figure 1. Galactic Absorption only, free normalization and Γ. PG 2303+029: $\phi_E(1 \text{ keV}) = 6.88 \times 10^{-6}$ photon s^{-1} cm^{-2} keV^{-1}, $\Gamma = 1.071$, $\chi^2_\nu = 2.57$ for 12 degrees of freedon (d.o.f.). PG 1254+047: $\phi_E(1 \text{ keV}) = 6.63 \times 10^{-7}$ photon s^{-1} cm^{-2} keV^{-1}, $\Gamma = 0.3787$, $\chi^2_\nu = 0.45$ for 4 d.o.f.

see Sabra & Hamann (in preparation) and Sabra et al. (in preparation) for a treatment where energy bins only above 0.5 keV are included.

3. Data Analysis and Results

Our procedure for spectral fitting is the following. We start by fitting a power law continuum ($\phi_E \propto E^{-\Gamma}$) absorbed only by the appropriate Galactic column density (Lockman & Savage 1995). Both the normalization of this continuum and its X-ray photon index, Γ, are left as free parameters. The results are shown in Figure 1. The slopes are rather flat for QSOs, where usually $1.3 < \Gamma < 2.3$ (e.g., Laor et al. 1997; Reeves et al. 1997). The X-ray fluxes are low, leading to the steep α_{ox} ($\alpha_{\text{ox}} = 2.1$ for PG2302+029; 2.5 for PG 1254+047). These values are consistent with the correlation between α_{ox} and the equivalent width (EW) of C IV $\lambda\lambda 1548, 1550$ shown in Figure 2. The correlation is indicative of intrinsic absorption: a large EW(C IV) results from an absorber that, in turn, is accompanied by an X-ray absorber, thus steepening α_{ox}.

To test for absorption, we adopt a "normal" QSO continuum, specified by $\alpha_{\text{ox}} = 1.6$ and $\Gamma = 1.9$ (Laor et al. 1997), attenuated through a neutral absorber. The choice of α_{ox} determines the normalization of the powerlaw at 2 keV/$(1+z)$, $\phi_E(1 \text{ keV}) \propto f_\nu^{obs}(1 \text{ keV})$, which we derive in the following way. We first calculate the rest-frame $f_\nu(2500 \text{ Å})$ from the B-magnitude, including the appropriate Galactic de-reddening and the k-correction (see Green 1996). The rest-frame $f_\nu(2 \text{ keV})$ and $f_\nu(2500 \text{ Å})$ are related by $\alpha_{ox} = 0.384 \log(f_\nu(2500 \text{ Å})/f_\nu(2 \text{ keV}))$. Therefore, $f_\nu^{obs}(1 \text{ keV}) \approx f_\nu^{rest}(2 \text{ keV})/2)$, for $z \approx 1$. We experimented by adding neutral absorbers at the redshift of the QSO (z_{em}) and at the redshift of the UV absorption lines(z_{abs}). For PG 2302+029, the fits did not favour any particular redshift. The redshifts of the absorber and the QSO are too close to be resolved in PG 1254+047. In both cases, however, the X-ray absorption at the systemic

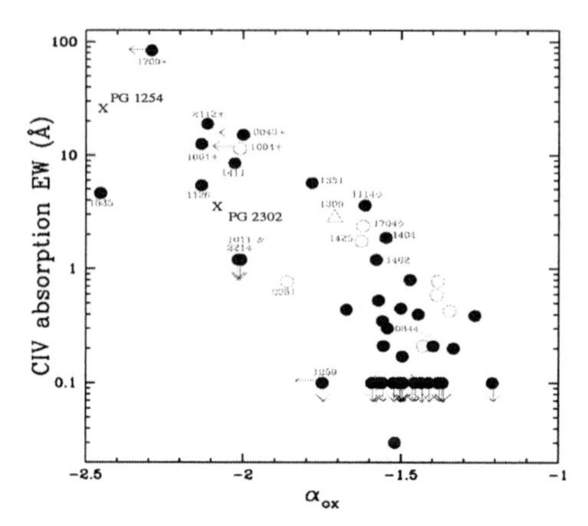

Figure 2. EW(C IV) vs. α_{ox}. Solid dots: radio-quiet QSOs, open triangles: core-dominated radio-loud QSOs, open circles: lobe-dominated radio-loud QSOs. Soft X-ray Weak QSOs are labeled their RAs, while an asterisk indicates a BAL QSO, and a diamond signifies a warm absorber. We overplot our quantities for PG 2302+029 and PG 1254+047 as X's. We have recalculated our α_{ox} to be in agreement with the convention in this plot, i.e. between 3000 Å and 2 keV, instead of what we have used so (2500 Å and 2 keV). (Adapted from Brandt et al. 2000.)

velocity of the QSO leads to higher column densities (by a factor $\lesssim 2$). We hereafter fix the redshifts of the X-ray absorbers at $z_{\rm em}$. The results are shown in Figure 3. The normalizations, based on high energies where absorption has little effect, hint at intrinsic X-ray weakness; the poor fits, especially at lower energies, hint at intrinsic absorption. The bad overall fit indicates the observed X-ray weakness cannot be explained by absorption alone.

To study the possibility of both intrinsic X-ray weakness and absorption, we remove the constraint that $\alpha_{\rm ox} = 1.6$ and hence allow the normalization of the power law to vary. The fits improve drastically, though not to the extent of giving an acceptable fit (i.e., such that $\chi^2_\nu \approx 1$). The intrinsic power law flux density at 1 keV with these improved fits decreased by about an order of magnitude. We show the results in Figure 4.

The results do not strongly support neutral absorption due to the large discrepancy between the data and models at low energies. Also, we know that there is not a neutral absorber with the above quoted column densities because the UV spectra do not contain low-ionization metal lines. To improve the fits, we experimented with absorption by ionized gas, and neutral and ionized partial coverage. We fix the ionization parameter, U, the ratio of ionizing (above 13.6 eV) flux density to hydrogen density. Experiments showed that $\log U = 0.1$ is consistent with UV data (Hamann 1998) and leads to $\chi^2_\nu < 2$. The partial covering fraction, $f_{\rm c}$, was fixed at 0.8 for PG 1254+047 because we found that it improved the quality of the fits. These values are consistent with absorption studies for such objects (e.g., Hamann 1998). Such models (shown in Figure

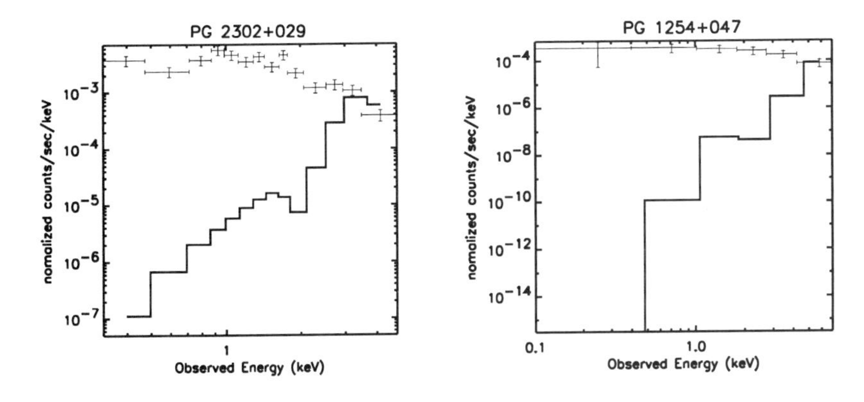

Figure 3. Galactic Absorption + Neutral Intrinsic Absorber, frozen normalization. PG 2302+029: $N_{\rm H} = 149.0 \times 10^{22}$ cm^{-2}, $\chi^2_\nu = 23.98$ for 13 d.o.f. PG 1254+047: $N_{\rm H} = 501.1 \times 10^{22}$cm^{-2}, $\chi^2_\nu = 6.28$ for 5 d.o.f.

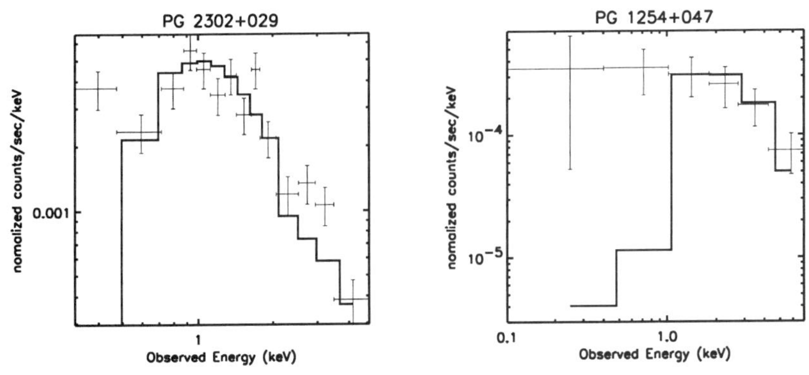

Figure 4. Galactic Absorption + Neutral Absorber, free normalization. PG 2302+029: $N_H = 1.12 \times 10^{22}$ cm^{-2}, ϕ_E(1 keV)= 1.36×10^{-5} photon s^{-1} cm^{-2} keV^{-1}, $\chi^2_\nu = 3.57$ for 12 d.o.f. PG 1254+047: $N_H = 10.8 \times 10^{22}$ cm^{-2}, ϕ_E(1 keV)= 5.87×10^{-5} photon s^{-1} cm^{-2} keV^{-1}, $\chi^2_\nu = 2.05$ for 4 d.o.f.

5) allow us to place limits on the column densities of the absorbers. We find that for PG 2302+029, $2.39 < \frac{N_H}{10^{22}\ {\rm cm}^{-2}} < 3.66$, while for PG 1254+047, $6.73 < \frac{N_H}{10^{22}\ {\rm cm}^{-2}} < 41.6$, both at the 90% confidence level.

4. Conclusions

We have presented *CXO* observations of PG 2302+029, a QSO that shows ultra high-velocity UV absorption, and PG 1254+029, a BAL QSO. The following points can be made:

1) The data suggest that there is evidence the both objects are intrinsically X-ray weak, though the X-ray slope is normal. The evidence is somewhat stronger

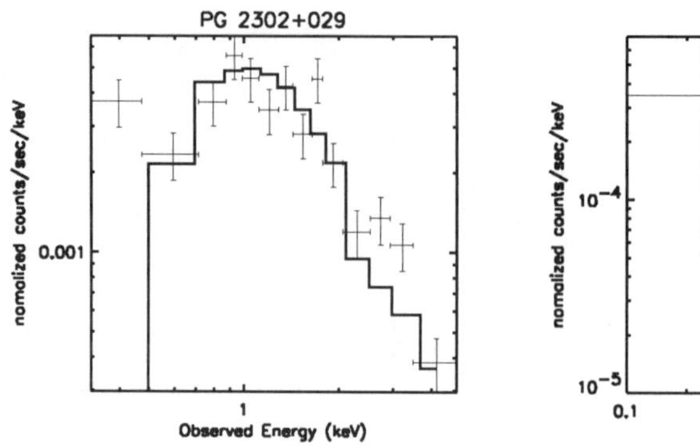

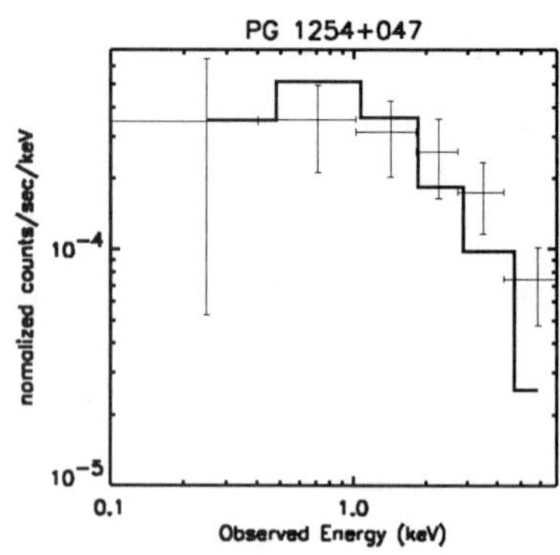

Figure 5. Galactic Absorption + Ionized Absorber, free normalization, partial coverage. PG 2302+029: $N_{\rm H} = 2.98 \times 10^{22}$ cm^{-2}, $\log U = 0.1$ (frozen), $f_{\rm c} = 1.0$ (frozen), $\phi_E(1$ keV$)= 1.54 \times 10^{-5}$ photon s^{-1} cm^{-2} keV^{-1}, $\chi^2_\nu = 1.66$ for 12 d.o.f. PG 1254+047: $N_{\rm H} = 17.3 \times 10^{22}$cm^{-2}, $\log U = 0.1$ (frozen), $f_{\rm c} = 0.8$ (frozen), $\phi_E(1$ keV$)= 3.05 \times 10^{-6}$ photon s^{-1} cm^{-2} keV^{-1}, $\chi^2_\nu = 1.76$ for 4 d.of. The normalizations correspond to intrinsic $\alpha_{\rm ox}$=2.0, 2.3, for PG 2302+029, PG 1254+047 respectively.

for PG 2302+029 given the higher number of counts. No amount of absorbing column density was able to suppress the X-ray flux down to the observed values while at the same time reproduce the overall X-ray spectral shape. The intrinsic $\alpha_{\rm ox}$'s are steep: $\alpha_{\rm ox}^{\rm PG\ 2302} = 2.0$, $\alpha_{\rm ox}^{\rm PG\ 1254} = 2.3$.
2) There is intrinsic X-ray absorption, most probably ionized with $logU = 0.1$ and $N_{\rm H}^{\rm PG\ 2302} \gtrsim 2.98 \times 10^{22}$ cm^{-2}, $N_{\rm H}^{\rm PG\ 1254} \gtrsim 17.3 \times 10^{22}$ cm^{-2} for 80% partial coverage. The derived column densities are consistent with results from the UV data, if the UV lines are very saturated.
3) We were not able to determine the X-ray absorber's redshift in PG 2302+029.

Acknowledgments. FH and BMS wish to acknowledge support through *CXO* grants GO 0-1123X and GO 0-1157X.

References

Brandt, W. N., Laor, A., & Wills, B. J. 2000, ApJ, 528, 637
Green, P.J., 1996, ApJ, 467, 61
Hamann, F. 1998, ApJ, 500, 798
Jannuzi, B. T., et al. 1996, ApJ, 470, L11
Laor, A., et al. 1997, ApJ, 477, 93
Lockman, F.J. & Savage, B.D. 1995, ApJS, 97, 1
Mathur, S., Elvis, M., & Singh, K. 1995, ApJ, 455, L9
Murray, N., Chiang, J., Grossman, J. A., & Voit, G. M. 1995, ApJ, 451, 498
Reeves, J. N., et al. 1997, MNRAS, 292, 468

Oral Presentations
Room 108

Mass Outflow in Active Galactic Nuclei: New Perspectives
ASP Conference Series, Vol. 255, 2002
D.M. Crenshaw, S.B. Kraemer, and I.M. George

FUSE Observations of Warm Absorbers in AGN

Gerard A. Kriss

Space Telescope Science Institute, 3800 San Martin Drive, Baltimore, MD 21218

Abstract. In a survey of the UV-brightest AGN using the *Far Ultraviolet Spectroscopic Explorer (FUSE)*, we commonly find associated absorption in the O VI $\lambda\lambda 1032, 1038$ resonance doublet. Of 34 Type I AGN observed to date with $z < 0.15$, 16 show detectable O VI absorption. Most absorption systems show multiple blueshifted components with intrinsic widths of $\sim$100 km s^{-1}spread over a velocity range less than 1000 km s^{-1}in width. With the exception of three galaxies (Ton S180, Mrk 478, and Mrk 279), those galaxies in our sample with O VI absorption and existing X-ray or longer wavelength UV observations also show C IV absorption and evidence for a soft X-ray warm absorber. In some cases, a UV absorption component has physical properties similar to the X-ray absorbing gas, but in others there is no clear physical correspondence between the UV and X-ray absorbing components.

1. Introduction

Roughly 50% of all Seyfert galaxies show UV absorption lines, most commonly seen in C IV and Lyα (Crenshaw et al. 1999). X-ray "warm absorbers" are equally common in Seyferts (Reynolds 1997; George et al. 1998). Crenshaw et al. (1999) note that all instances of X-ray absorption also exhibit UV absorption. While Mathur et al. (1994; 1995) have suggested that the same gas gives rise to both the X-ray and UV absorption, the spectral complexity of the UV and X-ray absorbers indicates that a wide range of physical conditions are present. Multiple kinematic components with differing physical conditions are seen in both the UV (Crenshaw et al. 1999; Kriss et al. 2000) and in the X-ray (Kriss et al. 1996a; Reynolds 1997; Kaspi et al. 2001).

The short wavelength response (912–1187 Å) of the *Far Ultraviolet Spectroscopic Explorer (FUSE)* (Moos et al. 2000; Sahnow et al. 2000) enables us to make high-resolution spectral measurements ($R \sim 20,000$) of the high-ionization ion O VI lines and the high-order Lyman lines of neutral hydrogen. The O VI doublet is a crucial link for establishing a connection between the higher ionization absorption edges seen in the X-ray and the lower ionization absorption lines seen in earlier UV observations. The high-order Lyman lines provide a better constraint on the total neutral hydrogen column density than Lyα alone. Lower ionization species such as C III and N III also have strong resonance lines in the *FUSE* band, and these often are useful for setting constraints on the ionization level of any detected absorption. The Lyman and Werner bands of molecular hydrogen also fall in the *FUSE* band, and we have serached for intrinsic H_2 absorption that may be associated with the obscuring torus.

We have been conducting a survey of the $\sim$ 80 brightest AGN using *FUSE*. To date (March 1, 2001) we have observed a total of 57; of these, 34 have $z <$ 0.15, so that the O VI doublet is visible in the FUSE band. In this presentation, I discuss the UV absorption properties of this sub-sample. A more extensive discussion of the full survey can be found in Kriss (2000). Results on *FUSE* observations of individual interesting objects such as NGC 3516 (Hutchings et al. 2002), NGC 3783 (Kaiser et al. 2002), and NGC 5548 (Brotherton et al. 2002) can also be found in these proceedings.

2. Survey Results

Roughly 50% (16 of 34) of the low-redshift AGN observed using *FUSE* show detectable O VI absorption. None show H_2 absorption. I'd first like to review the spectral morphology of the O VI absorption features. We see three basic morphologies. (1) **Blend**: multiple O VI absorption components that are blended together. This is the most common morphology (8 of 16 objects fall in this class), and the spectrum of Mrk 509 shown in Figure 1 illustrates the typical appearance. (2) **Single**: single, narrow, isolated O VI absorption lines, as illustrated by the spectrum of Ton S180 in Figure 2. (3) **Smooth**: this is an extreme expression of the "blend" class, where the O VI absorption is so broad and blended that individual O VI components cannot be identified. The spec-

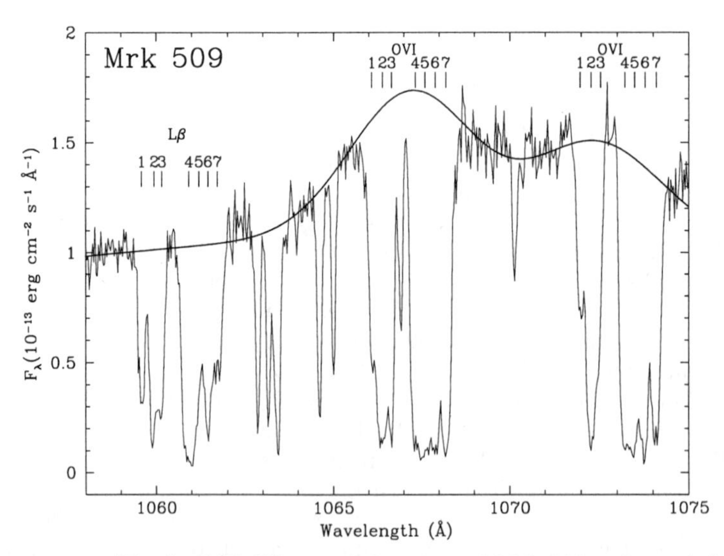

Figure 1. The Lyβ/O VI spectral region of Mrk 509, adapted from Kriss et al. (2000), illustrates the "blend" spectral morphology discussed in the text. The smooth, heavy line shows the local continuum comprised of a powerlaw continuum plus O VI emission from Mrk 509. Seven blended kinematic components are marked.

trum of NGC 4151 shown in Figure 3 is typical of this class. (Note, however, that both broad smooth absorption as well as discrete components are often visible in this class.)

As shown in the summary of characteristics presented in Table 1, individual O VI absorption components have FWHM of 50–750 km s^{-1}, with most objects having an average FWHM < 100 km s^{-1}. The multiple components that are typically present are almost always blue shifted, and they span a velocity range (total velocity coverage for all components) of 200–4000 km s^{-1}; half the objects span a range of < 1000 km s^{-1}.

Table 1. Properties of O VI Absorbers in AGN Observed with FUSE

Name	# Comp	Type	Avg. Line Widths (km s^{-1})	Velocity Range (km s^{-1})	Other UV Abs.[a]	X-ray Abs.[b]
I Zw 1	2	single	50	850	Y	...
NGC 985	1	smooth	100	1000	...	...
NGC 3516	5	blend	100	1000	Y	Y
NGC 3783	4	blend	300	1600	Y	Y
NGC 4151	3	smooth	100	1600	Y	Y
NGC 5548	6	blend	50	1300	Y	Y
NGC 7469	3	single	50	1000	Y	Y
Mrk 279	5	blend	100	1600	Y	N
Mrk 290	1	single	200	400	...	Y
Mrk 304	5	smooth	750	1500	...	...
Mrk 478	5	blend	50	2700	N	...
Mrk 509	5	blend	50	700	Y	Y
Mrk 817	2	blend	250	4000	...	...
PG1351+64	4	blend	100	2000	Y	...
Ton 951	1	single	40	200	...	...
Ton S180	3	single	50	300	N	N

[a]UV reference: Crenshaw et al. (1999).
[b]X-ray references: George et al. (1998); Reynolds (1997).

The 50% O VI absorption fraction is comparable to those Seyferts that show longer-wavelength UV (Crenshaw et al. 1999) or X-ray (Reynolds 1997; George et al. 1998) absorption. In Table 1 we make a detailed comparison to these UV and X-ray absorption studies. We can summarize this comparison in two ways: (1) all objects that have shown previous evidence of either UV or X-ray absorption also show O VI absorption in our *FUSE* observations; (2) of those objects showing detectable O VI absoption that also have previous UV or X-ray observations available, only 3 do not show X-ray absorption or longer-wavelength UV absorption. These three exceptions deserve some attention:

Ton S180 was observed simultaneously with *FUSE*, *Chandra*, and *HST* (Turner et al. 2001). As shown in Figure 2, the absorption features observed with *FUSE* are quite weak. At the resolution of the *HST* observations, which used the STIS low-resolution gratings, corresponding C IV and Lα absorption wo· not be expected to be detectable. No X-ray absorption features are seen

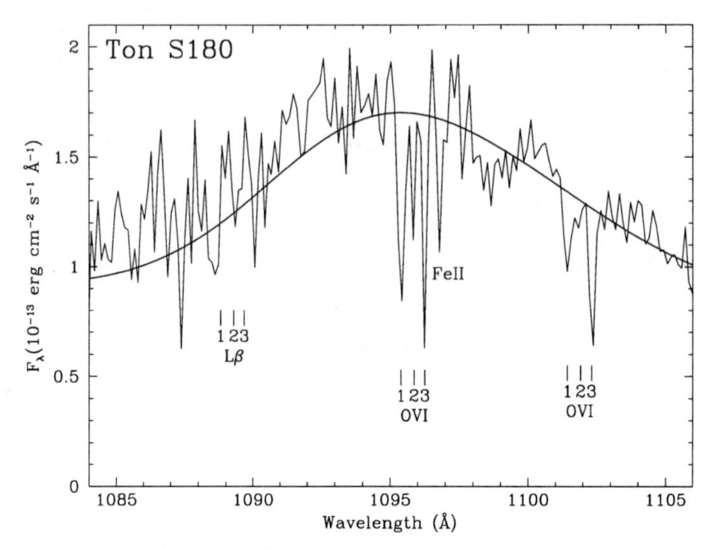

Figure 2. The Lyβ/O VI spectral region of Ton S180 shows the "single" absorption line spectral morphology discussed in the text. As in Fig. 1, the heavy smooth line shows the sum of a powerlaw continuum plus broad O VI emission. Three, narrow, isolated absorption components are visible in O VI. These are marked for both O VI and Lyβ, even though no absorption is detected in Lyβ. Ton S180 is also one of the objects in our sample for which we do not detect longer wavelength UV absorption or X-ray absorption (Turner et al. 2001).

in the *Chandra* grating observation, which implies that either the total column density of the absorbing gas is very low, and/or that its ionization state is lower than that necessary to create X-ray absorbing species.

Mrk 478 was observed previously with *HST* using the FOS. As for Ton S180, these observations were low spectral resolution, so that weak C IV or Lα absorption at the equivalent widths observed for O VI would not be detectable.

Mrk 279 shows strong Lyα absorption in an archival *HST* spectrum. However, the *ASCA* observation of Mrk 279 does not show evidence of a warm absorber (Weaver et al. 2001). This was a short observation, however, and total column densities of 10^{21} cm^{-2} could easily be present, which would be detectable in *Chandra* grating observations.

3. Discussion

The multiple kinematic components frequently seen in the UV absorption spectra of AGN clearly show that the absorbing medium is complex, with separate UV and X-ray dominant zones. In some cases, the UV absorption component corresponding to the X-ray warm absorber can be clearly identified (e.g., Mrk 509, Kriss et al. 2000). In others, however, *no* UV absorption component shows physical conditions characteristic of those seen in the X-ray absorber

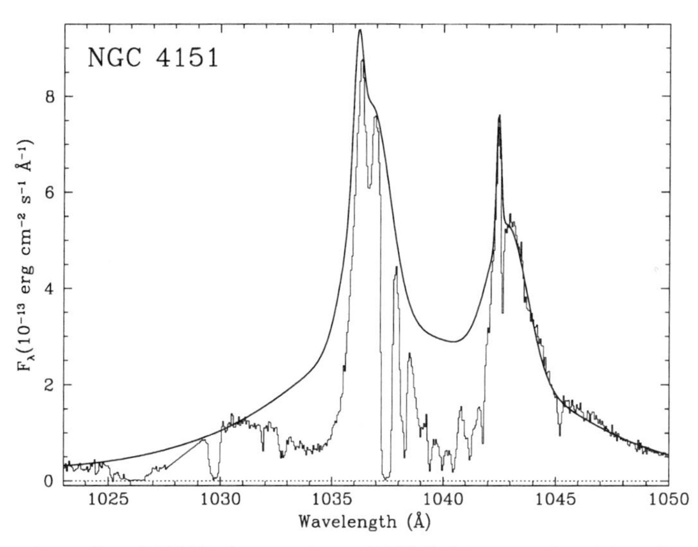

Figure 3. Our FUSE observation of NGC 4151 caught this galaxy in a low-luminosity state. The O VI region is dominated by narrow-line emission, but smooth, broad absorption illustrating the "smooth" morphology discussed in the text. The heavy, smooth solid line shows the modeled total emission from a powerlaw continuum plus weak broad-line and strong narrow-line emission.

(NGC 3783: Kaiser et al. 2002; NGC 5548: Brotherton et al. 2002). One potential geometry for this complex absorbing structure is high density, low column UV-absorbing clouds embedded in a low density, high ionization medium that dominates the X-ray absorption. As discussed by Krolik & Kriss (1995; 2001), this is possibly a wind driven off the obscuring torus. At the critical ionization parameter for evaporation, there is a broad range of temperatures that can coexist in equilibrium at nearly constant pressure; for this reason, the flow is expected to be strongly inhomogeneous. What would this look like in reality? We cannot spatially resolve this region, so it is instructive to look at nearby analogies. The HST images of the pillars of gas in the Eagle Nebula, M16, show the wealth of detailed structure in gas evaporated from a molecular cloud by the UV radiation of nearby newly formed stars (Hester et al. 1996). In an AGN one might expect a dense molecular torus to be surrounded by blobs, wisps, and filaments of gas at various densities. It is plausible that the multiple UV absorption lines seen in AGN with warm absorbers are caused by high-density blobs of gas embedded in a hotter, more tenuous, surrounding medium, which is itself responsible for the X-ray absorption. Higher density blobs would have lower ionization parameters, and their small size would account for the low overall column densities.

In summary, we find that O VI absorption is common in low-redshift ($z < 0.15$) AGN: 16 of 34 AGN observed with *FUSE* show multiple, blended O VI absorption lines with typical widths of $\sim 100\ \mathrm{km\,s^{-1}}$ that are blueshifted over a velocity range of $\sim 1000\ \mathrm{km\,s^{-1}}$. With three exceptions, those galaxies in our sample with existing X-ray or longer wavelength UV observations also show C IV absorption and evidence of a soft X-ray warm absorber. In some cases, a UV absorption component has physical properties similar to the X-ray absorbing gas, but in others there is no clear physical correspondence between the UV and X-ray absorbing components.

References

Brotherton, M., et al. 2002, in Mass Outflow in Active Galactic Nuclei: New Perspectives, eds. D.M. Crenshaw, S.B. Kraemer, & I.M. George (San Francisco: ASP), 127

Crenshaw, D. M., et al. 1999, ApJ, 516, 750

George, I. M., et al. 1998, ApJS, 114, 73

Hester, J. J., et al. 1996, AJ, 111, 2349

Hutchings, J., et al. 2002, in Mass Outflow in Active Galactic Nuclei: New Perspectives, eds. D.M. Crenshaw, S.B. Kraemer, & I.M. George (San Francisco: ASP), p. 267

Kaiser, M. E., et al. 2002, in Mass Outflow in Active Galactic Nuclei: New Perspectives, eds. D.M. Crenshaw, S.B. Kraemer, & I.M. George (San Francisco: ASP), p. 75

Kaspi, S., et al. 2001, ApJ, 467, 622

Kriss, G.A. 2000, in ASP Conf. Proc. 224, "Probing the Physics of Active Galactic Nuclei by Multiwavelength Monitoring", ed. B. Peterson, R. Polidan, & R. Pogge, (San Francisco: ASP), 224, 45

Kriss G. A., et al. 1996, ApJ, 467, 622

Kriss, G. A., et al. 2000, ApJ, 538, L17

Krolik, J. H., & Kriss, G. A. 1995, ApJ, 447, 512

Krolik, J. H., & Kriss, G. A. 2001, ApJ, in press

Mathur, S., Wilkes, B., & Elvis, M. 1995, ApJ, 452, 230

Mathur, S., Wilkes, B., Elvis, M., & Fiore, F. 1994, ApJ, 434, 493

Moos, H. W., et al. 2000, ApJ, 538, L1

Reynolds, C. S. 1997, MNRAS, 286, 513

Sahnow, D., et al. 2000, ApJ, 538, L7

Turner, T. J., et al. 2001, ApJ, 548, L13

Weaver, K. A., Gelbord, J., & Yaqoob, T. 2001, ApJ, 550, 261

Mass Outflow in Active Galactic Nuclei: New Perspectives
ASP Conference Series, Vol. 255, 2002
D.M. Crenshaw, S.B. Kraemer, and I.M. George

O VI Intrinsic Absorption in NGC 3783: *FUSE* Observations

Mary Elizabeth Kaiser

Department of Physics and Astronomy, Johns Hopkins University, 3400 North Charles Street, Baltimore, MD 21218-2686

Gerard A. Kriss

Space Telescope Science Institute, 3800 San Martin Drive, Baltimore, MD 21218

Kenneth R. Sembach

Department of Physics and Astronomy, Johns Hopkins University, 3400 North Charles Street, Baltimore, MD 21218-2686

Abstract. High resolution *FUSE* spectroscopy of NCG 3783 reveals 4 intrinsic absorption systems, 3 of which have been detected previously. The new component has an absorption velocity of -1029 km s^{-1} which is bracketed by the velocities of the other absorbers. O VI column densities for the four systems range from $1.7 - 6.0 \times 10^{15}$cm^{-2} with covering factors ranging from ~ 0.5 to unity.

1. Introduction

Observations of the circumnuclear regions of AGN in the UV and X-ray indicate that approximately half of all Seyfert 1 galaxies show significant amounts of ionized gas in absorption against the central continuum source, with an apparent one-to-one correspondence between the presence of a warm X-ray absorber and intrinsic absorption in the UV (Crenshaw et al. 1999; Reynolds 1997; George et al. 1998). The higher resolution UV observations determined that the absorption lines are blue-shifted with respect to the systemic velocity of the host galaxy, indicating that the gas is outflowing (Crenshaw et al. 1999; Kriss 2002). Recent high resolution X-ray spectra have corroborated this result (e.g. NGC 3783: Kaspi et al. 2001). Although this suggests that the UV and X-ray absorbers may represent the same gas column, the UV absorption often arises in multiple components, some of which present a column density of higher ionization species that is too low and a column density of lower ionization species that is too high to be spatially coexistent with the X-ray warm absorber.

The far-ultraviolet bandpass (900 - 1200Å) of the Far Ultraviolet Spectroscopic Explorer (*FUSE*) (Moos et al. 2000; Sahnow et al. 2000) is a critical tool for understanding the nature and origin of this gas. The high spectral resolution of *FUSE* (R $\sim$ 20,000) permits the isolation of multiple kinematic components in the absorption spectrum of O VI facilitating the association of an UV absorber

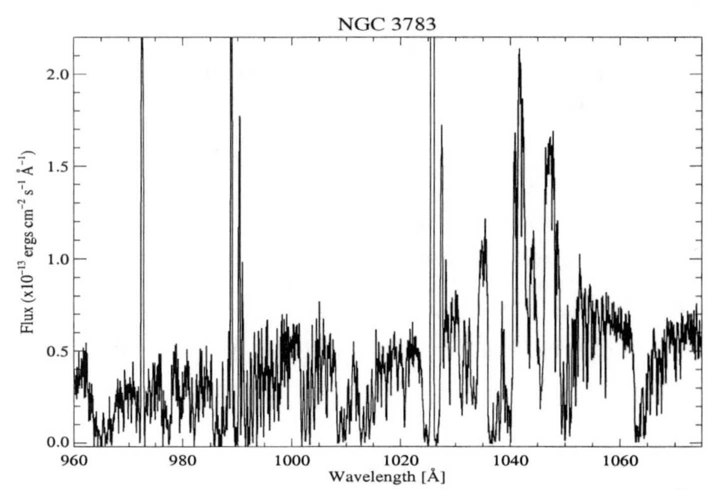

Figure 1. The *FUSE* spectrum of NGC 3783 binned to 0.05Å resolution.

with its X-ray counterpart. O VI is critical for establishing the connection between the lower ionization lines of Si IV, CIV and NV seen in the UV with the high ionization absorption edges of O VII and O VIII seen in the X-ray. In addition, Ly β and the higher order Lyman lines present in the *FUSE* bandpass provide a constraint on the neutral hydrogen column density.

NGC 3783, a bright Seyfert 1 galaxy with one of the strongest X-ray warm absorbers known, is part of the *FUSE* AGN survey (Kriss et al. 2000; Kriss 2002). Its extremely variable absorption and rapid continuum variability in both the UV and X-ray have made it a cornerstone of reverberation mapping programs, and as such it has been well-monitored in the UV (Reichert et al. 1994) and X-ray. However, high resolution far-UV spectra have not been available prior to these *FUSE* data.

In this paper we present preliminary O VI column densities, covering factors, and velocities for the intrinsic absorbers. We then discuss our results in the context of recent (non-simultaneous) observations in the X-ray (Kaspi et al. 2001) and UV (Kraemer et al. 2001).

2. Observations

Our *FUSE* data (Figure 1) were obtained on 2 February 2000 using all four channels of the *FUSE* spectrograph. We obtained 13 exposures totaling 37 ksec on target. The data were reduced with the *FUSE* pipeline and subsequently binned to 0.05 Å. The resolution of the spectra are $\sim$20 km s^{-1} (0.07 Å).

NGC 3783 resides at low galactic latitude and illuminates high-velocity cloud HVC 287.5+22.5+240, thus presenting a complex spectrum which is of interest for both AGN and ISM physics. Before analyzing the intrinsic absorption in NGC 3783, it is necessary to model the ubiquitous atomic and molecular hydrogen absorption lines present in the Ly β - O VI region due to absorption by the foreground HVC and our galaxy. Strong C II and Ar lines are also present,

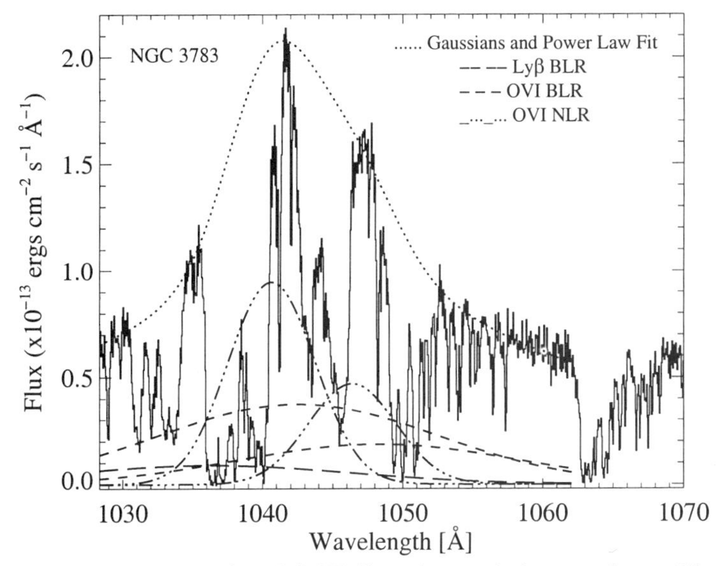

Figure 2. The Ly β and O VI Gaussian emission envelope. The composite emission envelope is shown (dotted line) overplotted on the *FUSE* spectrum, along with the individual BLR and NLR emission components.

with the C II lines in particular obscuring the O VI λ1032 intrinsic absorption region. We modeled the foreground absorption using the velocities and hydrogen column densities identified by Sembach et al. (2001) for these extrinsic absorbing systems.

Obtaining accurate column densities for the Ly β and O VI intrinsic absorption lines requires accurate modeling of both the foreground ISM absorption systems and the emission from both the BLR and NLR. To determine the local continuum we fit the Ly β – O VI region with a power law continuum and five Gaussian emission components. We assumed BLR emission for Ly β and the O VI doublet. We constrained the fit by fixing the relative wavelengths, requiring the BLR and NLR emission components to have the same FWHM, and fixing the O VI doublet line ratio at 2:1. Thus, 15 free parameters were reduced to 7. Although we tested the robustness of this model, uncertainties in the emission envelope are our largest systematic error in obtaining accurate column densities. The resulting fit to the continuum and emission structure in the Ly β – O VI region is shown in Figure 2. This fit yields a FWHM of 6692 km s^{-1} and 2216 km s^{-1} for the O VI BLR and NLR emission respectively. Measurements of the [OIII] line profile indicate a FWHM of 220 km s^{-1} (Whittle 1985), suggesting that the narrower of the two O VI emission components resides in a region between the canonical BLR and NLR.

The ISM model was scaled by the emission envelope (Figure 3) to permit the incorporation of the ISM absorption in the fit of the intrinsic absorption lines. First, we fit the scaled ISM model in regions without intrinsic absorption

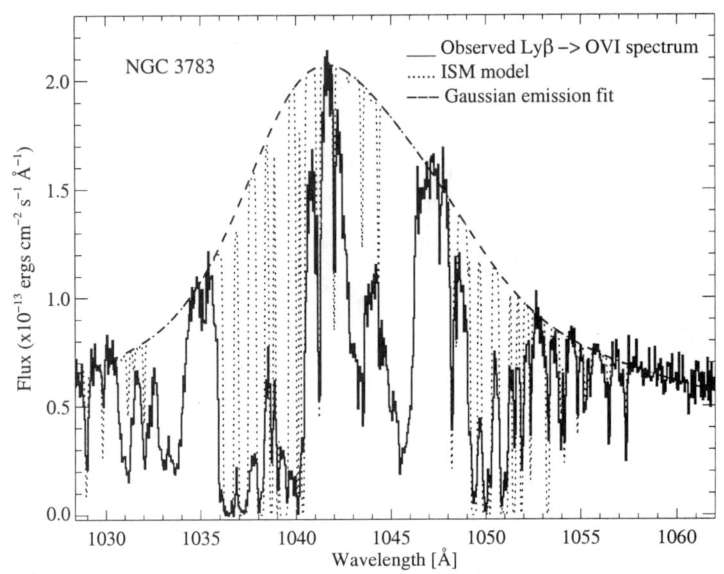

Figure 3. The ISM model (dotted line), consisting of the HVC and three lower velocity galactic absorption systems, is scaled by the Gaussian emission envelope (dashed line) and overplotted on the data.

(Figure 4) and then included this ISM fit as part of the continuum. Next, the intrinsic and extrinsic absorption were simultaneously fit. The model fit to the data assumes that all the emission components are absorbed by the intrinsic absorbers and that the ISM lines are foreground absorption against the intrinsically absorbed spectrum.

We identify four absorbing systems (Figure 5, Table 1) with covering factors ranging from ~ 0.5 to unity. Three of these system have been observed previously (e.g. Kraemer et al. 2001) and are known to be variable. Our velocity measurements for these components are consistent with the earlier measurements. The *FUSE* spectra provide the first detection of the fourth absorbing component (see also Gabel et al. 2002), which occurs at a relative velocity of $-1029\,\mathrm{km\,s^{-1}}$. We find O VI column densities ranging from $1.6 - 6\times 10^{15}\mathrm{cm}^{-2}$, which are an order of magnitude lower than predicted by the modeling of Kraemer et al. (2001).

Although these results are preliminary and we have not yet run extensive models to determine which, if any, UV absorption system is consistent with the X-ray warm absorber, we expect that the intrinsic absorber at $-599\ \mathrm{km\,s^{-1}}$ (component 2) is most likely to be identified with the warm absorber. The X-ray, FUV, and UV velocities are in agreement for this component. We do not see absorption by lower ionization species such as Si IV (Kraemer et al. 2001), and it possess a relatively large column of O VI. Fits to the FUV data indicate a covering factor of 0.9 for this component, while the X-ray data indicates a unity line-of-sight covering factor (Kaspi et al. 2001). Furthermore, this component may harbor two or more systems, only one of which is likely to be associated with

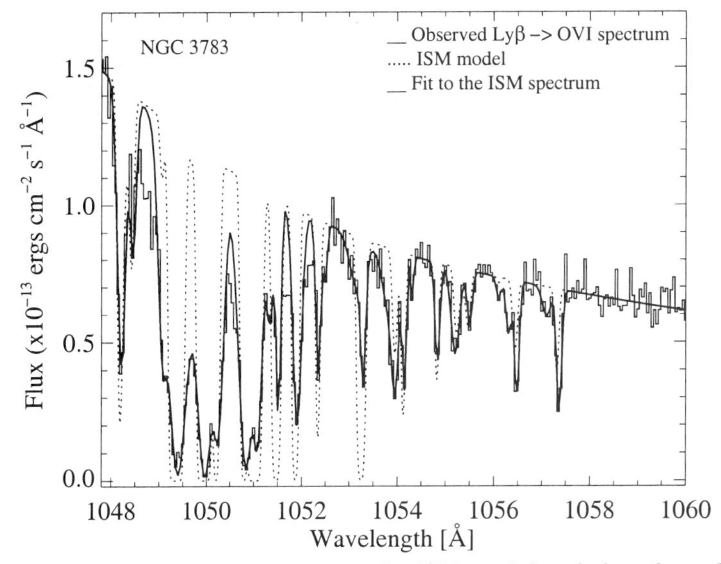

Figure 4. Comparison between the ISM model scaled to the red wing of the O VI emission and the explicit fit to the ISM absorption lines in this region.

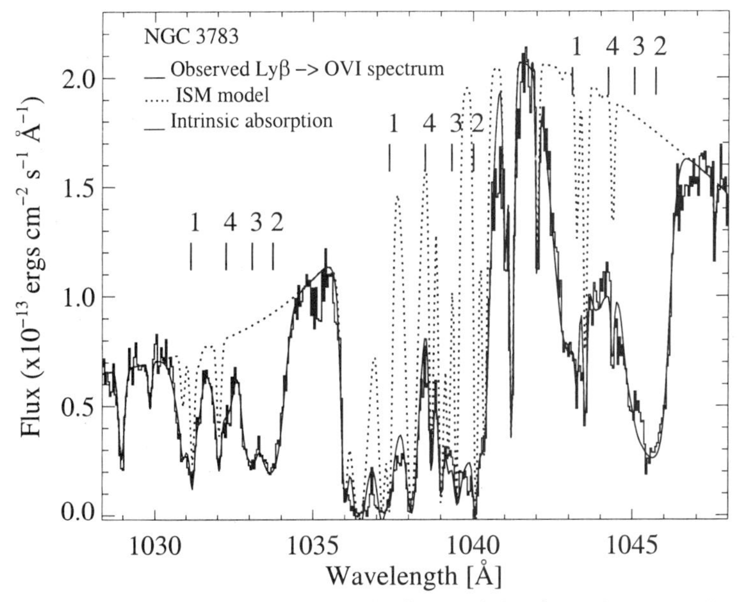

Figure 5. The preliminary best fit (smooth line) to the intrinsic absorption in the O VI region is shown overplotted on the spectrum. The dotted line depicts the ISM model in this region.

Table 1. Fit results for the four intrinsic absorption systems detected in Ly β and O VI $\lambda\lambda$1032,1038. The components are numbered according to their chronological appearance in the literature. Their velocities, O VI and hydrogen column densities, and line-of-sight covering factors are tabulated here.

Component	Velocity [km s^{-1}]	N_{OVI} [cm^{-2}]	N_H [cm^{-2}]	Covering Factor
1	– 1352	1.7×10^{15}	9.0×10^{14}	1.0
2	– 599	4.1×10^{15}	2.1×10^{15}	0.9
3	– 792	2.1×10^{15}	1.7×10^{15}	0.7
4	– 1029	6.0×10^{15}	2.3×10^{15}	0.5

the X-ray warm absorber. Modeling of this component as multiple absorbing systems is in progress.

Acknowledgments. We thank the *FUSE* AGN working group for their support of this program. This work is based on data obtained for the Guaranteed Time Team by the NASA-CNES-CSA *FUSE* mission operated by the Johns Hopkins University. Financial support has been provided by NASA contract NAS5-32985.

References

Crenshaw, D.M., et al. 1999, ApJ, 516, 750

George, I.M., et al. 1998, ApJS, 114, 73

Gabel, J.R., Kraemer, S.B., & Crenshaw, D.M. 2002, in Mass Outflow in Active Galactic Nuclei: New Perspectives, eds. D.M. Crenshaw, S.B. Kraemer, & I.M. George (San Francisco: ASP), p. 81

Kaspi, S., et al. 2001, ApJ, 554, 216

Kaspi, S., et al. 2001, astro-ph/0101540

Kraemer, S.B., Crenshaw, D.M., & Gabel, J.R. 2001, astro-ph/0104303

Kriss, G.A., et al. 2000, ApJ, 538, L17

Kriss, G.A. 2002, in Mass Outflow in Active Galactic Nuclei: New Perspectives, eds. D.M. Crenshaw, S.B. Kraemer, and I.M. George (San Francisco: ASP), p. 69

Moos H.W., et al. 2000, ApJ 538, L1

Reichert G. A., et al. 1994, ApJ, 425, 582

Reynolds, C.S. 1997, MNRAS, 286, 513

Sahnow, D., et al. 2000, ApJ 538, L7

Sembach K. R., et al. 2001, ApJ, 121, 992

Whittle, M., 1985, MNRAS, 213, 1

Mass Outflow in Active Galactic Nuclei: New Perspectives
ASP Conference Series, Vol. 255, 2002
D.M. Crenshaw, S.B. Kraemer, and I.M. George

The Physical Conditions in the UV Absorbers in the Seyfert 1 Galaxy NGC 3783

Jack R. Gabel & Steven B. Kraemer

Catholic University of America and Laboratory for Astronomy and Solar Physics, NASA's Goddard Space Flight Center, Code 681, Greenbelt, MD 20771

D. Michael Crenshaw

Department of Physics and Astronomy, Georgia State University, Atlanta, GA 30303

Abstract.
We present observations and analysis of the intrinsic UV absorption lines in the Seyfert 1 Galaxy NGC 3783 with the Space Telescope Imaging Spectrograph (STIS) on the *Hubble Space Telescope (HST)*. Lα, C IV, and N V appear in three distinct kinematic components with radial velocities ranging between -500 and -1400 km s^{-1} with respect to the systemic velocity of the host galaxy, indicating net radial outflow. The highest velocity component, component 1, also shows absorption in Si IV, indicating that a relatively low ionization zone exists in this component. We have calculated photoionization models to match the measured column densities. Each component is characterized by a zone with high ionization parameter (U $\geq$ 0.65); a second, low ionization zone (U = 0.0018) having a low column density (4.9 x 10^{18} cm^{-2}) is required for component 1. The O VIII column densities predicted by our models are smaller than the X-ray columns measured in *Chandra* spectra. We also present preliminary results of a *Far Ultraviolet Spectroscopic Explorer (FUSE)* observation of NGC 3783. The *FUSE* spectrum reveals strong O VI and Lyβ absorption, consistent with our model predictions. The covering factors derived from these features are larger than those deduced from the STIS data. This provides insight into the geometry of the absorber and emission sources.

1. Introduction

NGC 3783 is a bright, nearby (z = 0.00976) Seyfert 1 galaxy that exhibits variable intrinsic absorption in its UV spectrum. Intrinsic Lyα and C IV absorption were first detected in the 1992 July 27 *HST*/FOS spectrum at $\sim$$-550$ km s^{-1} with respect to the systemic velocity of the galaxy by Reichert et al. (1994). Subsequent *HST*/GHRS observations revealed the absorption to be highly variable; C IV was undetectable in a spectrum obtained 11 months later and then reappeared in a 1994 Jan 16 observation (Maran et al. 1996; Crenshaw et al. 1999). Additionally, a 1995 Apr 11 GHRS spectrum revealed Lyα, C IV, and

N V absorption in a second kinematic component, at v = -1370 km s^{-1}. X-ray observations of NGC 3783 also reveal absorption (George et al. 1998 and references therein), although the relation between these "warm absorbers" and the UV absorbers is uncertain. Using the results of an analysis of *ASCA* spectra by George et al. (1995), Shields & Hamann (1997) demonstrated that the UV and X-ray absorption could plausibly arise from the same zone. Further, the mean velocity of the absorption lines detected in *Chandra* spectra by Kaspi et al. (2000; 2001) is consistent with the strong UV absorption component at -550 km s^{-1}. However, the ionization state derived for the X-ray absorber may be too high to produce the UV lines. Clearly, the exact nature of the absorption in NGC 3783 is, as yet, undetermined.

2. STIS Echelle Observations of UV Absorption in NGC 3783

We have obtained medium-resolution STIS echelle spectra of the nucleus of NGC 3783 through a 0".2 x 0".2 aperture on 2000 February 27. In Figure 1, the STIS E140M spectrum is plotted in velocity space showing absorption in Lyα,

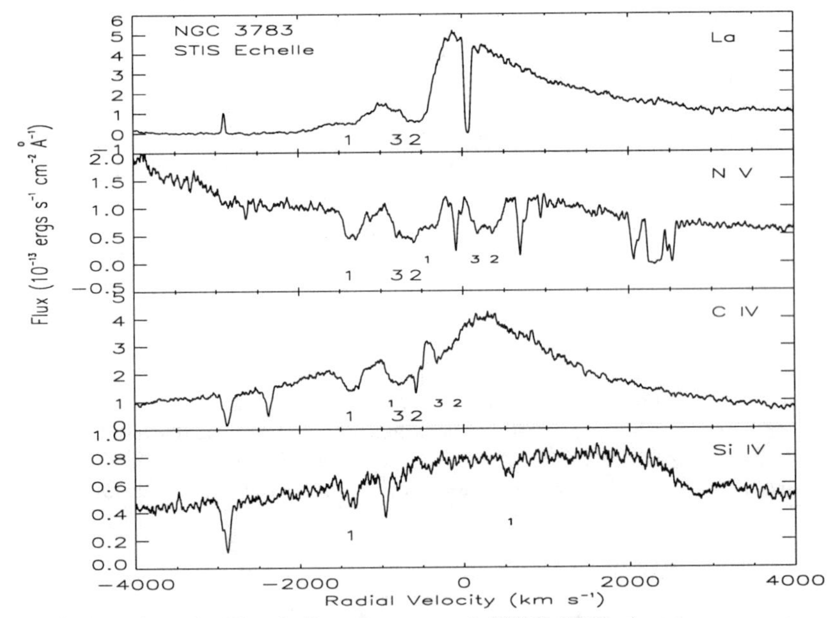

Figure 1. STIS echelle spectrum of NGC 3783 showing resonance line intrinsic absorption. Fluxes are plotted vs radial velocity (of the strongest member for the doublets) relative to the systemic velocity of 2919 km s^{-1}. The kinematic components are identified with large and small numbers for the strong and weak members of the doublets, respectively. Unmarked absorption lines are Galactic, except the feature in the Si IV profile near 2800 km s^{-1}, which is an artifact.

N V, C IV, and Si IV. In addition to the two kinematic components detected in the GHRS spectra (component 1 at v = -1365 km s^{-1} and component 2 at v = −550 km s^{-1}, following the notation of Crenshaw et al. 1999), we detect a third absorption component at v = −725 km s^{-1} in Lyα, N V, and C IV. Si IV is only present in component 1 in the STIS spectrum, indicating that a lower ionization region exists in this component. We find no detectable absorption from other low ionization lines that lie in the STIS bandpass, such as Mg II, C II, and Si II.

We used the techniques described in Crenshaw et al. (1999) to measure the intrinsic absorption lines in the STIS spectrum. The continuum plus line emission was fit with a cublic spline on each side of the absorption features and divided from the observed spectrum to give the normalized absorption profiles. To correct for blending, which affects several of the absorption features as seen in Figure 1, uncontaminated lines were used as template profiles. In many AGNs with intrinsic absorption, the line of sight covering factor of the absorbing material, C_{los}, is less than unity and, if not corrected for, results in an underestimate of the total column density (Hamann et al. 1997; Crenshaw et al. 1999). We estimated C_{los} following the method of Hamann et al., using the N V doublet line ratio because it is the least blended. For all three kinematic components, we find $C_{los} \approx 0.6$. Ionic column densities were measured by converting the normalized flux to optical depth and integrating over the line profiles, after correcting for C_{los} (Hamann et al.). In Table 1, we give the radial velocity centroid, FWHM, C_{los} and column densities measured for each kinematic component.

Table 1. Intrinsic Absorption Measured in STIS Spectrum

Comp.	v_r km s^{-1}	FWHM km s^{-1}	C_{los}	$N_{Ly\alpha}$	N_{NV} 10^{14} cm^{-2}	N_{CIV}	N_{SiIV}
1	−1365	193	0.61	$\geq$0.84	7.6($\pm$1.5)	3.5($\pm$1.5)	0.6($\pm$0.2)
2	−548	170	0.60	$\geq$0.77	3.7($\pm$1.8)	0.6($\pm$0.2)	–
3	−724	280	0.65	$\geq$3.86	16($\pm$2)	3.2($\pm$0.5)	–

3. Photoionization Modeling of the Intrinsic UV Absorption Observed with STIS

Photoionization models were generated to match the measured ionic columns using the code CLOUDY90 (Ferland et al. 1998). To model the ionizing continuum emission, we began with the fit to the hard X-ray (0.6 - 10 keV) *ASCA* data by George et al. (1998), $F_\nu \propto \nu^{-\alpha}$ with $\alpha = 0.8$. Extrapolating this power-law flux distribution to UV wavelengths underpredicts the observed UV luminosity (corrected for reddening) by over two orders of magnitude and thus, the SED steepens at some point below 0.6 keV. We arbitrarily chose this break in the continuum emission to occur at 0.6 keV and interpolated between the observed UV and X-ray fluxes to derive a power-law index of $\alpha = 1.4$ for $h\nu \leq 0.6$ keV. It is important to note that the uncertainty in the spectral shape in the EUV adds an uncertainty in the photoionization models.

In Figure 1, only component 1 is seen to possess detectable Si IV absorption, indicating low ionization. However, N(N V)/N(C IV) $\geq$ 2 in this same component, which is indicative of highly ionized gas (Shields & Hamann 1997).

This situation may arise if the ionizing continuum is modified by an optically thick, intervening absorber in our line-of-sight, as demonstrated by Kraemer et al. (2001) for the case of NGC 4151 However, there is no evidence for such absorbing material in the nucleus of NGC 3783 (see discussion below). Alternatively, component 1 may consist of two zones of absorbing gas with similar velocity profiles, but characterized by different ionization parameters. Physically, this may correspond to denser, low ionization gas co-located with a more tenuous, and thus more highly ionized, absorber. Accordingly, we modeled component 1 using two separate zones with the following parameters: 1) U = 0.78, N_H = 1.2 x 10^{21} cm^{-2}, and 2) U = 0.0018, N_H = 4.9 x 10^{18} cm^{-2}. If the two zones are at the same radial distance, the density of the lower ionization zone is $\sim$ 430 times higher. The sum of the predicted ionic columns from the two zones provides a good fit to the observed columns. Furthermore, the low ionization model predictions for N(C II) and N(Si II) are below their respective 2σ upper limits for component 1.

Since modeling components 2 and 3 involves matching only two ionic columns (N V and C IV), single-zoned models are sufficient for these components. Based on the ratio of the N V and C IV column densities, component 2 is the most highly ionized. We were able to fit the observed ionic columns with U = 0.80 and N_H = 6.4 x 10^{20} cm^{-2}. Component 3, which has a lower N(N V)/N(C IV) ratio and, thus, a lower ionization parameter, is matched well with U = 0.65 and N_H = 1.5 x 10^{21} cm^{-2}. None of the absorbers has sufficient optical depth in the H or He Lyman limits to affect the ionization state of gas further from the nucleus, hence, we see no effects due to the screening of one absorbing component by another.

Our calculations predict that all three kinematic components should possess significant O VII and O VIII column densities, with summed values of $N_{O\ VII}$ = 1.3 x 10^{18} cm^{-2} and $N_{O\ VIII}$ = 7.7 x 10^{17} cm^{-2}. In an analysis of roughly contemporary *Chandra* spectra, Kaspi et al. (2001) found an X-ray absorption component with a mean velocity roughly coincident with our components 2 and 3; however, their measured O VII and O VIII columns are larger than our predicted values by factors of 2.6 and 5, respectively (Kaspi et al. 2000). If component 2, which has the largest uncertainty (Table 1), is more highly ionized than indicated by the mean N V/C IV ratio, its O VIII column may be significantly larger than our model predictions. However, this would worsen the fit to O VII, and the columns of other high ionization species measured in the X-ray by Kaspi et al. (2000), e.g. Si XIV, Fe XVIII, and Fe XIX, would still be underpredicted. Due to the small errors in our measurements of the ionic columns in component 3, the ionization parameter in this absorber is well constrained. Thus, this component cannot produce the observed O VIII column. Based on these results, we conclude that the UV absorbers cannot fully account for the observed X-ray absorption.

4. Preliminary Results from *FUSE* Observations

In Figure 2, we show the *FUSE* spectrum of NGC 3783, obtained three weeks prior to the STIS observations, plotted in velocity space to show intrinsic absorption in Lyβ and the O VI resonance doublet lines. In addition to the three

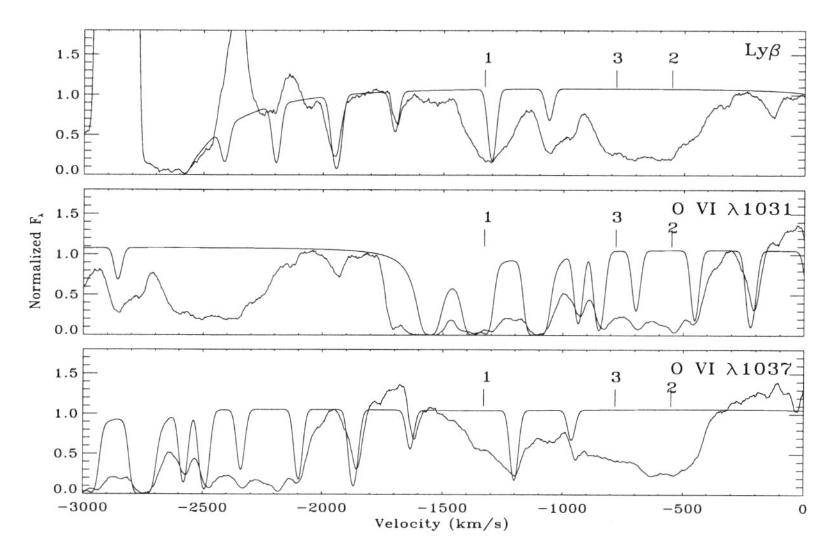

Figure 2. *FUSE* spectrum of NGC 3783 showing intrinsic absorption in Lyβ and the O VI resonance doublet lines. Normalized fluxes are plotted vs radial velocity for each line relative to the systemic velocity. The four kinematic components are identified on each plot. A model of the Galactic H_2 absorption, which consists of two kinematic components (Sembach et al. 2001), is plotted as a smooth line.

kinematic components identified in the STIS data, we find another absorption component in O VI and Lyβ, at v ≈ -1000 km s^{-1}, which we call component 4 (see also Kaiser et al. 2002). Reexamination of the STIS spectrum in Figure 1 reveals component 4 as a small notch between components 1 and 3 in N V, C IV, and Lyα. Although we have not modeled this absorption component, qualitatively, the high N V/C IV ratio and lack of Si IV indicates a high ionization state.

One peculiarity in the *FUSE* spectrum is that the depths of the O VI features indicate a minimum covering factor for the absorbing gas, $C_{los} \geq 0.8$, that is larger than the value derived from the N V doublet lines in the STIS data, $C_{los} \approx 0.6$. This may indicate that a more complicated geometry exists for the absorbers and emission sources than assumed in our models - we assumed the continuum and BLR emission sources were covered by equal percentages in our previous analysis. For example, the O VI zone in the broad emission line region may be more occulted by the absorber than the lower ionization BLR gas (i.e., N V and C IV). This would arise, for example, if the central core of the AGN (i.e., continuum source and inner BLR gas) is more occulted than the outer BLR, and the ionization in the BLR gas decreases with radial distance from the ionizing source. Alternatively, the descrepancy in covering fractions may be a result of the ionization structure within the absorber. A more detailed analysis of all absorption lines in the combined *FUSE* and STIS spectra is required to better constrain the geometry of the absorption gas with respect to the emission sources.

5. Conclusions

In summary, we have measured the intrinsic absorption observed in a recent STIS spectrum of NGC 3783. The measured ionic columns were matched with one-zoned or, in the case of component 1, two-zoned photoionization models. We find that the UV absorbers cannot fully account for the highly ionized columns measured in X-ray spectra. The O VI absorption seen in a near-simultaneous *FUSE* spectrum indicates that the line-of-sight covering fraction of the absorbing gas may be a function of ionization.

We are currently undertaking an intensive, multiwavelength monitoring campaign of NGC 3783 with *HST*, *FUSE*, and *Chandra*. Simultaneous data from these observatories will provide unprecedented constraints on the intrinsic absorption in a Seyfert 1 galaxy.

References

Crenshaw, D.M., et al. 1999, ApJ, 516, 750

Crenshaw, D.M., et al. 2002, in Mass Outflow in Active Galactic Nuclei: New Perspectives, eds. D.M. Crenshaw, S.B. Kraemer, & I.M. George (San Francisco: ASP), p. 87

George, I.M., Turner, T.J., & Netzer, H. 1995, ApJ, 438, L67

George, I.M., et al. 1998, ApJS, 114, 73

Hamann et al. 1997, ApJ, 478, 78

Kaiser, M.E., Kriss, G.A., & Sembach, K.R. 2002, in Mass Outflow in Active Galactic Nuclei: New Perspectives, eds. D.M. Crenshaw, S.B. Kraemer, & I.M. George (San Francisco: ASP), p 75

Kaspi, S., Brandt, W.N., Netzer, H., Sambruna, R., Chartas, G., Garmire, G.P., & Nousek, J.A. 2000, ApJ, 535, L17

Kaspi, S. et al. 2001, ApJ, in press

Kraemer, S.B., et al. 2001, ApJ, 551, 671

Krolik, J.H., & Kriss, G.A. 1997, ASP Conference Series, 143, 271

Maran, S.P, Crenshaw, D.M., Mushotzky, R.F., et al. 1996, ApJ, 465, 733

Reichert, G.A., et al. 1994, ApJ, 425, 582

Sembach, K.R., Howk, J.C., Savage, B.D., & Shull, J.M. 2001, ApJ, 121, 992

Shields, J.C., & Hamann, F. 1997, ApJ, 481, 752

Mass Outflow in Active Galactic Nuclei: New Perspectives
ASP Conference Series, Vol. 255, 2002
D.M. Crenshaw, S.B. Kraemer, and I.M. George

Variable UV Absorption in NGC 3783 and NGC 4151

D. Michael Crenshaw

Department of Physics and Astronomy, Georgia State University, Atlanta, GA 30303

Steven B. Kraemer and Jack R. Gabel

Catholic University of America and Laboratory for Astronomy and Solar Physics, NASA's Goddard Space Flight Center, Code 681, Greenbelt, MD 20771

Abstract. We present new observations of the intrinsic UV absorption lines in the Seyfert 1 galaxies NGC 3783 and NGC 4151, which we obtained from the Space Telescope Imaging Spectrograph (STIS) on the Hubble Space Telescope (*HST*). We show that the observed variability in the column densities of the absorption lines can be attributed primarily to variable ionization in NGC 4151 and transverse motion in NGC 3783. In general, both sources of variability are likely present and operating on different time scales in Seyfert galaxies with intrinsic absorption. Intensive multiwavelength monitoring is needed to disentangle these effects and derive the parameters (densities, distances, transverse velocities) needed to test dynamical models of the outflowing UV and X-ray absorbers.

1. Introduction

Since the launch of the *International Ultraviolet Explorer (IUE)*, it has been known that the UV spectra of Seyfert 1 galaxies show absorption lines intrinsic to their nuclei (Ulrich 1988). With the advent of the *HST*, it is now understood that intrinsic absorption is a common phenomenon, present in more than half of the well-studied Seyfert 1 galaxies (Crenshaw et al. 1999). Among those Seyferts that show absorption, high ionization lines such as N V $\lambda\lambda$1238.8, 1242.8 and C IV $\lambda\lambda$1548.2, 1550.8 are always present, along with Lyα, while lower ionization lines, such as Si IV $\lambda\lambda$1393.8, 1402.8 and Mg II $\lambda\lambda$2796.3, 2803.5, are less common. The absorption lines are blueshifted (by up to 2500 km s^{-1}) with respect to the systemic velocities of the host galaxies, indicating net radial outflow. The ionic columns are highly variable, which may be the result of changes in the ionizing continuum or transverse motion of the absorbers. In either case, the variability indicates the close proximity of the absorbers to the central active nuclei of these galaxies.

The identification of the source of variability and the determination of the time scales of variability are important, because they provide information on physicial conditions that cannot be easily obtained by other means. If the variations are attributed to changing ionization, the time lag between the continuum and absorption variations provides the ionization and recombination time scales, and hence the radial location and density of the gas (Krolik & Kriss 1997; Shields & Hamann 1997). If the variations are attributed to motion of gas across the line of sight, then the transverse velocities can be obtained (cf., Maran et al. 1996). These quantities are crucial for testing the applicability of current dynamical models (cf., Proga et al. 2000; Bottorff et al. 2000) to the outflowing UV and X-ray absorbers.

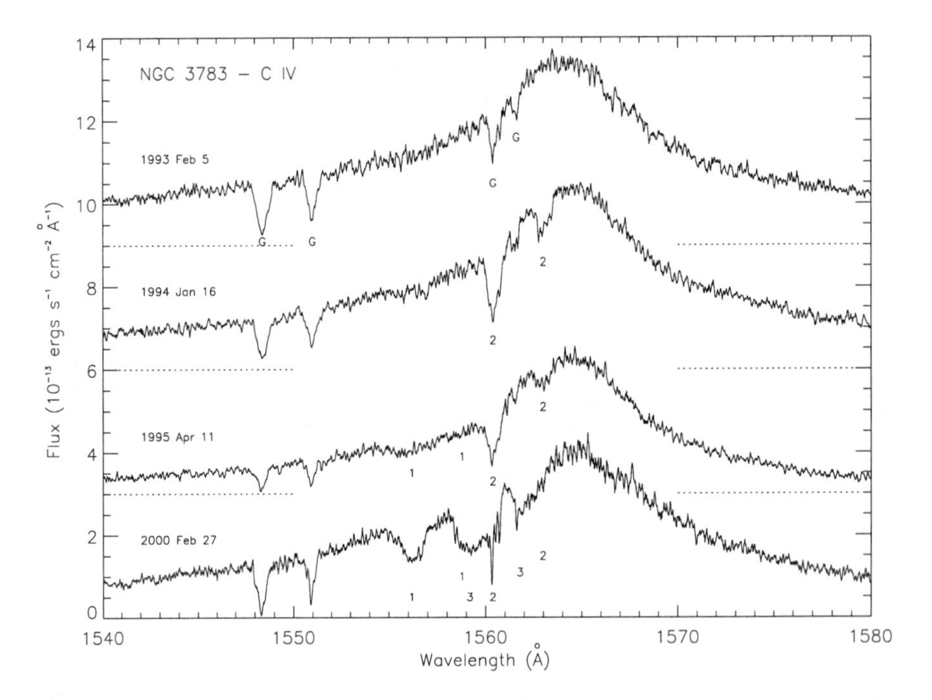

Figure 1. GHRS spectra (top three plots) and STIS spectra (lowest plot) of NGC 3783. The zero flux levels are given by dotted lines. Galactic absorption lines are indicated with a"G", and the three kinematic components of the intrinsic C IV absorption doublets are numbered.

2. Transverse Motion in NGC 3783

Figure 1 shows the extreme variability of the UV absorption in NGC 3783. The top three spectra are from GHRS observations obtained about one year apart, and the lowest spectrum is from a STIS echelle observation obtained about 5 years after the last GHRS spectrum. In the GHRS spectra, only Galactic absorption was present on 1993 February 5, but a C IV doublet (component 2) was present 11 months later at -550 km s^{-1} (relative to the systemic velocity). Another component (1) appeared $\sim$15 months later at -1370 km s^{-1}. In the

STIS spectrum, we see that component 1 became much stronger, and component 2 is weaker (confirmed in the corresponding N V absorption). Meanwhile, a new broad component (3) appeared at a radial velocity of -720 km s^{-1}. Clearly, we are undersampling the absorption variations in this object. However, it is interesting that the continuum flux levels for these spectra are all nearly the same, except for the 1995 spectrum. The appearance and disappearance of absorption components appears to be independent of continuum flux, which indicates that the total columns are changing on time scales of years due to transverse motion of gas across the line of sight.

We have measured the radial velocities, covering factors, and ionic column densities (or upper limits) for the kinematic components at each epoch and have calculated photoionization models to match the ionic columns (Kraemer et al. 2001a; Gabel et al. 2001). We find no evidence for variations in the radial velocity coverage for any kinematic component, consistent with previous findings for this and other Seyfert galaxies (Crenshaw et al. 1999). An unexpected result is that each component has a covering factor (of the continuum source plus broad emission-line region) of $\sim$0.6. The models predict a zone characterized by high ionization parameter (U = 0.65 – 0.80) and hydrogen column density (6.4 x 10^{20}– 1.5 x 10^{21} cm^{-2}) for each component, and a second, low ionization (U = 0.0018) and low hydrogen column density (4.9 x 10^{18} cm^{-2}) zone for Component 1 (Kraemer et al. 2001a; Gabel et al. 2001).

To quantify our conclusion that the absorption variations seen in Figure 1 are primarily due to transverse motion, Figure 2 shows the changes in continuum flux and C IV column density over time. There is no obvious correlation between the continuum and C IV variations. In the bottom plot of Figure 2, we show the ionization parameters needed to produce the observed C IV columns assuming the total hydrogen column densities determined from the STIS observations did not vary (i.e., assuming no tranverse motion). There is no correlation between the ionization parameter and observed continuum flux, which indicates that the source of variability is not likely to be variable ionization. The only reasonable alternative is transverse motion. Given the size of the C IV emitting region ($\sim$8 light days, Reichert et al. 1994), the covering factor, and the yearly variations, we estimate that the transverse velocities of these components are $\geq$3000 km s^{-1} (see Maran et al. 1996).

We note that the case for variable ionization as the primary source of these absorption variations is not completely dead if 1) the EUV continuum variations responsible for the ionization of C IV are not correlated with the observed UV continuum variations or 2) the time delay between continuum and absorption variations is sufficient to yield a complete lack of apparent correlation, given the undersampling of the variations. We are in the process of testing these scenarios with an intensive *HST*, *Chandra*, and *FUSE* monitoring campaign on this object.

3. Variable Ionization in NGC 4151

We have obtained a number of STIS spectra of NGC 4151 as part of an Instrument Definition Team key project (J. Hutchings, PI). In Crenshaw et al. (2001), we demonstrated that the broad absorption component (identified as D+E in the GHRS spectra of Weymann et al. 1997) showed a dramatic increase in the

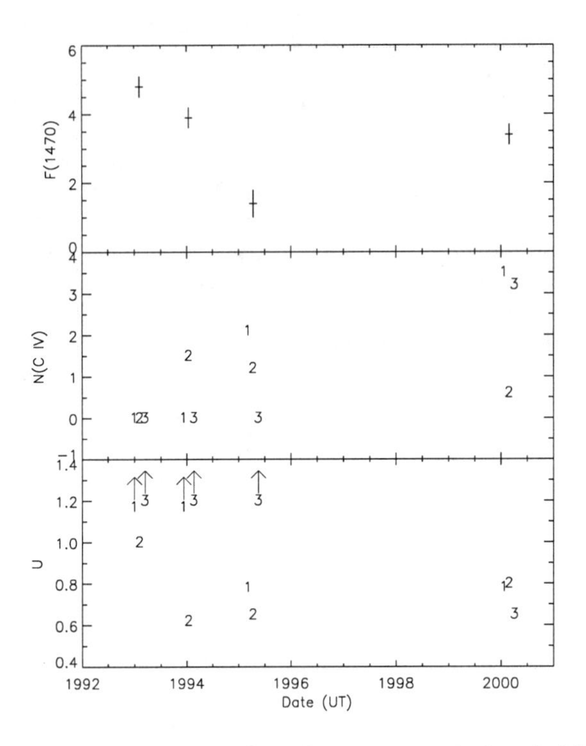

Figure 2. Continuum and C IV absorption variations in NGC 3783 as a function of Julian date. The top plot shows the continuum variations at 1470 Å in units of 10^{-14} ergs s^{-1} cm^{-2}Å^{-1}. The middle plot shows variations in the C IV column density in units of 10^{14} cm^{-2} for each component (1, 2, and 3). The lower plot shows ionization parameters (U) derived from photoionization models assuming constant hydrogen column density.

column densities of the low ionization lines (e.g., Si II, Fe II, and Al II) after a factor of ~4 decrease in the UV continuum over two years. We concluded that this occurence provides direct evidence for variable ionization of this component. We also showed that metastable C III and Fe II absorption are associated with this component, indicating very high densities ($> 10^{9.5}$ cm^{-3}). From photoionization models of this component (Kraemer et al. 2001b, 2001c), we have determined that it has a large total column (N_H = 2.75 x 10^{21} cm^{-2} and is very close to the central continuum source ($\leq$ 0.03 pc).

In Figure 3, we show the first STIS echelle spectrum obtained on 1999 July 19 and two new echelle spectra obtained on 2000 June 15 and 2000 November 5. In the 2000 June spectrum, the continuum flux decreased to an extremely low level, and the absorption lines are difficult to detect due to the low constrast. By 2000 November, however, the continuum and broad-line fluxes were up again, though not as high as in 1997 when the low-ionization lines were absent. The

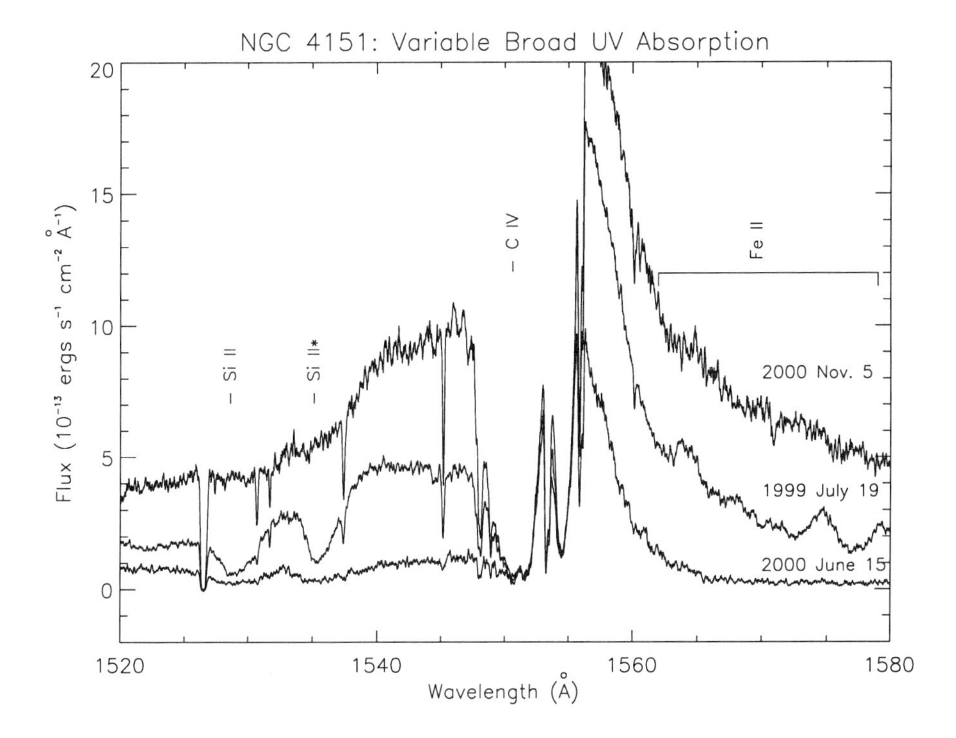

Figure 3. STIS echelle spectra of the nucleus of NGC 4151 in the region around the broad C IV emission line at three epochs. These spectra show the extreme variability of the broad absorption component (D+E, see Kraemer et al. 2001b) in the low-ionization species of Si II and Fe II.

weakness of the low-ionization lines in the 2000 November spectrum confirms the strong correlation between continuum and absorption variations in this object. Additional photoionization models of these data should be helpful in placing tighter constraints on the physical conditions in this broad component.

4. Conclusions

Unfortunately, the techniques we used for determining densities and distances of the absorbers in NGC 4151 (Kraemer et al. 2001b) cannot be applied to absorbers in most other Seyfert galaxies. The reason is that most Seyfert galaxies show only high-ionization lines, and nearly all of the fine-structure and metastable lines for high-ionization species are in the extreme ultraviolet ($\lambda < 912$ Å). Thus, intensive monitoring is probably the only way to determine the densities and distances of most of the UV absorbers in Seyfert galaxies.

We have shown that variations in the ionizing continuum and total column of gas in the line of sight are both sources of absorption-line variability. It is likely that both sources are operable on different time scales in a given Seyfert galaxy. To disentangle these effects and provide valuable constraints (e.g., densities, distances, and transverse velocities) on dynamical models of the absorbers, we have initiated a multi-wavelength monitoring campaign on NGC 3783, which will be completed by 2001 July.

The authors acknowledge support from NASA grant NAG5-4103. We thank John Hutchings for making the NGC 4151 spectra available for this paper.

References

Bottorff, M. 2000, ApJ, 537, 134

Crenshaw, D.M., et al. 1999, ApJ, 516, 750

Crenshaw, D.M., et al. 2000, ApJ, 545, L27

Gabel, J.R., et al. 2002, in Mass Outflow in Active Galactic Nuclei: New Perspectives, eds. D.M. Crenshaw, S.B. Kraemer, & I.M. George (San Francisco: ASP), p. 81

Kraemer, S.B., et al. 2001a, ApJ, in press

Kraemer, S.B., et al. 2001b, ApJ, 551, 671

Kraemer, S.B., et al. 2002, in Mass Outflow in Active Galactic Nuclei: New Perspectives, eds. D.M. Crenshaw, S.B. Kraemer, & I.M. George (San Francisco: ASP), p. 93

Krolik, J.H., & Kriss, G.A. 1997, ASP Conference Series, 143, 271

Maran, S.P, Crenshaw, D.M., Mushotzky, R.F., et al. 1996, ApJ, 465, 733

Proga, D., Stone, J.M., & Kallman, T.R. 2000, ApJ, 543, 686

Reichert, G.A., et al. 1994, ApJ, 425, 582

Shields, J.C., & Hamann, F. 1997, ApJ, 481, 752

Ulrich, M.-H. 1988, MNRAS, 230, 121

Weymann. R.J., Morris, S.L., Gray, M.E., & Hutchings, J.B. 1997, ApJ, 483, 717

Mass Outflow in Active Galactic Nuclei: New Perspectives
ASP Conference Series, Vol. 255, 2002
D.M. Crenshaw, S.B. Kraemer, and I.M. George

Physical Conditions in the UV Absorbers in NGC 4151

S.B. Kraemer

Catholic University of America and Laboratory for Astronomy and Solar Physics, NASA's Goddard Space Flight Center, Code 681, Greenbelt, MD 20771

D. M. Crenshaw

Department of Physics and Astronomy, Georgia State University, Atlanta, GA 30303

J.B. Hutchings

Dominion Astrophysical Observatory, National Research Council of Canada, Victoria, BC V8X 4M6, Canada

I.M. George

UMBC and Laboratory for High Energy Astrophysics, NASA's Goddard Space Flight Center, Code 662, Greenbelt, MD 20771

T.R. Gull

Laboratory for Astronomy and Solar Physics, NASA's Goddard Space Flight Center, Code 681, Greenbelt, MD 20771

M.E. Kaiser

Department of Physics and Astronomy, Johns Hopkins University, 3400 North Charles Street, Baltimore, MD 21218

C.H. Nelson and D. Weistrop

Department of Physics, University of Nevada, Las Vegas, 4505 Maryland Parkway, Las Vegas, NV 89154

Abstract. We have examined the physical conditions in intrinsic UV-absorbing gas in the Seyfert galaxy NGC 4151, using echelle spectra obtained with the Space Telescope Imaging Spectrograph (STIS) on the *Hubble Space Telescope (HST)*. The UV continuum of NGC 4151 was a factor of about 4 lower than in observations taken over the previous two years, and we argue the changes in the column density of the low ionization absorption lines associated with the broad component at -490 km s^{-1} reflect the decrease in the ionizing flux. Most of the strong absorption lines (e.g., N V, C IV, Si IV, etc.) from this component are saturated, but show substantial residual flux in their cores, indicating that the absorber does not fully cover the source of emission. For the first time in such a study, we have been able to constrain the densities for this kinematic component and several others based on the strength of absorption lines from metastable states of C III and Fe II, and/or the ratios of ground and fine structure lines of O I, C II, and Si II. Based on photoionization model predictions, we have been able to map the relative radial positions of the absorbers. None of the UV absorbers is of sufficiently large column density or high enough ionization state to account for the observed X-ray absorption, hence the X-ray absorption must arise in a separate component of circumnuclear gas.

1. Introduction

NGC 4151 is the first Seyfert galaxy known to show intrinsic absorption that could be attributed to the active nucleus. Ultraviolet observations of NGC 4151 by the *IUE* (Boksenberg et al. 1978) and subsequent far-ultraviolet observations by the *Hopkins Ultraviolet Telescope* (*HUT*, Kriss et al. 1992) revealed a number of absorption lines from species that span a wide range in ionization potential (e.g., O I to O VI), as well as fine-structure and metastable absorption lines. The intrinsic UV absorption was found to be variable in ionic column density, but no variations in radial velocities were detected (Bromage et al. 1985). The ionization state of the UV absorbers are generally well correlated with the UV continuum flux level; the equivalent widths of low ionization species are negatively correlated with the flux level, while the those of the high ionization species are positively correlated (Bromage et al. 1985). Goddard High Resolution Spectrograph (GHRS) spectra, obtained by Weymann et al. (1997), revealed that the C IV and Mg II absorption lines, detected in six major kinematic components, were remarkably stable over the time period 1992 – 1996. X-ray spectra of NGC 4151 reveal the presence of a large intrinsic column of atomic gas, which is most likely ionized, as suggested by Yaqoob, Warwick, & Pounds (1989).

2. Observations

We observed the nucleus of NGC 4151 with a $0''.2 \times 0''.2$ aperture and the STIS E140M and E230M gratings to obtain a spectral coverage of 1150 – 3100 Å at a velocity resolution of 7 – 10 km s^{-1}. We obtained the STIS echelle spectra on 1999 July 19, when the continuum and broad emission lines were in a low state compared to previous STIS and GHRS observations (see Crenshaw et al. 2000; hereafter Paper I). We found that the UV continuum decreased by a factor of $\sim$4 over the previous two years, which resulted in a dramatic increase in the column densities of the broad components of the low-ionization absorption lines (e.g., Si II, Fe II, and Al II).

Figure 1 shows the kinematic components in the Si IV λ1393.8 and S II λ1253.8 lines. The broad absorption features (FWHM = 435 km s^{-1}) appear at a heliocentric redshift of 0.0017 (velocity centroid = -490 km s^{-1} with respect to the nucleus), and correspond to a blend of kinematic components D and E in Weymann et al. (1997). In addition, Component D+E is present in the resonance lines of C IV, N V, Mg II, excited fine-structure lines of C II, O I, and Si II, and the metastable C III* λ1175 blend. The narrow kinematic components of absorption identified by Weymann et al. (1997) are also shown in Figure 1. For the relatively high ionization line of Si IV, components A, B (Galactic), C, D+E (the broad feature), F, and F$'$ are evident (F and F$'$ likely arise in the interstellar medium or halo of NGC 4151, Weymann et al. 1997). For the low ionization line of S II, components B, D+E, E$'$, and F are present (E$'$ is not seen in the high ionization lines). In addition, we discovered a new component, which we call D$'$; this component is shallow, broad (FWHM $\approx$ 940 km s^{-1}), highly blueshifted ($-$1680 km s^{-1}), and only seen in the high ionization lines (N V, C IV, and Si IV). The D$'$ component is not detected in previous STIS or GHRS spectra obtained at higher states but it is clearly seen as a flattening of the C IV absorption profile around 1544 Å in the STIS echelle spectra (Paper I). Figure 2 shows the D$'$ component in the N V λ1238.8 absorption.

The measurement of the absorption components, the details of the covering factor calculations and evidence for heavy saturation of Component D+E and a strong unocculted component of UV light are discussed in Kraemer et al. (2001, hereafter Paper II). The unocculted spectrum must consist of continuum, broad emission lines, and broad absorption lines similar to those of the nuclear spectrum itself. The most straightforward explanation is that there is a scattering region that is outside of the region responsible for the D+E component, which reflects a fraction of the nuclear spectrum into our line of sight.

3. Modeling The Absorbers

The details of the photoionization models (e.g., the element abundances, spectral energy distribution (SED) of the ionizing continuum, and the criteria for a "successful" model) are given in Paper II. As discussed in Paper I (and references therein), the UV continuum of NGC 4151 is highly variable. In several cases, we have compared our models of the current ("low" state) absorbers with the high state absorbers observed with the GHRS; in order to do so, we increased the ionizing flux by a factor of 4, while, for simplicity, using the same SED, although there is some evidence for variations in the latter as a function of continuum flux.

As discussed in Paper II, the predictions of our photoionization models succesfully matched the measured ionic columns densities for the narrow absorption components A, C, and E$'$. In addition, we were able to use the fine structure lines to constrain the densities of these components. For Component A, n_e (electron density) is $\sim$ 100 cm^{-2} and we were able to fit the ionic columns densities of A in the STIS spectra with a single component photoionization mode assuming an ionization parameter U = 0.0012, and column density N_{eff} = 1.3 x 10^{18} cm^{-2}. For Component C, n_e is $\sim$ 10 cm^{-2}, U= 0.0012, and N_{eff} = 1.0 x 10^{18} cm^{-2}, and, for Component E$'$, n_e is $\geq$ 10^6 cm^{-3}, U = 0.0001, and N_{eff} = 1.5 x 10^{18} cm^{-2}.

The cumulative absorption component (D+E) possesses strong high ionization lines, e.g. N V, C IV, and Si IV. In our spectra, a number of lower ionization absorption lines are associated with Component D+E, the most striking of which are those from metastable levels of Fe II (Paper I), including those with lower levels up to 4.1 eV above the ground state. With the resolution of the STIS echelle, we have now established that the C III* λ1175 absorption, which arises from a level 6.5 eV above the ground state, is also associated with D+E, indicating densities of $n_e \geq 10^9$ cm^{-3}. The only isolated unsaturated lines associated with D+E are S II λ1254 (which has a weak oscillator strength), Ni II λ1317 (which has a low abundance), and some of the weaker Fe II metastable lines. We have, therefore, used the S II and Ni II columns to constrain the fraction of low ionization gas in the absorber. Although there are no weak high ionization lines in the STIS data, P V $\lambda\lambda$1117.98, 1128.01 has been detected in HUT Astro-1 and Astro-2 observations (Kriss et al. 1992; Kriss et al. 1995) and with the Berkeley sepctrometer aboard *Orbiting and Retrievable Far and Extreme Ultraviolet Spectrometers (ORFEUS) - Shuttle Pallet Satellite (SPAS) II* mission (Espey et al. 1998). Based on the equivalent widths determined by Espey et al., and the appearance that the lines are unsaturated, N(P V) is $\approx$ 1.2 x 10^{14} cm^{-2}. Since the ionization potentials of P IV and C III are similar (51.4 eV and 47.9 eV, respectively), it is reasonable to expect that the ratio of ionic columns of P V and C IV will be approximately the same as that of the elemental abundances of phosphorus and carbon (i.e, 0.001; cf. Grevesse & Anders 1989). Hence, at the time of *ORFEUS-SPAS II* observations, during which NGC 4151 was in a high flux state, N(C IV) must have been $\sim 10^{17}$ cm^{-2}. Based on these assumptions, we have successfully modeled the Component D+E as a single slab, with U = 0.015, n_H = 3.2 x 10^9 cm^{-2}, and N_{eff} = 2.75 x 10^{21} cm^{-3} (see Paper II). As an additional test of the model, we ran a high state version, increasing U by a factor of 4 (see above), while holding N_{eff} fixed. In this case, the Fe II and Si II columns are small enough that their absorption lines would be weak or below the limit of detectability, which is precisely the case for the earlier STIS (Paper I) and GHRS (Weymann et al. 1997) observations. The higher ionization lines, such as N V $\lambda\lambda$1238, 1241 and C IV $\lambda\lambda$ 1548, 1551, would remain heavily saturated, while N(P V) is close to the value derived from the *ORFEUS - SPAS II* observations. Notably, D+E does not have sufficient X-ray opacity to produce the observed X-ray absorption.

For Component D', N(Si IV) is = 4.5 x 10^{14} cm^{-2}, hence there must be a significant column of C III, although no corresponding C III* λ1175 is detected, indicating that the electron density is $< 10^{9.5}$cm^{-3}. In paper II, we argued that the coincidence of the appearance of D' at the same time as the low ionization lines from D+E can be best explained if the former is screened by the latter. To demonstrate this, we generated a single-component model, with N_{eff} = 1 x 10^{20} cm^{-2}, using the ionizing continuum filtered by D+E as the input spectrum. Assuming that these components are at the same approximate radial distance, the D' component must have U = 0.012, which yields a density of n_H = 2.6 x 10^7 cm^{-3}. As shown in Paper II, the model predictions provide a good fit to the N V, C IV, and Si IV columns, and indicate that Si II and Mg II lines should be weak or absent (due to the small column densities and broadness of this feature). A second model was generated, using the filtered continuum from the high state model for Component D+E as the input, while holding n_H and N_{eff} fixed, which results in U = 4.15. As required by the observations, the predicted column densities are quite small. Therefore, the appearance of D' while NGC 4151 was in its low state is the direct result of the higher EUV opacity of D+E.

4. Discussion

We estimate that the number of ionizing photons emitted by the central source is Q = 2 x 10^{53} s^{-1} (Paper II). From our determination of their densities and ionization parameters, we place the absorbers at the following radial distances: Component C, 2.15 kpc; Component A, 681 pc; Component E', 23.6 pc, and Component D+E, 0.03 pc. Based on our model results, the absorbers decrease in density with increasing radial distance, faster than distance^{-2} up to Component A. We estimate that at least 15% of the high-state continuum emission, broad emission lines, and absorption lines are scattered into our line of sight, from a plasma that extends outside the solid angle subtended by D+E. For a covering factor of unity, the scatterer must have a free electron column density of N_{eff} $\sim$ 2 x 10^{23} cm^{-2}. Based on the constraints on the total Hβ flux, the covering factor, and the EUV opacity of the scatterer, we argue that it must be highly ionized (U $\geq$ 4.5) and, hence, cannot produce the observed X-ray absorption. Instead, a smaller column of lower ionization gas (N_{eff} = 4 x 10^{22} cm^{-2}, U = 0.86) must also be present.

References

Boksenberg, A., et al. 1978, Nature, 275, 404

Bromage, G.E., et al. 1985, MNRAS, 215, 1

Crenshaw, D.M., et al. 2000, ApJ, 545, L27 (Paper I)

Espey, B.R., et al. 1998, ApJ, 500, L13

Grevesse, N., & Anders, E. 1989, in Cosmic Abundances of Matter, ed. C.J. Waddington (New York: AIP), 1

Kraemer, S.B., et al. 2000, ApJ, 531, 278 (Paper II)

Kriss, G.A., Davidsen, A.F., Zheng, W., Kruk, J.W., & Espey, B.R. 1995, ApJ, 454, L7

Kriss, G.A., et al. 1992, ApJ, 392, 485

Weymann, R.J., Morris, S.L., Gray, M.E., & Hutchings, J.B. 1997, ApJ, 483, 717

Yaqoob, T., Warwick, R.S., & Pounds, K.A. 1989, MNRAS, 236, 153

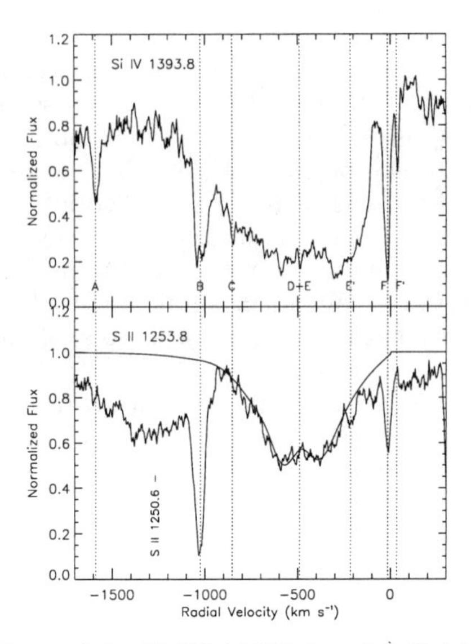

Figure 1. Plots of the Si IV λ1393.8 and S II λ1253.8 lines as functions of radial velocity. The expected positions of the absorption lines are plotted as vertical dotted lines. The smooth line in the lower plot is our fit to the D+E component of S II.

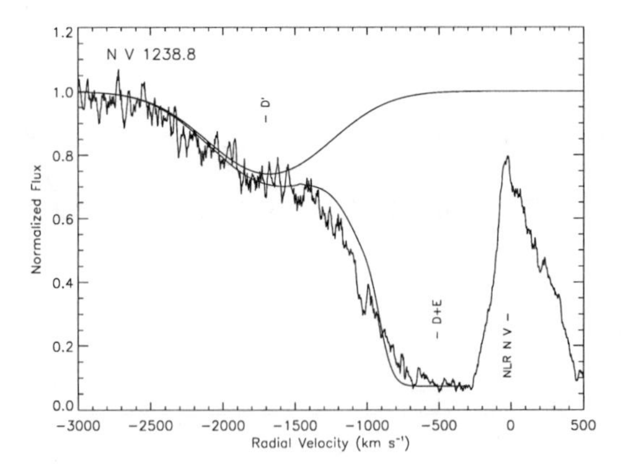

Figure 2. Plot of the N V λ1238.8 absorption as a function of radial velocity. The upper plot gives our fit to the D′ and D+E components.

Mass Outflow in Active Galactic Nuclei: New Perspectives
ASP Conference Series, Vol. 255, 2002
D.M. Crenshaw, S.B. Kraemer, and I.M. George

The Luminosity Dependence of UV Absorption in AGNs

Ari Laor

Technion, Physics Dept., Haifa 32000, Israel

Niel Brandt

Penn State University, Dept. of Astronomy & Astrophysics, 525 Davey Lab, University Park, PA 16802

Abstract. We describe the results of a survey of the UV absorption properties of the Boroson & Green AGN sample, which extends from the Seyfert ($M_V = -22$) to the quasar ($M_V = -27$) level, based on data available from the *HST* archives. The absorption equivalent width EW(abs), and the maximum outflow velocity $v_{\rm max}$, in the soft X-ray weak quasars (SXWQs), increase with increasing optical luminosity and decreasing EW([O III]). Thus, BALQs are the high-luminosity subset of SXWQs. Non SXWQs have lower EW(abs) and $v_{\rm max}$ and do not show the above strong correlations. This, and the low EW([O III]) of SXWQs, suggests that strong UV and X-ray absorption (without optical continuum absorption) occurs in a physically distinct group of AGN, and is not just an inclination effect. The correlation of $v_{\rm max}$ and continuum luminosity is expected for radiation pressure driven outflows, but the strong correlation of $v_{\rm max}$ and EW([O III]) remains a puzzle. Some further outstanding questions are briefly outlined.

1. Introduction

Fast outflows are common in AGNs. The outflow properties are very different in high and low-luminosity AGNs. Whereas luminous high z quasars display outflows reaching a few 10^4 km s^{-1}, Seyfert galaxies display typical outflow velocities up to only 10^3 km s^{-1}. *Why are the outflow properties so different at low and high-luminosity? How do the outflow properties vary with luminosity? Are AGNs with outflows identical to other AGNs, or are they physically distinct objects? How is the UV absorption related to X-ray absorption?* The answers bear important clues to the origin of AGN outflows, and their acceleration mechanism. In this study we make a first step toward a uniform and systematic survey of the UV absorption properties of AGN from the Seyfert to the luminous quasar level. We use archival *HST* observations of the Boroson & Green (1992; hereafter BG92) sample, which includes the 87 $z < 0.5$ Palomar-Green (PG) AGNs, and extends from Seyferts at $M_V = -21$ to luminous quasars at $M_V = -27$. The high-quality optical data set available from BG92 allows us to explore relations between quasar absorption and emission properties, in particular the BG92 eigenvector 1, which may be an indicator of $L/L_{Eddington}$.

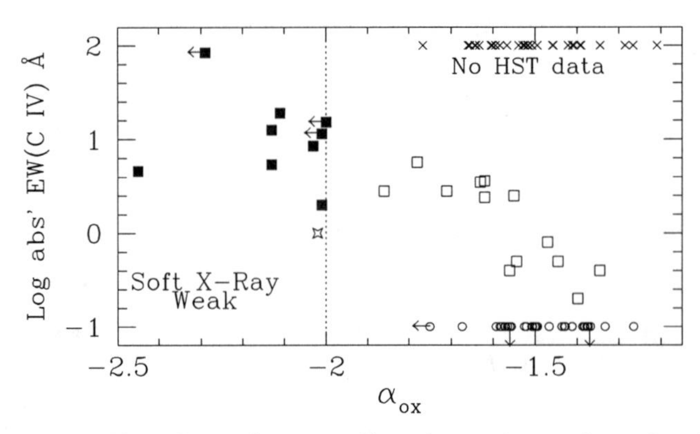

Figure 1. The relation between C IV absorption and α_{ox} for the BG92 AGNs (see also Fig. 4 in BLW). SXWQs are marked as filled squares, other absorbed AGNs by open squares. Objects with *HST* upper limits are marked by open circles, and objects with no *HST* data by crosses. The star marks PG 2214+139 for which C IV absorption is questionable.

2. Method

We retrieved all public spectra of the BG92 AGNs which are available in the *HST* archives (nearly all from the Faint Object Spectrograph). We measure the absorption for the C IV line, since this is the most commonly observed line. The absorption is parametrized here using two numbers, the absorption equivalent width, EW(C IV), and maximum velocity of absorption, $v_{\rm max}$. Since the C IV absorption is rather weak in some objects, we verified that it is also present in either N V or Lyα, which are available for most objects. Absorption was searched for from 0 to 20,000 km s^{-1} in blueshift with respect to the redshift determined from [O III] λ5007. The *HST* spectra were complemented by four UV spectra from the *IUE* archives (see Brandt, Laor & Wills 2000; hereafter BLW). The 3000 Å to 2 keV spectral slopes, α_{ox}, for all objects are taken from BLW.

3. Results & Discussion

Line profiles were available for C IV in 48 of the objects, and for N V+Lyα only in five other objects. Blueshifted absorption with respect to [O III] λ5007 was detected in 15 objects, and very weak or marginal systemic (or slightly redshifted) absorption was detected in four objects, which are not further discussed here.

BLW have shown that EW(C IV) is strongly correlated with α_{ox}. A somewhat revised version of their Figure 4 is shown here in Figure 1. The correlation between UV absorption and X-ray weakness strongly suggests that the X-ray weakness is due to absorption, and is not an intrinsic property. This complements the discovery of Green & Mathur (1996) that nearly all BALQs are soft X-ray weak. Hard X-ray observations of some of the steep α_{ox} quasars revealed a high-energy recovery (e.g. Gallagher et al. 2001), which provides a clear proof they are highly absorbed.

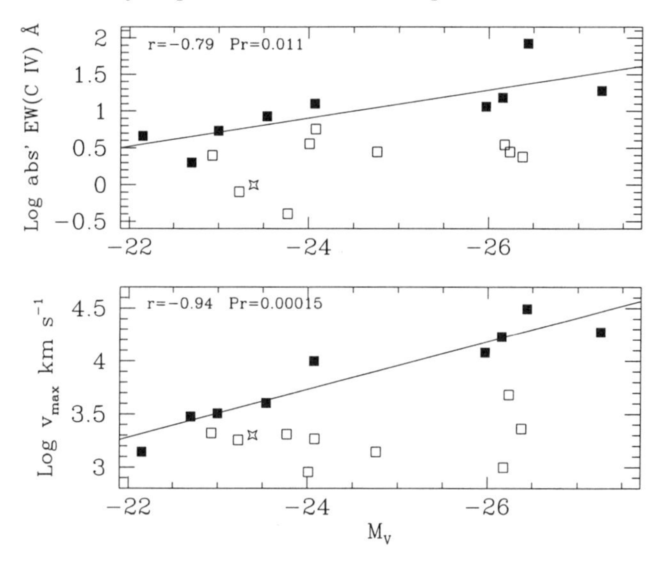

Figure 2. The luminosity dependence of EW(C IV) and $v_{\rm max}$ for SXWQs (filled squares) and non SXWQs (open squares). SXWQs have a higher EW(C IV) and $v_{\rm max}$ than non SXWQs at any given luminosity. Both parameters are significantly correlated with luminosity for the SXWQs. The Spearman correlation coefficient, and its significance level are indicated in each panel.

BLW defined a complete sample of Soft X-ray Weak quasars (SXWQs), which includes all 10 AGNs from BG92 with $\alpha_{ox} \leq -2$. The EW(C IV) in SXWQs covers a wide range ($1 - 100$ Å), and although practically all BALQs are SXWQs, the reverse is not true as only some of the SXWQs can be defined as BALQs. *What is it that determines whether a SXWQ is a BALQ?*

The answer appears to be remarkably simple. Figure 2 shows that the EW(C IV) and $v_{\rm max}$ of SXWQs are strongly correlated with the continuum luminosity. A least squares fit to the SXWQs gives EW(C IV)$\propto L^{0.48\pm0.14}$, and $v_{\rm max} \propto L^{0.56\pm0.07}$. BALQs are just the more luminous SXWQs. The absorption in non SXWQs has both a lower EW(C IV) and a lower $v_{\rm max}$ than that of the SXWQs with the same continuum luminosity. Both EW(C IV) and $v_{\rm max}$ appear to be independent of luminosity for the non SXWQs. The X-ray weakness criterion thus appears to differentiate between two qualitatively different UV absorbers in AGN.

The strong luminosity dependence of $v_{\rm max}$ is consistent with some radiation pressure acceleration scenarios. For example, if the BAL outflow is driven by radiation pressure acceleration on dust grains residing in clouds moving at some Keplerian velocity $v_{\rm Kep}$, then for optically thin gas $v_{\rm max} \propto v_{\rm Kep} \times \sqrt{\Gamma L/L_{\rm Edd}}$, where the force multiplier, Γ, depends on the dust properties and dust/gas ratio. The maximum Keplerian velocity $v_{\rm Kep,max}$ is obtained at the minimum radius where grains survive, $R_{\rm min} \propto L^{1/2}$, and thus $v_{\rm Kep,max} \propto M_{\rm BH}^{1/2}/L^{1/4}$. Since $L/L_{\rm Edd} \propto L/M_{\rm BH}$ we obtain that $v_{\rm max} \propto M_{\rm BH}^{1/2}/L^{1/4} \times \sqrt{L/M_{\rm BH}}$, or simply

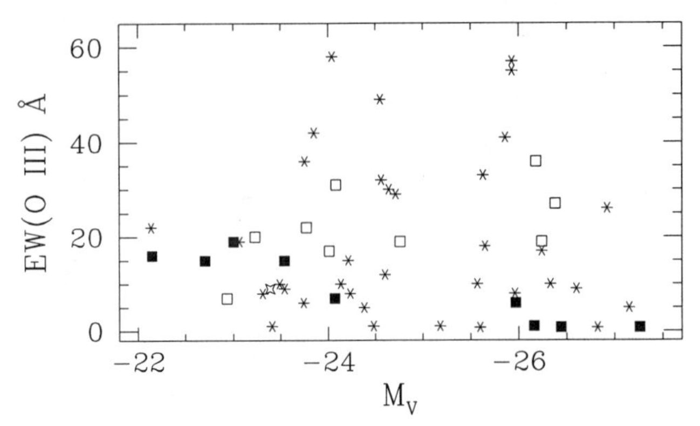

Figure 3. The EW([O III]) distribution for the BG92 AGNs with UV spectra. SXWQs (filled squares) have systematically low EW([O III]), while the EW([O III]) distribution of the other UV absorbed AGNs (open squares) is similar to that of the unabsorbed AGNs (stars).

$v_{\rm max} \propto L^{1/4}$, independent of $M_{\rm BH}$. The observed slope 0.56 ± 0.07 is steeper than this simplified model prediction of 0.25, but the difference may not be significant given the small sample size.

As pointed out by BLW (see their Fig. 3), SXWQs tend to lie toward the weak [O III] end of the BG92 eigenvector 1. Figure 3 shows the positions of all the AGNs studied here in the EW([O III]) vs. M_V plane. The SXWQs occupy only the lower portion of this plane, where they form $\sim 1/3$ of the AGNs. None of the AGNs with a high EW([O III]) are SXWQs. Since [O III] is most likely emitted isotropically (e.g. Kuraszkiewicz et al. 2000 and references therein), the weakness of [O III] strongly suggests SXWQs are physically distinct from the high-EW([O III]) AGNs (see discussion in BLW §8). This is contrary to the common interpretation in which BALQs, a subset of SXWQs, are normal quasars seen edge on (e.g. Weymann et al. 1991). However, SXWQs may form a closer to edge on sub-population of the low-EW([O III]) AGNs. This conclusion agrees with the one reached by Boroson & Meyers (1992) for low-ionization BALQs, and by Turnshek et al. (1997) for normal BALQs.

Are the UV absorption properties related to the set of optical emission-line correlations (EV1) of BG92? Figure 4 shows that EW(C IV), and in particular $v_{\rm max}$, of the SXWQs are significantly correlated with the EW([O III]). However, we do not find any significant relation with the Fe II strength, or with the Hβ FWHM, although both are major components of EV1, which suggests the correlations in Fig. 4 are not just due to a correlation with EV1.

Is there a direct physical link between $v_{\rm max}$ *and* EW([O III])? Possible direct links may be the destruction of NLR clouds by very fast outflows, or maybe an increased obscuration of the NLR due to interactions of the fast outflow with nearby dense clouds which increases their scale height.

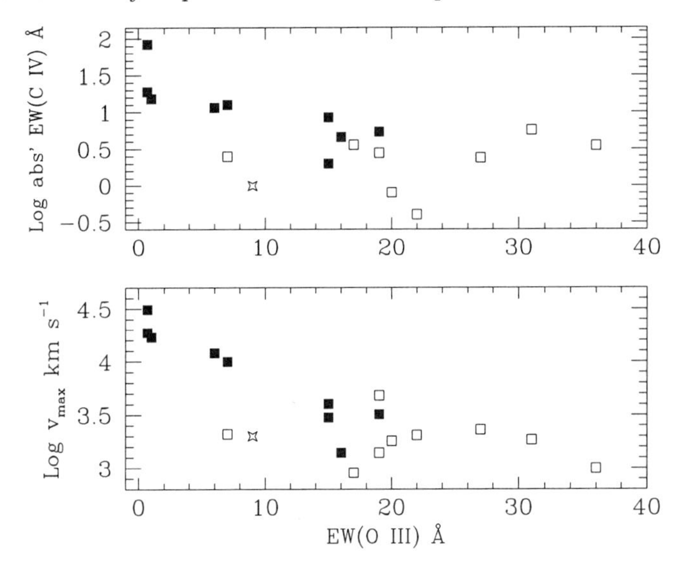

Figure 4. The dependence of absorption EW(C IV) and $v_{\rm max}$ on EW([O III]). The upper panel is consistent with the known tendency of BALQs to have a low EW([O III]). The strong dependence of $v_{\rm max}$ on EW([O III]) for the SXWQs is remarkable and puzzling.

4. Some Further Questions

Do any of the remaining 34 PG AGNs show significant UV absorption? If the answer is no, and the upper right part of Fig. 1 remains empty, that would imply that it is apparently impossible to get strong UV absorption without corresponding X-ray absorption. This result is not trivial as one can imagine a low column absorber ($N_{\rm HI} \sim 10^{20}$ cm^{-2}) which can produce very strong UV absorption, but no X-ray absorption. The complete absence of such absorbers would have important theoretical implications.

Are the X-ray and UV absorbers the same? The strong link found in BLW (Fig. 1) suggests they are. However, simple estimates of the UV columns appear to be highly discrepant with the X-ray columns, and such discrepancies appear to remain even in the two best studied low and high-luminosity AGNs (Kraemer et al. 2001; Arav, these proceedings). If they are the same, they should share the same column and the same kinematics. The available UV spectroscopy allows an accurate handle on the outflow kinematics, but only weak constraints on the column, while the reverse is true for the available low resolution X-ray spectroscopy. This problem can be alleviated, at least in part, by getting high-quality UV spectroscopy of low-luminosity SXWQs, where the low outflow velocities allow one to resolve various absorption doublets, and thus better constrain the absorbing column. In addition, one can also obtain X-ray spectra of high-luminosity SXWQs, where the high outflow velocities would be easier to resolve in X-ray spectroscopy, even if the S/N is not very high (e.g. PG 2112+059 in Gallagher et al. 2001).

What determines the outflowing column? The results presented here suggest that the maximum outflow velocity is set by the luminosity, as some radiation pressure scenarios suggest. Radiation pressure can just overcome gravity and start a radially accelerating outflow for a column of $1.5 \times 10^{24} L/L_{\rm Edd}$ cm^{-2}, assuming the outflow is optically thick. For example, objects with $L/L_{\rm Edd} \lesssim 0.01$ should not be able to drive an outflow with $\tau > 1$ at 2 keV. Combined X-ray and UV spectroscopy may allow us to obtain the absorbing column as a function of outflow velocity, and thus test directly the predictions of radiation pressure acceleration models.

Why are there no SXWQs with a strong [O III]? All the SXWQs in BG92 sample have EW([O III])$<$ 19Å. There are 35/87 AGNs in the BG92 sample (i.e. 40%) with EW([O III])$\geq$ 20Å, and if the frequency of X-ray weakness (10/87) were independent of EW([O III]), then 4 of these 35 AGNs should have been SXWQs. Yet, none of them are (0.018 significance level). Since the strength of [O III] is not just an inclination effect, it implies the existence of "high-EW([O III]) type" AGNs which for some reason do not show X-ray absorption. Do these AGNs just lack high column density clouds ($N_{\rm H\ I} \geq 10^{22}$ cm^{-2}) capable of X-ray absorption? This seems rather implausible as models of the BLR strongly suggest larger column clouds are present. A possible alternative explanation is that the X-ray absorbing gas in high-EW([O III]) AGNs has, for some reason, a normal dust/gas ratio, and this would make these AGNs "optically weak" as well. This can be tested in hard X-ray surveys which may reveal the missing population of high-EW([O III]) AGNs with strong X-ray absorption.

The first step toward answering some of the above questions is to make a complete survey of the UV absorption properties of the BG92 AGNs. Together with the BLW study, this will establish complete UV + soft X-ray coverage of the sample. This can then be followed by detailed UV and X-ray spectroscopy of all the "interesting" AGNs, which may lead to a solution of some of the above questions. A complete survey is a resource intensive approach, but unlike the study of individual AGNs, will allow one to draw conclusions about the general AGN population. In addition, it is important to study the new relations described here using a larger complete sample of SXWQs.

References

Boroson, T. A., & Green, F. G. 1992, ApJS, 80, 109 (BG92)

Boroson, T. A., & Meyers, K. A. 1992, ApJ, 397, 442

Brandt, W. N., Laor, A., & Wills, B. J. 2000, ApJ, 528, 637 (BLW)

Gallagher, S. C., Brandt, W. N., Laor, A., Elvis, M., Mathur, S., Wills, B. J., & Iyomoto, N. 2001, ApJ 546, 795

Green, P. J. & Mathur, S. 1996, ApJ, 462, 637

Kraemer, S. B., et al. 2001, ApJ, 551, 671

Kuraszkiewicz, J., Wilkes, B. J., Brandt, W. N., & Vestergaard, M. 2000, ApJ, 542, 631

Turnshek, D. A., Monier, E., Sirola, C. J., & Espey, B. R. 1997, ApJ, 476, 40

Weymann, R. J., Morris, S. L., Foltz, C. B., & Hewett, P. C. 1991, ApJ, 373, 23

Mass Outflow in Active Galactic Nuclei: New Perspectives
ASP Conference Series, Vol. 255, 2002
D.M. Crenshaw, S.B. Kraemer, and I.M. George

Ultraviolet Absorption in LINERs

Joseph C. Shields[1], Bassem M. Sabra[2], Luis C. Ho[3], Aaron J. Barth[4], and Alexei V. Filippenko[5]

[1]*Ohio University, Physics & Astronomy Dept, Athens, OH 45701*

[2]*University of Florida, Astronomy Dept, Gainesville, FL 32611*

[3]*Obs. of the Carnegie Institution of Washington, Pasadena, CA 91101*

[4]*Harvard-Smithsonian Center for Astrophysics, Cambridge, MA 01238*

[5]*University of California, Astronomy Dept, Berkeley, CA 94720*

Abstract. LINERs are a very common form of AGN, and it is natural in the context of this meeting to ask whether they show evidence of outflows of the type seen in Seyferts and QSOs. Ultraviolet spectroscopy suitable for measuring resonance line absorption is available for only a small number of LINERs. These data indicate that a significant fraction of LINERs may not be accretion-powered systems at all, but instead are small-scale starburst phenomena. Low-ionization line absorption is seen in the majority of observed LINERs, including both AGN-like and starburst-like systems. In most if not all of these objects, however, the absorption may arise from the normal interstellar medium of the host galaxy, rather than in material residing near or outflowing from the central source of activity. The lack of strong outflow signatures in accretion-powered LINERs, as compared with more powerful systems, may reflect their low luminosities, or differences in the underlying accretion structure.

1. Introduction

Low-Ionization Nuclear Emission-Line Regions, or LINERs (Heckman 1980), are a class of low-luminosity nebular sources that have been the subject of growing attention over the past 20 years. These objects are classified on the basis of optical emission-line ratios, and show emission in low-ionization features (e.g. O I, N II, S II) that is enhanced beyond that typically seen in H II regions. The Hα luminosities of LINERs are, however, often comparable to those of giant H II regions. One reason for the growing interest in LINERs is the recognition that they are common, with surveys revealing their existence in $\sim 40\%$ of bright galaxies (Ho et al. 1997).

A question central to the understanding of LINERs is the nature of their power source(s). In some cases these objects are clearly weak versions of classical AGN, as revealed by broad Hα emission, hard X-rays, and high brightness-temperature radio cores. In many sources, however, the evidence is more ambiguous, and stellar phenomena remain a possibility for generating the observed nebulosity. The addition of information on outflows associated with LINERs is one further means for investigating the physical nature of these objects.

2. LINERs as AGN

For those LINERs that are clearly AGN, increasing attention has been devoted to understanding the accretion physics that describes these systems. If the underlying black holes have masses comparable to those inferred for luminous AGN and some nearby quiescent galaxies, the low luminosities of LINERs imply that they are accreting at well below Eddington rates. As a result, we might expect the accretion mode in LINERs to differ from that of luminous AGN, and several observational findings support this idea. A growing number of LINERs are now known to exhibit double-peaked or double-shouldered Hα emission (Barth et al. 2001 and references therein), which resembles line profiles in some broad-line radio galaxies that have been interpreted as emission from an accretion disk (Eracleous & Halpern 1994 and references therein). The spectral energy distribution of LINERs is also noteworthy for the absence of the "Big Blue Bump," and the presence of radio emission sufficient to label these objects as radio-loud (Ho 1999).

The emerging paradigm that may account for the distinct properties of LINERs posits that in the low-$\dot{M}$ regime, accretion no longer proceeds through a thin disk, but instead occurs through an Advection-Dominated Accretion Flow (ADAF), Adiabatic Inflow-Outflow Solution (ADIOS), or similar structure (e.g., Narayan & Yi 1994; Blandford & Begelman 1999). In this picture, the inner part of the accretion disk is replaced by a hot torus undergoing quasi-spherical inflow. The lack of an inner thin disk explains the missing Big Blue Bump, while X-rays from the torus irradiating the outer thin disk account for the double-peaked Balmer emission. Material entrained in hot outflows from this accretion structure might be detectable in UV line absorption.

3. UV Properties of LINERs

Imaging surveys with *HST* have found that only $\sim$ 20% of LINERs are detectable as UV sources (Maoz et al. 1995; Barth et al. 1998). Extinction and reddening effects are clearly significant in many cases, and even when detected, most LINERs are relatively weak UV emitters. As a result, UV spectra from *HST* are available for only $\sim$ 10 sources.

Efforts to study AGN-related outflow phenomena with the available UV spectra are subject to some significant complications. In luminous AGN there are suspicions that the emission and absorption line properties may be linked in some way, for example, with both tracing an outflowing disk wind. For LINERs, however, the UV emission properties of LINERs are very heterogeneous, in terms of line strengths and widths. In addition, a number of LINERs show evidence that the UV spectrum is dominated by radiation from hot *stars*, rather than an AGN. A survey of the available spectra indicates, in fact, that LINERs can be sorted into two broad categories that can be characterized as "AGN-like" and "starburst-like." The available spectra provide roughly equal numbers of examples for each class, although this ratio should not be adopted as the intrinsic value, due to potentially strong selection effects. In the following sections we describe two case studies exemplifying these types.

3.1. NGC 4569

Ultraviolet spectra of the LINER in NGC 4569 were published by Maoz et al. (1998). The data show no emission lines, but unambiguous absorption features attributable to hot O stars, notably C IV and Si IV absorption, are present; the overall continuum structure is well matched by a starburst template with an age of 4 – 5 Myr. In addition to photospheric features, the data show narrow interstellar absorption lines from low-ionization species: C I, C II, O I, Si II, Fe II. If solar abundances are assumed, the total column density implied by these measurements is $N_H \gtrsim 10^{19}$ cm^{-2}, or $\gtrsim 10^{21}$ cm^{-2} if iron is depleted onto grains at a level typical of the Milky Way. These values are consistent with extinction estimates of $A_V \approx 2$ obtained from Balmer line ratios, assuming a standard dust/gas ratio and extinction curve (see Maoz et al. 1998 for further discussion). Finally, the interstellar features show no indication of being offset in velocity from the stars ($\Delta v \lesssim 40$ km s^{-1}).

The picture that emerges is that the nucleus of NGC 4569 is essentially a small-scale starburst phenomenon, despite its LINER-like spectrum. Efforts to understand the underlying ionization and excitation processes in this situation are described by Barth & Shields (2000) and references therein. The interstellar absorption that is observed is probably produced by the normal interstellar medium of this galaxy. Little evidence is present to suggest that the absorbing matter is closely linked to an AGN (if one is present at all) or the region of star formation. NGC 4569 is not unique; to varying degrees (limited mostly by signal-to-noise), a similar pattern is seen in a number of other type 2 LINERs (Maoz et al. 1998; Sabra et al., in preparation).

3.2. M87

The nucleus of M87, with its famous synchrotron jet, is clearly an AGN. *HST* spectra of its nucleus show weak but nonnegligible UV emission lines, with stronger emission in low-ionization transitions (Sabra et al. in preparation). The continuum is dominated by a power-law component throughout the UV/optical bandpass, with no evidence of hot stars. The UV spectrum shows absorption attributable to M87 in low-ionization features including C I, Fe II, Si II, and Mg II; absorption is also seen in Si IV but *not* in C IV or N V (see Figure 1). The implied total column density in this case is $N_H \gtrsim 10^{18}$ cm^{-2}. A limit on H I 21 cm absorption reported by Van Gorkom et al. (1989) suggests a neutral column density of $N_H < 5 \times 10^{19}$ cm^{-2}, and reddening estimates as well as X-ray studies indicate that $N_H \lesssim 10^{21}$ (Böhringer et al. 2001). The UV absorption lines show a small velocity offset (outflowing) from the systemic velocity of M87, with $\Delta v = -127 \pm 26$ km s^{-1} (Sabra et al., in preparation; see also Tsvetanov et al. 1999).

4. Discussion

In the accretion-powered LINER M87, we thus have some evidence for an AGN-related outflow. Among AGN-like LINERs, M87 actually provides the best

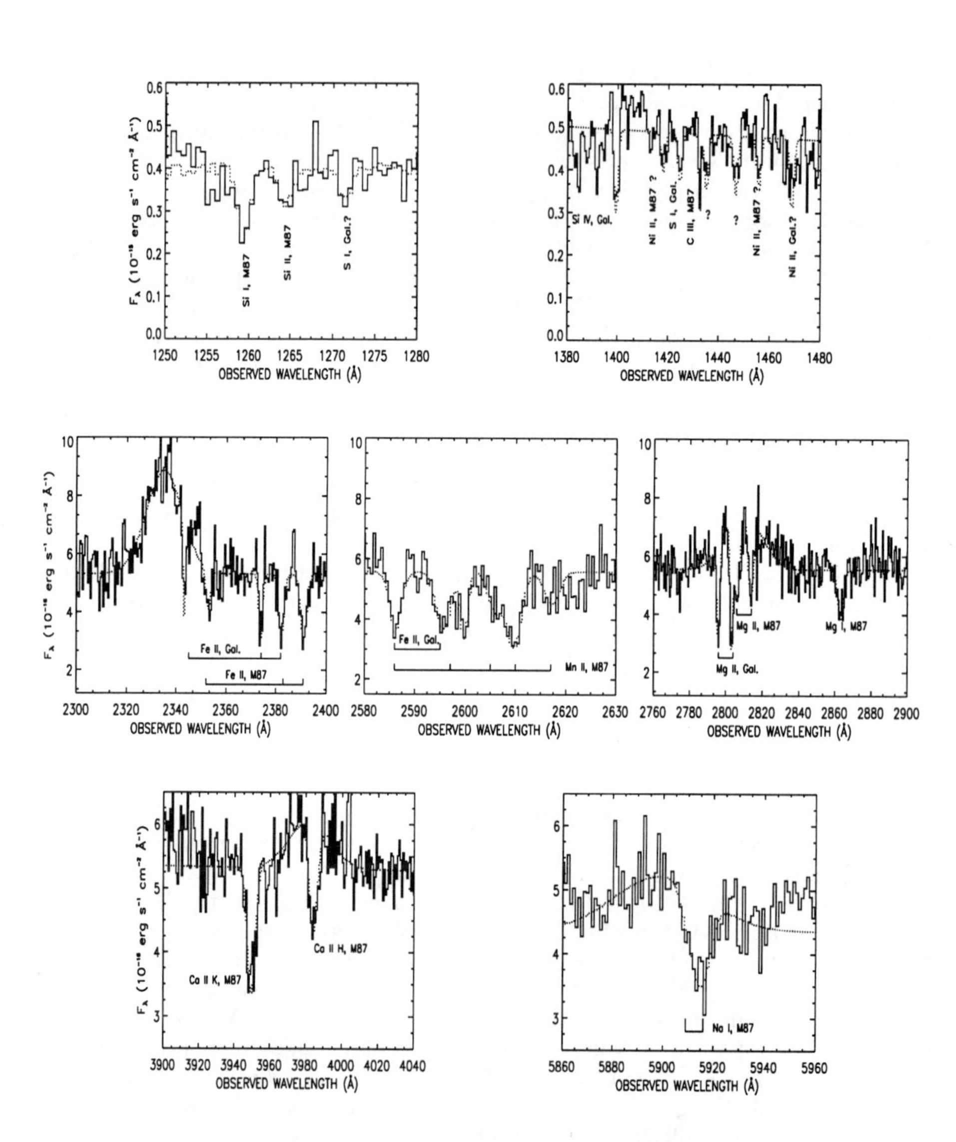

Figure 1. Absorption spectra for the nucleus of M87, obtained with FOS and STIS (Sabra et al, in preparation). Prominent interstellar absorption features are labeled. The dotted curves represent fits to the observed spectrum.

case for an outflow from the nucleus. The absorbing matter, like the nebular plasma, is described by a low-ionization state; LINERs are thus different from luminous AGN in both their emission and absorption properties. The velocity offset in M87 is comparatively low, however, leaving some ambiguity as to the interpretation of the absorber. Its column density and ionization properties resemble those of the absorber in NGC 4569; are we seeing merely a signature of the normal ISM in both cases?

Although the majority of observed LINERs show UV resonance line absorption, the detected features provide only weak indications that these systems (whether powered by starbursts or accretion) produce significant outflows. The lack of such outflows may be another distinction between LINERs and more luminous AGN. The difference may arise from simply the reduced power output of LINERs, or perhaps from an inability of ADAFs to drive outflows of low-ionization plasma. Unfortunately the small number of UV-bright LINERs suitable for *HST* spectroscopy will limit our ability to investigate this question further in the near term. Broad-line radio galaxies, which appear to be close cousins to LINERs, may provide an alternative class of object that can be observed in order to clarify this matter.

References

Barth, A. J., Ho, L. C., Filippenko, A. V., & Sargent, W. L. W. 1998, ApJ, 496, 133

Barth, A. J., Ho, L. C., Filippenko, A. V., Rix, H.-W., & Sargent, W. L. W. 2001, ApJ, 546, 205

Barth, A. J., & Shields, J. C. 2000, PASP, 112, 753

Blandford, R. D., & Begelman, M. C. 1999, MNRAS, 303, L1

Böhringer, H., et al. 2001, 365, L181

Eracleous, M., & Halpern, J. P. 1994, ApJS, 90, 1

Heckman, T. M. 1980, A&A, 87, 152

Ho, L. C. 1999, ApJ, 516, 672

Ho, L. C., Filippenko, A. V., & Sargent, W. L. W. 1997, ApJ, 487, 568

Maoz, D., et al. 1995, ApJ, 440, 91

Maoz, D., et al. 1998, AJ, 116, 55

Narayan, R., & Yi, I. 1994, ApJ, 428, L13

Tsvetanov, Z. I., et al. 1999, in the Radio Galaxy Messier 87, ed. H.-J. Röser & K. Meisenheimer (Berlin: Springer), 307

Van Gorkom, J. H., et al. 1989, AJ, 97, 708

EXIT

Mass Outflow in Active Galactic Nuclei: New Perspectives
ASP Conference Series, Vol. 255, 2002
D.M. Crenshaw, S.B. Kraemer, and I.M. George

Properties of QSO-Intrinsic Narrow Ultraviolet Absorption

Rajib Ganguly

Department of Astronomy & Astrophysics, University Park, The Pennsylvania State University, State College, PA 16802

Abstract. I present the current state of knowledge about narrow ($\lesssim 500$ km s^{-1}) ultraviolet absorption that is intrinsic to QSOs. I consider interpretations in the context of the accretion-disk/wind scenario of QSOs.

1. Introduction

In the past four years, the study of truly intrinsic narrow absorption has exploded both as a result of new insights about how to identify these systems and because of the advent of high resolution spectroscopy with large ground-based telescopes. Before summarizing the work and attempting to understand it in a coherent scenario, I start with a few conventions regarding nomenclature. A narrow absorption line nominally is one in which resonance UV doublets are well separated (Hamann & Ferland 1999). This implies widths less than about 500 km s^{-1} in order to resolve the C IV$\lambda\lambda1548, 1550$ doublet. I use the term NALs to refer to narrow absorption lines (in general) and NALQSO to refer to any QSO that has truly intrinsic narrow absorption (by analogy with the term BALQSO). I also use the term "associated" absorption lines as signifying that the absorption lies within 5000 km s^{-1} of the QSO emission redshift. (Note that an associated absorption line need not be intrinsic to the QSO.) Furthermore, a strong system is one whose C IV rest-frame equivalent width is larger than 1–2 Å. First, I will consider how we can identify intrinsic systems – that is, how we separate them from intervening gas. I will then discuss separately what we know from studies at high redshift and at low redshift. Finally, I will unify observed properties in the context of the accretion-disk/wind scenario.

2. Identifying Intrinsic NALs

There are two basic methods which one uses to infer the properties of intrinsic narrow absorbers. One can identify large populations of absorbers in a sample and consider if there is a relationship with the host QSO. Identifying populations of intrinsic absorbers typically involves a demonstration that there is an excess of absorbing systems over what is expected in a given redshift or velocity path. As a consequence, it is known which systems are actually intrinsic.

Alternatively, one can identify a specific absorber as intrinsic and decipher the physical conditions of the gas. Identifying specific intrinsic absorbers requires either multiple epochs of observation to look for time variability and/or high resolution spectroscopy to show that the absorbing gas only partly occults the QSO central engine.

Thus far, only fifteen QSOs have been shown conclusively to have intrinsic NALs through time variability and/or partial coverage. Of these, nine are radio-quiet, five are radio-loud, and only one is at low redshift. Six have been shown to be time variable (TV) while ten have been shown to exhibit the signature of partial coverage (PC). These are listed in Table 1 along with the citation to the work that showed the intrinsic origin.

Table 1. Known NALQSOs

Name	z_{em}	Radio	Method	Reference
Q 0123 + 257	2.358	Loud	PC	Barlow & Sargent (1997)
Q 0150 − 203	2.139	Loud	PC, TV	Hamann et al. (1997a)
PKS 0424 − 131	2.166	Loud	PC	Petitjean, Rauch, & Carswell (1994)
Q 0449 − 134	3.093	Quiet	PC	Barlow, Hamann, & Sargent (1997)
Q 0450 − 132	2.253	Quiet	PC	Ganguly et al. (1999)
Q 0835 + 580	1.534	Loud	TV	Aldcroft, Bechtold, & Foltz (1997)
Q 0935 + 417	1.980	Quiet	TV	Hamann et al. (1997b)
PKS 1157 + 014	1.986	Loud	TV	Aldcroft, Bechtold, & Foltz (1997)
PG 1222 + 228	2.038	Quiet	PC	Ganguly et al. (1999)
PG 1329 + 412	1.930	Quiet	PC	Ganguly et al. (1999)
HS 1700 + 6416	2.722	Quiet	PC, TV	Barlow, Hamann, & Sargent (1997)
Q 2116 − 358	2.341	Quiet	PC	Wampler, Bergeron, & Petitjean (1993)
QSO J2233 − 606	2.238	...	PC	Petitjean & Srianand (1999)
MRC 2251 − 178	0.066	Loud	TV	Ganguly, Charlton, & Eracleous (2001b)
Q 2343 + 125	2.515	Quiet	PC, TV	Hamann et al. (1997c)

3. Intrinsic NALs at High Redshift

At high redshift, since the rest-frame ultraviolet transitions are shifted into the optical, we can take advantage of optical spectroscopy with large ground-based telescopes to identify specific absorbing systems as intrinsic. Unfortunately, since the QSOs are at higher redshift, it is generally more difficult to obtain detailed multiwavelength information about the QSOs themselves.

Strong systems seem to prefer optically-faint (OF), radio-loud QSOs with steep radio spectra (Foltz et al. 1986; Anderson et al. 1987; Møller & Jakobsen 1987; Foltz et al. 1988). In addition, the equivalent width of strong systems seems to correlate with orientation, with stronger systems existing in more radio-lobe dominated (that is, edge-on) QSOs (Barthel, Tytler, & Vestergaard 1997; Baker et al. 2002). A recent study of QSO absorption systems down to a 0.15 Å limiting equivalent width also found that intrinsic systems can appear at very large "ejection" velocities (Richards et al. 1999, Richards 2001). This happens more so in radio-quiet QSOs than radio-loud QSOs and more so in flat-spectrum, radio-loud (FSRL) QSOs than steep-spectrum, radio-loud (SSRL) QSOs.

4. Intrinsic NALs at Low Redshift

At low redshift, we have the advantage of knowing very well the QSO properties. However, it is harder to identify intrinsic systems since we must rely on smaller space-based telescopes (*HST* and *FUSE*). High-resolution spectra can be obtained for only the brightest targets.

The first remarkable property of low-redshift NALQSOs is that none host strong systems (Ganguly et al. 2001a). As a result, the correlations with QSO properties seen at high redshift, which were driven by strong absorption, are largely absent. [Strong systems seem to exist in compact, steep-spectrum (CSS) radio-loud QSOs down to $z_{\rm em} \sim 0.7$ (Baker et al. 2002).] The equivalent widths of weak systems do not correlate with any single QSOs property but their velocity distribution seems to peak at the same velocity as the broad emission lines. This seems to indicate a relationship between the line-of-sight velocity of the absorbers and the velocity of maximum emissivity of the broad line region. A multivariate analysis of associated absorption indicates a combination of QSO properties that seem to prohibit the detection of associated NAL gas. Associated NALs are absent in FSRL QSOs that have mediocre C IV emission FWHM ($\lesssim 6000$ km s^{-1}), but are present in a finite fraction of FSRL QSOs with large C IV FWHM.

Table 2. Properties of QSO-Intrinsic NALs

$z_{\rm em} \gtrsim 1$	$z_{\rm em} \lesssim 1$
Strong systems	
• Prefer OF SSRL QSOs • EW correlated with orientation • Velocity distribution peaks with BEL, not $z_{\rm sys}$ • Exist at high $v_{\rm ej}$	• Largely absent @ $z \lesssim 0.7$ but exist in CSS RL QSOs @ $z \gtrsim 0.7$
Weak systems	
• No preference for RL • Excess of high $v_{\rm ej}$ NALs in RQs compared to RLs, and FSRLs compared to SSRLs	• Absent in FSRL QSOs with C IV BEL FWHM $\lesssim 6000$ km s^{-1} • Velocity distribution peaks with BEL, not systemic velocity • Enhanced probability of NALs in BALQSOs

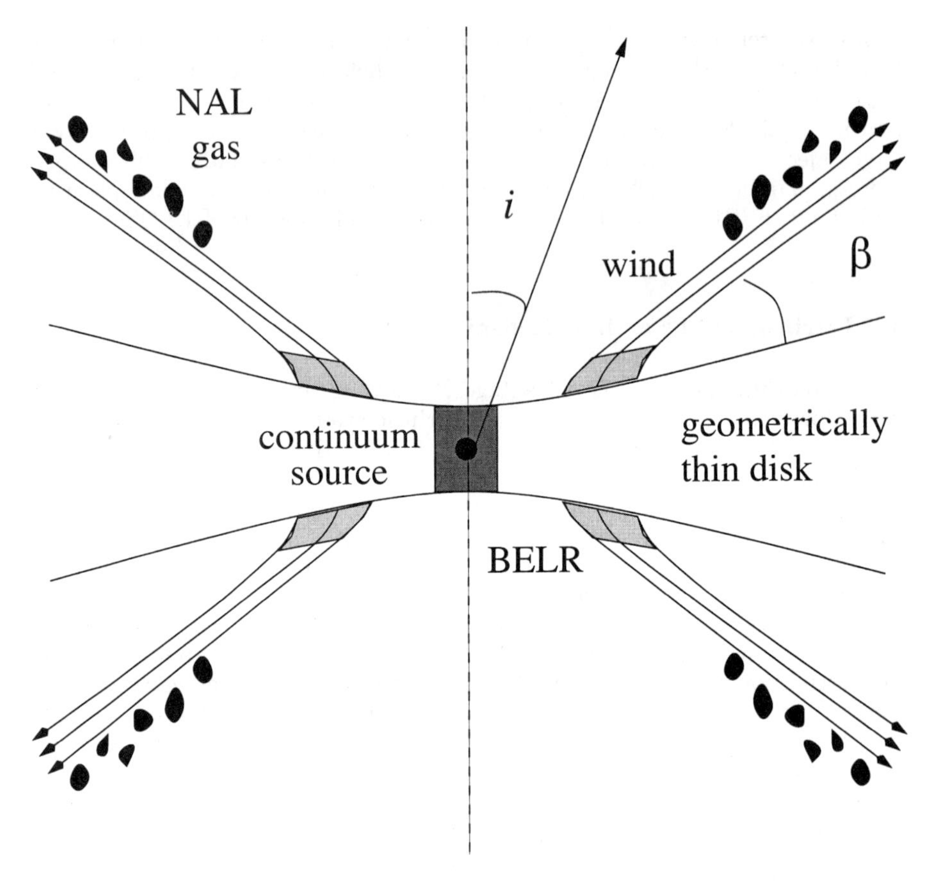

Figure 1. Disk-wind model for QSOs from Ganguly et al. (2001a). The inclination angle, i, and the wind opening angle, β, are shown.

5. Putting It All Together

These properties are summarized in Table 2, which is broken up according to both redshift and absorption strength. If we postulate that a unified model exists, there are three basic conclusions to draw from the table. First, there has been evolution such that strong systems are largely absent at low redshift. Second, the properties of weak absorbers at both, high and low redshift complement each other so that no evolution in their population is required. That is, the properties at both high and low redshift can be considered together as governing an unevolving population of absorbers. Similarly, the properties of high redshift strong systems and low redshift weak systems do not contradict each other and can be considered simultaneously.

We can understand these properties in the context of the disk-wind scenario. A cartoon of this scenario, from Ganguly et al. (2001a), is shown in Fig. 1. The scenario, a modification of the Murray et al. (1995) model originally employed to explain broad absorption lines, is that of a radiatively driven, outflowing wind in which clumps of gas hug the wind at "large" distances from the black hole. Hydrodynamic simulations by Proga, Stone, & Kallman (2000) show that such clumps do arise from Kelvin-Helmholtz shearing instabilities.

This scenario has, essentially three fundamental parameters: the black hole mass, the mass fueling rate, and the inclination with respect to the observer. The black hole mass and mass fueling rate can also be translated into a wind opening angle and wind density. A given black hole mass implies a maximum mass accretion rate (that is, the Eddington rate). For a larger mass fueling rate, this implies a larger wind density. Moreover, the luminosity of the QSO will be larger and thus the acceleration of the wind will be even more dominated by its radial component. This will result in a smaller wind opening angle.

If we hypothesize that the wind in radio-loud QSOs is less dense than in radio-quiet QSOs, then strong NALs can be thought of as the analogues of BAL, where the wind itself is the source of absorption. Reversing the above reasoning, a sparser wind implies a less luminous QSO, explaining the preference for "optically faint" QSOs. The projected velocity dispersion of the wind along sightlines is small (i.e. narrow). Likewise, the optical depth of the wind is largest when the QSO is viewed at higher inclination angle (similar to BALQSOs).

The evolution of strong systems can be viewed as a change in the mass outflow rate (or wind density). This change can result either from a decrease in the mass fueling rate or an increase in the mass accretion rate. Either case seems natural since (1) there is only a finite amount of gas to fuel the engine and (2) over the duty cycle of the QSOs life, the accretion process will increase the black hole mass, and therefore the maximum allowed accretion rate.

The absence of weak associated systems in FSRL QSOs w/ average CIV FWHM can be seen as mostly an inclination effect. If the population of weak absorbers is due the clumps produced by the shear, then NALs will only be detected when the line of sight intercept these clumps. The flat-spectrum radio-loud QSOs in the Ganguly et al. (2001) sample were also strongly radio core dominated. So, there is little doubt that the QSOs are viewed at small inclination angles (i.e. face-on geometries). In addition, the width of the C IV emission line implies that the velocity dispersion of the wind along the line of sight is not large. Thus, the wind opening angle is "small." In this case, neither photons from the compact continuum, nor photons from the broad emission line region intercept the weak NAL clouds.

Acknowledgments. This work was funded by NASA through grants NAG 5-6399, HST-GO-08681.01-A, and through an archival award from the Space Telescope Science Institute (STSI AR-08763.01-A), which is operated by AURA, Inc., under NASA contract NAS 5-26555. Travel expenses to the meeting were provided by the Zaccheus Daniel Foundation. R.G. acknowledges Ray Weymann for a thoughtful meeting summary, and a stimulating discussion after the meeting.

References

Aldcroft, T., Bechtold, J., & Foltz, C. 1997, in Mass Ejection from Active Galactic Nuclei, ed. N. Arav, I. Shlosman, & R. Weymann (San Francisco: ASP), 25

Anderson, S. F., Weymann, R. J., Foltz, C. B., & Chaffee, F. H. 1987, AJ, 94, 278

Baker, J.C. 2002, in Mass Outflow in Active Galactic Nuclei: New Perspectives, eds. D.M. Crenshaw, S.B. Kraemer, & I.M. George (San Francisco: ASP), p. 121

Barlow, T. A., & Sargent, W. L. W. 1997, AJ, 113, 136

Barlow, T. A., Hamann, F., & Sargent, W. L. W., 1997, in Mass Ejection from Active Galactic Nuclei, ed. N. Arav, I. Shlosman, & R. Weymann (San Francisco: ASP), 13

Barthel, P. D., Tytler, D. R., & Vestergaard, M. 1997, in Mass Ejection from Active Galactic Nuclei, ed. N. Arav, I. Shlosman, & R. Weymann (San Francisco: ASP), 48

Foltz, C. B., Weymann, R. J., Peterson, B. P., Sun, L., Malkan, M. A., & Chaffee, F. H. 1986, ApJ, 307, 504

Foltz, C. B., Chaffee, Jr., F. H., Weymann, R. J., & Anderson, S. F. 1988, in *QSO Absorption Lines*, J. C. Blades, D. A. Turnshek, & C. A. Norman, Cambridge: Cambridge Univ. Press, 53

Ganguly, R., Eracleous, M., Charlton, J. C., & Churchill, C. W. 1999, AJ, 117, 2594

Ganguly, R., Bond, N. A., Charlton, J. C., Eracleous, M., Brandt, W. N., & Churchill, C. W. 2001a, ApJ, 549, 133

Ganguly, R., Charlton, J. C., Eracleous, M. 2001b, ApJ, submitted

Hamann, F., Barlow, T. A., Junkkarinen, V., & Burbidge, E. M. 1997, ApJ, 478, 80

Hamann, F., Barlow, T. A., & Junkkarinen, V. 1997b, ApJ, 478, 87

Hamann, F., Barlow, T. A., Cohen, R. D., Junkkarinen, V., & Burbidge, E. M., 1997c, in Mass Ejection from Active Galactic Nuclei, ed. N. Arav, I. Shlosman, & R. Weymann (San Francisco: ASP), 19

Hamann, F., & Ferland, G. 1999, ARA&A, 37, 487

Møller, P., & Jakobsen, P. 1987, ApJ, 320, 75

Murray, N., Chiang, J., Grossmann, S. M., & Voit, G. M. 1995, ApJ, 454, 105

Proga, D., Stone, J. M., & Kallman, T. R. 2000, ApJ, 543, 686

Petitjean, P., Rauch, M., & Carswell, R. F. 1994, A&A, 291, 29

Petitjean, P., & Srianand, R. 1999, A&A, 345, 73

Richards, G. T., York, D. G., Yanny, B., Kollgaard, R. I., Laurent-Muehleisen, S. A., & vanden Berk, D. E. 1999, ApJ, 513, 576

Richards, G. T. 2001, ApJS, 133, 53

Wampler, E. J., Bergeron, J., & Petitjean, P. 1993, A&A, 273, 15

Mass Outflow in Active Galactic Nuclei: New Perspectives
ASP Conference Series, Vol. 255, 2002
D.M. Crenshaw, S.B. Kraemer, and I.M. George

The Outflow Column Density in NGC 5548

Nahum Arav

Astronomy Department, UC Berkeley, Berkeley, CA 94720, and Physics Department, University of California, Davis, CA 95616

Martijn de Kool

Research School of Astronomy and Astrophysics, ANU ACT, Australia

Kirk T. Korista

Western Michigan Univ.,Dept. of Physics,1120 Everett Tower, Kalamazoo, MI 49008

Abstract.
We re-analyze the HST high resolution spectroscopic data of the intrinsic C IV λ1549 absorber in NGC 5548. We find that the absorption column density is probably at least four times larger than previously determined. These results begin to bridge the gap between the high column densities measured in the X-ray and the low ones previously inferred for the UV lines. Combined with our findings for outflows in high luminosity quasars these results suggest that traditional techniques for measuring column densities; equivalent width, curve of growth and Gaussian modeling, are of limited value when applied to absorption associated with AGN outflows.

1. Introduction

The intrinsic absorber in NGC 5548 was studied in the UV using the *HST* Goddard High Resolution Spectrograph (GHRS, Mathur, Elvis & Wilkes 1999) and Space Telescope Imaging Spectrograph (STIS, Crenshaw & Kraemer 1999), the Far Ultraviolet Spectroscopic Explorer (*FUSE*, Brotherton et al. 2001, in preparation), and in X-rays with the *ASCA* (George et al. 1998) and *CXO* (Kaastra et al. 2000) satellites. These high quality observations combined with the relative simplicity of its intrinsic absorption features make NGC 5548 a prime target for understanding the nature of Seyfert outflows.

Our work on quasar outflows, led us to suspect that the current determination of column densities in Seyfert outflows is highly uncertain. In the last few years our group (Arav 1997; Arav et al. 1999a; Arav et al. 1999b; De Kool et al. 2001; Arav et al. 2001) and others (Barlow 1997, Telfer et al. 1998) have shown that in quasar outflows most lines are saturated even when not black. Therefore, a) the column densities inferred from the depth of the trough are only lower limits; b) the shape of the trough is almost entirely due to changing degree of covering at different velocities; and c) for a single line it is virtually impossible to decouple the effects of optical depth from those of covering factor, thus column density determination in such cases are not possible. In order to assess the importance of these effects in Seyfert outflows, we reanalyze the *HST* high resolution spectra of NGC 5548. Here we concentrate on analyzing the GHRS data of the C IV λ1549 (Mathur, Elvis & Wilkes 1999). These are of significantly higher signal to noise ratio than the STIS observations (Crenshaw & Kraemer 1999) and thus more suitable to our analysis. However, our results are consistent with the STIS data as well.

2. Analysis

Our method of extracting column densities relies upon using doublet or higher multiplets in order to have at least two related sets of data from which the two unknowns (optical depth and covering fraction) can be determined simultaneously (Barlow & Sargent 1997; Hamann et al. 1997, Arav et al 1999b).

In Figure 1, we show the results of the analysis, which assumes that the narrow emission line (NEL) is not covered by the flow. Such a scenario is plausible on physical grounds. In NGC 5548 the size of the continuum source is ~1 light day. Reverberation studies have shown that the C IV BEL region is a few light days across. In contrast the NEL region is larger than 10 light years (Wilson et al. 1989). The very large difference in sizes comfortably allows for a model where the outflow cover most (or all) of the relatively small continuum source and BEL region, while not covering the thousand times larger NEL region. A discussion on the applicability of this assumption to the outflow in NGC 5548, as well the full details for the analysis are found in Arav et al. (2001).

Figure 1 shows that the residual intensities in the doublet components do not adhere to the expected 1:2 optical depth ratio (i.e., $I_{\rm blue} = I^2_{\rm red}$), but are much closer to a 1:1. This has two effects: 1) To restore the 1:2 optical depth ratio the covering fraction has to be very close to the $I_{\rm blue}$ curve, which unequivocally demonstrate that the shape of the trough is determined almost solely by the covering fraction at each velocity. 2) Under these conditions, the optical depth is much higher than the apparent optical depth defined as $\tau_{ap} = -\ln[I_{\rm red}]$, which assumes complete coverage. We note that both previous analyses of these data sets derived column density for this component very similar in value to the one we extract from the apparent optical depth. Moreover, the integrated column density becomes very sensitive to the difference between the blue and red residual intensities when $(I_{\rm red} - I_{\rm blue})/(I_{\rm blue}) \ll 1$. Therefore, small changes in the NEL model can produce much higher optical depth in parts of the trough, and with that a large increase in the derived column density. We therefore consider the factor four increase over previous determination as rather conservative and note that the column density might be significantly higher.

What are the implications of the results presented here on our understanding of AGN outflows? Almost everything we deduce about the outflows depends upon the measured column densities. For example, the existing column density measurements in NGC 5548 coupled with photoionization modeling, suggests that nitrogen is significantly overabundent compared to carbon relative to their solar ratio (Srianand 2000). However, if the C IV column density if much higher than previously thought, the nitrogen overabundance may not be needed. Dynamical models of the winds are also strongly affected since the mass and en-

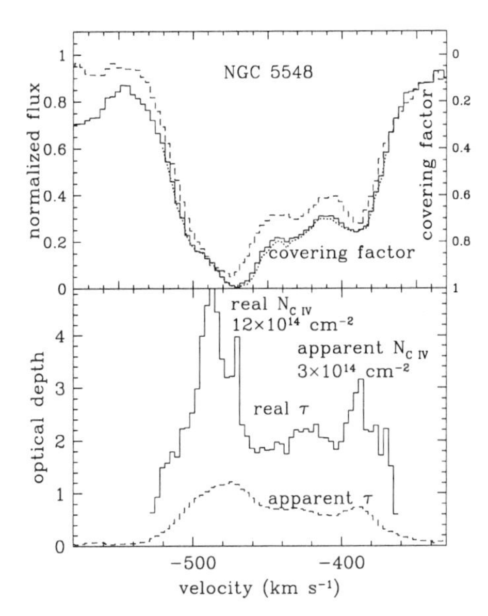

Figure 1. The histograms in the upper panel show the normalized data, for the deepest subtrough in the C IV absorption complex, where the solid and dashed histograms corresponds to the blue and red doublet components respectively. The dotted line is the covering factor solution, which demonstrates that the shape of the trough is determined almost entirely by the covering factor, which is a strong function of velocity. In the lower panel we show the real (solid) and apparent (dashed) optical depths.

ergy fluxes are proportional to the column densities (Arav, Li & Begelmen 1994; Proga, Stone & Kallman 2000). One of the most important implications is in regards to the connection between the UV and the warm X-ray absorber. Based on the existing UV column density measurements it appears that in NGC 5548 the two absorbers cannot rise from the same gas, since the X-ray column densities appear to be a hundred times higher (Kaastra et al. 2000) than those extracted in the UV. However, if the UV column densities are strongly underestimated, this situation might change.

References

Arav, N., 1997 in Mass Ejection from Active Galactic Nuclei, ed. N. Arav, I. Shlosman, & R. Weymann (San Francisco: ASP), p. 208

Arav, N., et al., 1999, ApJ, 524, 566

Arav, N., Li, Z. Y., & Begelman, M. C. 1994, ApJ, 432, 62

Arav, N., Korista & de Kool, 2001 in press

Barlow, T. A., 1997, in Mass Ejection from Active Galactic Nuclei, ed. N. Arav, I. Shlosman, & R. Weymann (San Francisco: ASP), p. 13

Barlow, Thomas A., Sargent, W. L. W., 1997, AJ, 113, 136

Crenshaw, D. M., Kraemer, S. B., 1999, ApJ, 521, 572

de Kool, M., et al., 2001, ApJ, in press

George, I. M., et al., 1998, ApJS, 114, 73

Hamann, F., et al. 1997, ApJ, 478, 80

Kaastra, J. S., et al. 2000, A&A, 354L, 83

Mathur, S., Elvis, M., & Wilkes, B. 1999, ApJ, 519, 605

Proga, D., Stone, J. M., Kallman, T. R., 2000, ApJ, 543, 686

Srianand, R., 2000, ApJ, 528, 617

Telfer, R.C., et al., 1998, ApJ, 509, 132

Mass Outflow in Active Galactic Nuclei: New Perspectives
ASP Conference Series, Vol. 255, 2002
D.M. Crenshaw, S.B. Kraemer, and I.M. George

Associated Absorption and Radio Source Growth

Joanne C. Baker
Astronomy Department, 601 Campbell Hall, University of California, Berkeley CA 94720, USA

Abstract.
Results are presented from a survey for C IV associated absorption in a complete low-frequency-selected sample of quasars. In agreement with previous work, associated absorbers are most common in steep-spectrum and lobe-dominated quasars, indicative of an anisotropic cloud distribution. Furthermore, we find the strongest C IV absorption occurs in sources of small radio size, suggesting that the absorbing clouds are destroyed or displaced as radio sources expand. Evidence for dust in the clouds is also found, such that quasars with strong absorption are systematically redder. Finally, we find no evidence for evolution in the frequency or properties of the absorbers from $z \sim 0.7$ to $z \sim 3$.

1. Introduction

Absorption lines occurring very close to the quasar redshift — associated absorbers — are potentially valuable probes of quasar environments. The precise origin of the absorbing material is unknown, but many interesting regions along the sightline may contribute. Absorption may arise, for instance, in gas near the quasar nucleus, in the ISM of the host galaxy or in neighbouring galaxies.

It seems clear that the majority of associated absorption systems (as defined by $|z_a - z_e| < 5000\,\mathrm{km\,s^{-1}}$) are related directly to the quasar phenomenon, rather than being due solely to cosmologically-distributed foreground galaxies. First, the density of associated absorption systems per redshift interval is greater than expected for intervening galactic systems alone (Foltz et al. 1988, Richards et al. 1999, 2001). Secondly, the characteristics of the absorbers depend on quasar type. Narrow absorption lines occur more frequently in radio-loud quasars, especially those with steep radio spectra (Anderson et al. 1987; Foltz et al. 1988; Richards et al. 1999, 2001). Alternatively, optically-selected samples might be biased against objects with absorption. Previous studies have interpreted the prevalence of $z_\mathrm{a} \approx z_\mathrm{e}$ systems in steep-spectrum and lobe-dominated quasars in terms of orientation (Barthel, Tytler & Vestergaard 1997). In addition, $z_\mathrm{a} \approx z_\mathrm{e}$ systems differ subtly in ionisation or velocity profiles from those seen along sightlines traversing normal galaxy halos, consistent with their proximity to the AGN (Hamann & Ferland 1999).

A major limitation of most previous studies of $z_\mathrm{a} \approx z_\mathrm{e}$ absorption is that they have used inhomogeneous samples which are prone to strong selection effects. To counter this, we have observed $z_\mathrm{a} \approx z_\mathrm{e}$ absorption in quasars drawn from a complete sample of 408-MHz selected quasars. Low-frequency radio selection and a high completeness level ensure that orientation bias is minimised and reddened sightlines are included. The initial results of this study, correlations between C IV absorption and quasar properties, are summarised here. Full results will be published shortly (Baker et al. 2001, in preparation). Cosmological parameters $H_0 = 50\ \mathrm{km\,s^{-1}Mpc^{-1}}$, $\Omega = 1.0$ and $\Lambda = 0$ are assumed for consistency with our earlier work.

2. The quasar sample and observations

We have obtained intermediate-resolution spectroscopy of the C IV to Ly α spectral region for 43 quasars drawn from the complete Molonglo Quasar Sample (MQS; Kapahi et al. 1998; Baker et al. 1999). Briefly, the MQS comprises all quasars in the $-30° < \delta < -20°$ strip of the 408-MHz Molonglo Reference Catalogue (MRC; Large et al. 1981) down to a limiting flux density of $S_{408} = 0.95$ Jy. The MQS contains 111 quasars with $0.1 < z < 3.0$.

MQS quasars were selected for the absorption study in two redshift ranges, $1.5 < z < 3.0$ where redshifted C IV is observable from the ground, and $0.7 < z < 1.0$ which was observed in the UV using STIS on HST. Ground-based spectra with spectral resolution 1–2.4Å (FWHM) were obtained for 22 out of a total of 27 MQS quasars with $z > 1.5$, using mostly the Anglo-Australian Telescope (AAT) and the ESO 3.6m telescope. In addition, four faint quasars were observed with FORS1 on the VLT (UT1). The STIS spectroscopy was carried out between May 1999 and February 2001 for 19 MQS quasars with redshifts $0.7 < z < 1.0$. The NUV-MAMA detector was used with the G230L grism, giving a spectral resolution of 3.0Å over the wavelength range 1570–3180Å.

Absorption systems were identified and measured using IRAF. The strongest absorption system was identified in each spectrum within $\pm 5000\mathrm{km\,s^{-1}}$ of the C IV emission-line redshift. The majority (50–70%) of the resulting systems lay within $\pm 500\mathrm{km\,s^{-1}}$ of the emission redshift, and were both blue- and red-shifted.

3. Results

3.1. Radio spectrum and morphology

Although not shown explicitly in this short contribution, we do confirm the trends for absorption to be most prevalent in steep-spectrum and lobe-dominated quasars (e.g. Anderson et al. 1997; Foltz et al. 1988; Barthel et al. 1997). In our MQS study, for example, strong absorption ($W_\lambda > 1$Å) was detected exclusively in steep-spectrum ($\alpha > 0.5$) quasars in both the high- and low-redshift subsamples.

3.2. Radio size

In the MQS data, the most striking result is that the strongest absorption occurs preferentially in the smallest radio sources. The equivalent widths of C IV absorption are plotted as a function of radio source size in Figure 1 for both high- and low-redshift datasets. Highly beamed core-dominated quasars are excluded as they are expected to be severely foreshortened. Compact, steep-spectrum

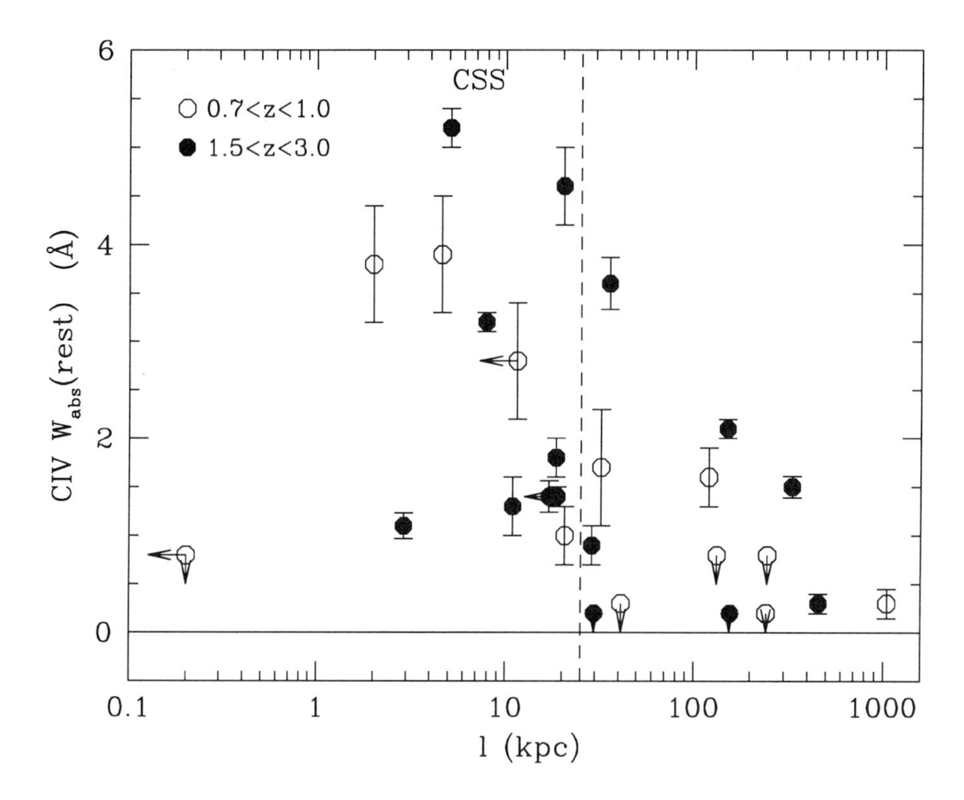

Figure 1. Equivalent width of C IV absorption as a function of radio source size, l (kpc). Quasars with $0.7 < z < 1.0$ are plotted as open circles, those with $1.5 < z < 3.0$ are plotted with filled symbols. The dotted line at $l = 25$ kpc illustrates our working definition of CSSs. Arrows indicate limits.

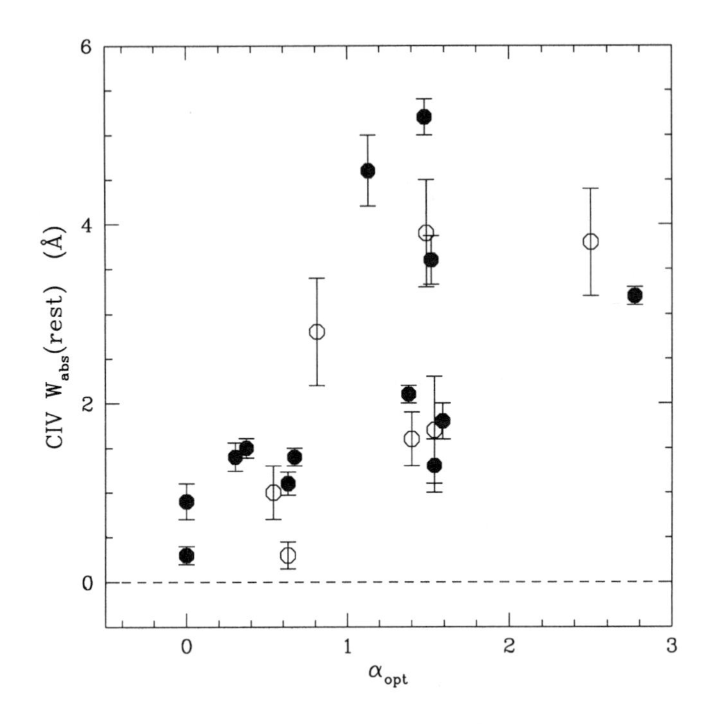

Figure 2. Equivalent width of C IV absorption as a function of optical spectral index, for all quasars with detected absorption. Quasars with $0.7 < z < 1.0$ are plotted as open circles, those with $1.5 < z < 3.0$ are plotted with filled symbols.

(CSS) sources are included on the plot. CSS sources (see review by O'Dea 1998) are intrinsically small (l < 25 kpc) with steep radio spectra ($\alpha > 0.5$ where $S_\nu \propto \nu^{-\alpha}$). The precise definition of CSSs is somewhat arbitrary, comprising essentially those sources whose radio emission is unbeamed (not core-dominated) yet unresolved with conventional arrays (arcsec resolution). Thus CSS sources do include intrinsically small, and perhaps young, sources. Radio sizes for the CSS quasars in the MQS study were measured from MERLIN images with $\sim 0.1''$ resolution (de Silva et al. 2001, in preparation). Notably, all the CSS quasars in our study (except an unusual GigaHertz-Peaked Spectrum source) show $z_a \approx z_e$ absorption stronger than $W_{abs} = 1$Å.

3.3. Reddening by absorbing clouds

For quasars where C IV absorption was detected, the equivalent width of the absorption is plotted in Figure 2 against the slope of the optical spectrum, α_{opt} (as observed between 3500 and 10000Å). There is a strong correlation between absorption-line strength and spectral slope — heavily absorbed quasars are systematically redder.

Baker & Hunstead (1995) and Baker (1997) presented evidence that the range in optical spectral slope observed in the MQS is due in part to reddening by an anisotropic dust screen lying outside the broad emission-line region. In this earlier study, the most direct evidence for dust reddening (as opposed to intrinsic spectral steepening) was the tight correlation between $\alpha_{\rm opt}$ and broad $\rm H\,\alpha/H\,\beta$ Balmer Decrement, at least in low-redshift quasars where it was measurable in the optical. The reader is referred to Baker (1997) for this result and a more detailed description of the dust-reddening hypothesis. By extension, the simplest explanation of the correlation of $W_{\rm abs}$ with $\alpha_{\rm opt}$ is that the absorbing gas clouds contain dust, and they lie outside the nuclear continuum source. Alternatively, if dust is not responsible for the red continuum slopes, then C IV absorption strength correlates with an intrinsically softer continuum shape.

4. Discussion

These results suggest that the distribution of associated absorbers in quasars is dependent on *both orientation and radio size*. Orientation explains the trends with radio spectral index and radio-core dominance, as described by Barthel et al. (1997). Orientation, however, cannot explain the stronger absorption in CSS sources and the global decrease in absorbing column density with radio size.

The radio-size dependence of associated absorption may be explained if the absorption column density either correlates with quasar environmental density, or changes with time. Currently, the first hypothesis is not supported by observations, which find that CSS host galaxies appear to be the same as those harbouring larger sources, and CSS quasars do not systematically reside in clusters more often than larger sources (de Vries et al. 2000; O'Dea 1998 and references therein). The alternative idea is that the absorbing column density decreases as the radio source grows. This could occur because of photoionisation of the clouds by the quasar over time, or by direct interaction of the radio jet and its cocoon on the absorbing clouds, or both.

The strong correlation between the C IV absorption strength and red continuum in the quasars is highly suggestive of dust in the absorbing clouds. However, a cospatial distribution of dust and gas is problematic, dust should be destroyed by sputtering in the hot gas where C IV absorption arises. De Young (1998) points out that the strong shocks in radio-source environments should destroy dust easily on timescales $\ll 10^6$ yrs, which is much shorter than the lifetime of the radio source.

Putting all the evidence together, we propose a consistent model whereby radio sources are born enshrouded in dust and gas, which is gradually destroyed and ionised (respectively) along the radio axis as the source expands.

In addition, we find no evidence for changes in the the frequency or strength of the absorbers with redshift from $z \sim 0.7$ to $z \sim 3$. This lack of evolution is perhaps unexpected given the absorbers are probably at kpc distances where they should be affected by quasar environmental and perhaps galactic evolution.

5. Conclusions

Initial results from a study of C IV associated absorption in a complete, homogeneous sample of radio-loud quasars are presented. The results confirm that the absorbing cloud distribution is anisotropic, such that absorption is more common in steep-spectrum and lobe-dominated quasars. Furthermore, we find new evidence that the strength of C IV absorption decreases with increasing radio source size. If we assume that the larger sources are older than the smaller (CSS) ones, then we can attribute the decrease in column density to the growth of the radio source envelope through the ISM. The absorbing clouds probably contain dust, which reddens the quasar light. Consequently we predict that absorbed quasars will be missed preferentially in optically-selected samples. Finally, these results appear to be independent of redshift, giving essentially the same picture at $z \approx 0.7$ and $z \sim 3$, epochs between which evolution of quasar environments should be discernible. Thus we are drawn to a picture where radio sources are born in gaseous and dusty cocoons, from which they emerge as their radio jets expand beyond the host galaxy.

Acknowledgments. JCB acknowledges support by NASA through Hubble Fellowship grant #HF-01103.01-98A from STScI, which is operated by the AURA, Inc., under NASA contract NAS5-26555.

References

Anderson S.F., Weymann R.J., Foltz C.B., Chaffee F.H. Jr. 1987, AJ, 94, 278

Baker J.C. 1997, MNRAS, 286, 23

Baker J.C., Hunstead R.W. 1995, ApJL, 452, L95

Baker J.C., Hunstead R., Kapahi V., Subrahmanya C. 1999, ApJS, 122, 29

Barthel P.D., Tytler D., Vestergaard M. 1997, in Mass Ejection from AGN, eds. N. Arav, I. Shlosman, & R. Weymann (San Francisco: ASP) 48

de Vries W.H., O'Dea C.P., Barthel P.D., Fanti C., Fanti R., Lehnert M.D. 2000, AJ, 120, 2300

De Young, D.S. 1998, ApJ, 507, 161

Foltz C.B., Chaffee F.H. ,Weymann R.J., Anderson, S.F. 1988, in 'QSO absorption lines: Probing the Universe', eds. J.C. Blades, D.A. Turnshek, C.A. Norman, CUP, p53

Hamann F., Ferland G. 1999, ARA&A, 37, 487

Kapahi V.K., Athreya R.M., Subrahmanya C.R., Baker J.C., Hunstead R.W., McCarthy P.J., van Breugel W. 1998, ApJS, 118, 327

Large M., Mills B., Little A., Crawford D., Sutton J. 1981, MNRAS, 194, 693

O'Dea C.P. 1998, PASP, 110, 4930

Richards G., Laurent-Muehleisen S., Becker R., York D. 2001, ApJ, 547, 635

Richards G., York D., Yanny B., Kollgaard R., Laurent-Muehleisen S. vanden Berk D. 1999, ApJ, 513, 576

Mass Outflow in Active Galactic Nuclei: New Perspectives
ASP Conference Series, Vol. 255, 2002
D.M. Crenshaw, S.B. Kraemer, and I.M. George

Far Ultraviolet Spectroscopic Explorer Observations of NGC 5548 in a Low State

Michael S. Brotherton
National Optical Astronomy Observatories, Kitt Peak National Observatory,950 N. Cherry Ave., Tucson, AZ 85719

Abstract. The far-UV spectrum of the Seyfert 1.5 galaxy NGC 5548 obtained with the Far Ultraviolet Spectroscopic Explorer (FUSE), shows a weak continuum and emission from O VI $\lambda\lambda$1032, 1038 and other species. In this low state, O VI shows strong narrow components in emission; interestingly the doublet ratio appears intermediate between the optically thin and optically thick cases. FUSE also resolves intrinsic, associated absorption lines of O VI and Lyman β. Several components are present, spanning velocities of -50 to -1200 km s^{-1}, consistent with what has been before seen for UV lines. I explore the relationships between the far-UV absorbers and those seen previously in the UV and X-rays.

1. Introduction to NGC 5548 and UV-X-ray Warm Absorbers

NGC 5548 is a bright, nearby AGN, often the target of monitoring campaigns (e.g., Peterson et al. 1999). The time-delayed response of emission lines to continuum changes indicates the size of the line-emitting regions. For NGC 5548 the Doppler widths of the broad lines and their sizes are consistent with Keplerian motion around a black hole of $5.9 \times 10^7 M_\odot$ (Peterson & Wandel 2000).

About half of Seyfert galaxies display intrinsic narrow UV absorption (Crenshaw et al. 1999), which so far always corresponds to the presence of an X-ray warm absorber. Warm absorbers are highly ionized (U = 0.1 to 10) with total column densities of 10^{21-23} cm^{-2} (George et al. 1998). The link between UV and X-ray absorption is not yet clear. Mathur et al. (1995) proposed a single zone model for a combined UV-X-ray absorber in NGC 5548. High-resolution UV spectroscopy shows that the UV absorber breaks into distinct components with outflow velocities ranging up to 1200 km s^{-1} (Mathur et al. 1999; Crenshaw & Kraemer 1999), with a range of ionization states and column densities.

The FUSE wavelength range features the O VI $\lambda\lambda$1032,1038 doublet, which is intermediate in ionization between the absorbed UV and X-ray species. Other FUSE results on AGNs are summarized by Kriss (2002).

2. Results from a Low-State FUSE Spectrum

NGC 5548 was observed on 7 June 2000 through the $30'' \times 30''$ aperture for 25 ksec, when it was in a very low state (Fig. 1, left). Coincidently, the low

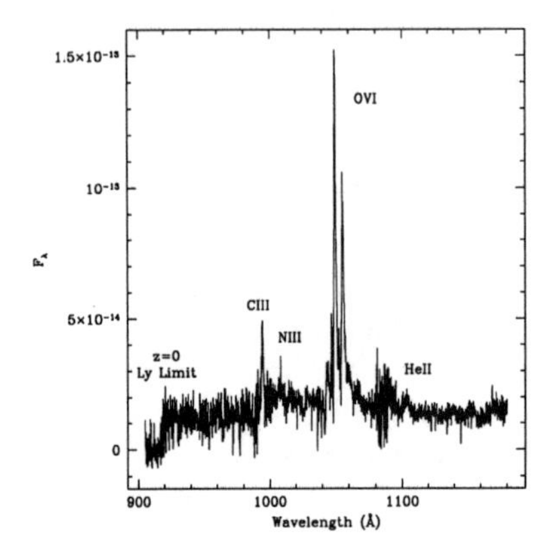

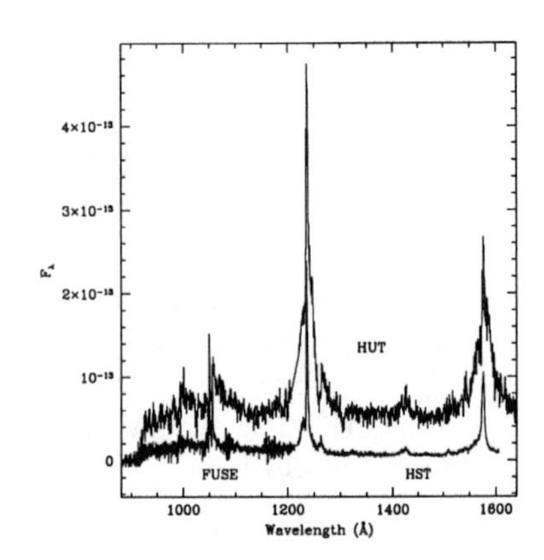

Figure 1. Left. NGC 5548 FUSE spectrum, 0.1Å bins. Right. A comparison between the FUSE spectrum, a low-state FOS spectrum from HST (Crenshaw et al. 1993), and a Hopkins Ultraviolet Telescope (HUT) spectrum at a more typical flux level.

flux level matches that seen when it was first observed with the Hubble Space Telescope (HST). This level is several times lower than typical (Fig. 1, right).

Analysis of the FUSE spectrum provides challenges. While the continuum and broad lines faded, narrow emission lines became prominent. In fitting the O VI region (Fig. 2), two extremes are feasible: a narrow line with FHWM = 700 km s^{-1} covered by the absorbers, or a narrow line with FWHM = 400 km s^{-1} not covered by the absorbers. Each has supporting arguments. The covered NLR model fits the blue wing better, other high-ionization narrow lines show FHWM $\sim$ 700+ km s^{-1} (Goad & Koratkar 1998), and the NLR of NGC 5548 is compact (Kraemer et al. 1998). On the other hand, the uncovered NLR is less of an extrapolation, and an extended NLR may not be expected to be covered.

The narrow emission line O VI doublet ratio does **not** appear consistent with an optically thin 2:1 as expected for a traditional low density NLR. Both models indicate a 1.5:1 doublet ratio. This is reminiscent of NGC 3516, which has a 1:1 narrow emission line ratio for O VI in a low state and may be evidence for an accelerating wind component (Hutchings 2002): the low flux state shows dense O VI emitting gas near the nucleus with low velocities, while in a high flux state the velocities are larger for more distant O VI emitting gas.

The current uncertainties (primarily involving the narrow emission lines), prevents robust conclusions about the absorbers with blueshifts of 200 to 700 km s^{-1} (components 2-5). These velocities correspond to the absorption features seen in O VII and O VIII in X-ray spectra (Kaastra et al. 2000). Minimum assumptions regarding the absorption suggests a lower limit to the O VI column density of about 10^{15} cm^{-2}. Taken together with the apparent O VII and O VIII column densities, photoionization modeling with CLOUDY (Ferland 2001) suggests an absorber with an ionization parameter of U = 3 and a total hydrogen column density of a few times 10^{21} cm^{-2}. Similar modeling of the FUV/UV column densities of the highest velocity component (1) indicates a low total col-

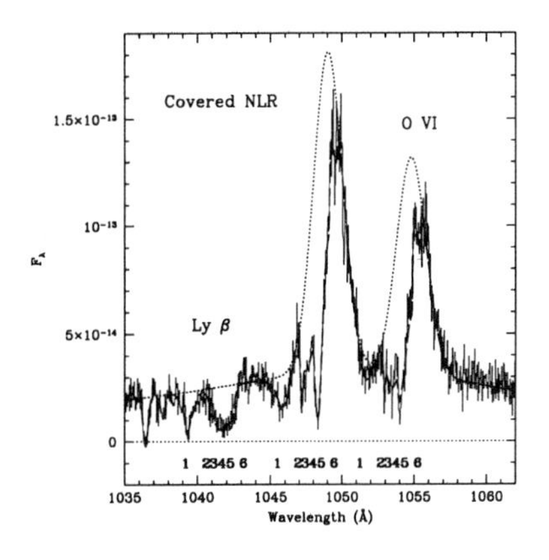

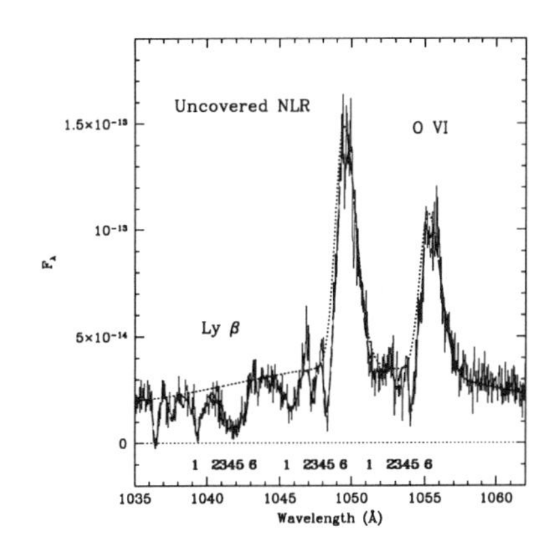

Figure 2. Two O VI fits. Marked components based on UV absorption redshifts (Crenshaw & Kraemer 1999). Left: a narrow emission line of 700 km s^{-1} covered by absorption. Right: a narrow emission line of 400 km s^{-1} uncovered by absorption.

umn and low ionization state contrary to previous reports (Crenshaw & Kraemer 1999). This reanalysis is consistent with the fact that no absorption is seen at these velocities in X-ray spectra (Kaastra et al. 2000).

Simultaneous high-resolution X-ray/UV spectroscopy is required to accurately characterize the emission and absorption properties of NGC 5548, permitting an understanding of the outflows from this much-studied AGN.

I thank the FUSE science team and the AGN working group. This work is based on data obtained for the Guaranteed Time Team by the NASA-CNES-CSA FUSE mission operated by the Johns Hopkins University. Financial support to U. S. participants has been provided by NASA contract NAS5-32985.

References

Crenshaw, D. M., Boggess, A., & Wu, C. 1993, ApJ, 416, L67

Crenshaw, D. M. & Kraemer, S. B. 1999, ApJ, 521, 572

Crenshaw, D. M., et al. 1999, ApJ, 516, 750

Ferland, G. J., 2001, Hazy, a brief introduction to Cloudy 94.00

George, I., et al. 1998, ApJS, 114, 73

Goad, M. & Koratkar, A. 1998, ApJ, 495, 718

Hutchings, J.B., 2002, in Mass Outflow in Active Galactic Nuclei: New Perspectives, eds. D.M. Crenshaw, S.B. Kraemer, & I.M. George (San Francisco: ASP), p. 267

Kaastra, J. S., et al. 2000, A&A, 354, L83

Kraemer, S. B., et al. 1998, ApJ, 499, 719

Kriss, G.A. 2002, in Mass Outflow in Active Galactic Nuclei: New Perspectives, eds. D.M. Crenshaw, S.B. Kraemer, & I.M. George (San Francisco: ASP), p. 69

Mathur, S., Elvis, M., & Wilkes, B. 1999, ApJ, 519, 605

Mathur, S., Elvis, M., & Wilkes, B. 1995, ApJ, 452, 230

Peterson, B. M., et al. 1999, ApJ, 510, 659

Peterson, B. M., & Wandel, A. 1999, ApJ, 521, 95

Mass Outflow in Active Galactic Nuclei: New Perspectives
ASP Conference Series, Vol. 255, 2002
D.M. Crenshaw, S.B. Kraemer, and I.M. George

Fe II Absorbers in Arp 102B and Other LINERs and BLRGs

Michael Eracleous

Department of Astronomy and Astrophysics, The Pennsylvania State University, 525 Davey Lab, University Park, PA 16802

Abstract.
The broad-line radio galaxy Arp 102B was found to have prominent Fe II absorption lines in a UV spectrum obtained with the *HST*. These line are particularly interesting because they arise from metastable levels of Fe II, up to 1 eV above the ground state. Here, we present new *ASCA* X-ray observations aimed at determining the intrinsic neutral gas column responsible for absorption at both soft X-ray and UV wavelengths. We determine a neutral Hydrogen column density of $N_{\mathrm{H}} = (2.8 \pm 0.3) \times 10^{21}\ \mathrm{cm}^{-2}$. The absorber must be mostly free of dust in order for this large column density to be consistent with the inferred optical extinction. Using this measurement, we are in the process of modelling the physical properties of the absorber and investigating whether the X-ray and UV absorption can arise in the same medium. Based on our preliminary results, we speculate that the absorber may be a "failed" accretion-disk, which would be consistent with its low dust content and with the fact that the broad emission lines of Arp 102B have disk-like, double-peaked profiles. In this context, it is rather tantalizing that two other double-peaked emitters, the LINER NGC 1097 and the broad-line radio galaxy 3C 332, show similar absorption lines in their UV spectra.

1. Introduction

The broad-line radio galaxy Arp 102B (which is also a certified LINER, based on the relative strengths of its *narrow* emission lines) is famous for the double-peaked profiles of its Balmer lines. We observed it with the *HST* in 1995 in order to study its UV emission lines and, much to our surprise, we discovered a forest of relatively strong Fe II *absorption* lines in the near UV. A segment of the *HST* spectrum is shown in Figure 1.

These lines constitute an associated absorber and appear to be intrinsic to the AGN. We identified many of the lines as originating from *excited* states of Fe II, namely from two other of the lowest states (terms a^4F and a^4D). Particularly surprising is the presence of multiplets UV62, UV 63, and UV 64, arising from the metastable a^4D term, which lies 0.99–1.10 eV above the ground state. Fig. 1 shows an expanded view of the near-UV spectrum with identified lines marked. When the original observations were made only very few quasars were known to show such absorption lines: Q 0059–2735 (e.g., Hazard et al. 1987; Wampler, Chugai, & Petitjean 1995), Hawaii 167 (e.g., Cowie et al. 1994),

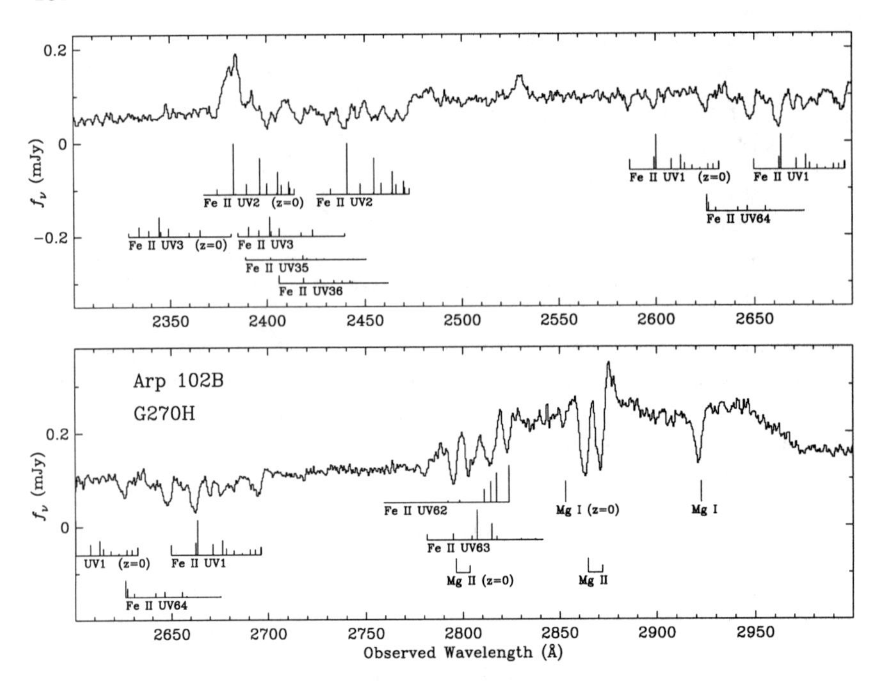

Figure 1. An expanded view on the near-UV spectrum of Arp 102B, with absorption lines identified. There are absorption lines of Fe II, Mg II, and Mg I from the ISM of the Milky Way. The same lines are also observed at the redshift of Arp 102B, but there are additional intrinsic Fe II lines from metastable levels up to 1.1 eV above the ground state.

and Mrk 231 (e.g., Boroson, Meyers, & Morris 1991), but more examples have been found since.

To understand the nature of the absorber we used archival *ROSAT* and new *ASCA* observations to determine the column density of neutral gas along the line of sight to the X-ray source. With the help of this measurement and photoionization models, we are attempting to constrain the nature of the absorber.

2. X-Ray Observations and Results

The X-ray spectrum of Arp 102B shows the signature of significant absorption by neutral gas. The *ROSAT* PSPC spectrum is so heavily attenuated that it places only loose limits on the absorbing column of $(2-8)\times 10^{21}$ cm^{-2} and hardly any constraints on the spectral index. The *ASCA* spectrum, shown in Figure 2, gave much better constraints, since it goes up to 10 keV and allows us to determine the spectral index fairly well. The spectrum is well-described by a simple, absorbed power law with the following parameters: $N_{\rm H} = (2.8\pm 0.3)\times 10^{21}$ cm^{-2}, $\Gamma = 1.58\pm 0.04$, and $L_{2-10\ {\rm keV}} = 3.1\times 10^{43}$ erg s^{-1}. The Galactic column density is an order of magnitude smaller and cannot account for the observed absorption, thus the measured column must be intrinsic. Even if we try other models for

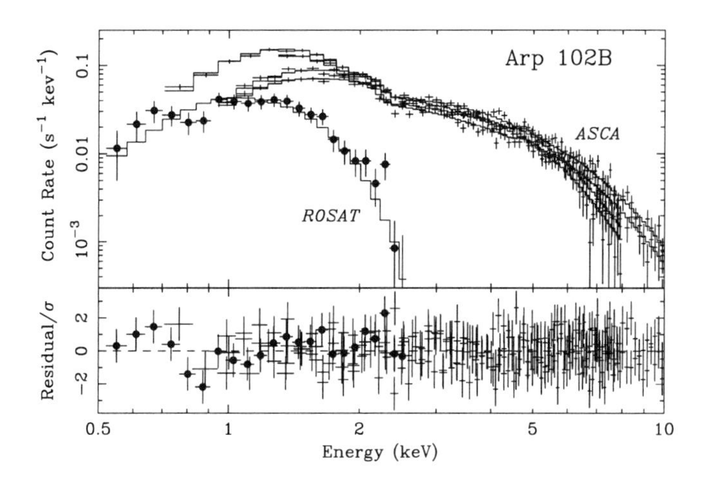

Figure 2. The *ASCA* and *ROSAT* spectra of Arp 102B, fitted (independently) with a simple power-law model modified by photoelectric absorption in neutral matter with solar abundances. The corresponding confidence contours are shown in Figure 3.

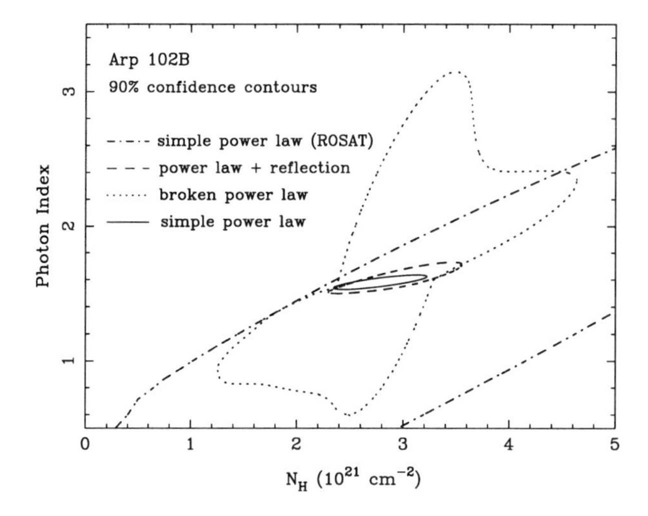

Figure 3. Confidence contours from the *ASCA* and *ROSAT* spectra of Arp 102B for different assumed continuum models. The simple power-law model (shown in Figure 2) gives an acceptable description of the spectrum and yields $\Gamma = 1.58$ and $N_{\rm H} = 2.8 \times 10^{21}$ cm^{-2}.

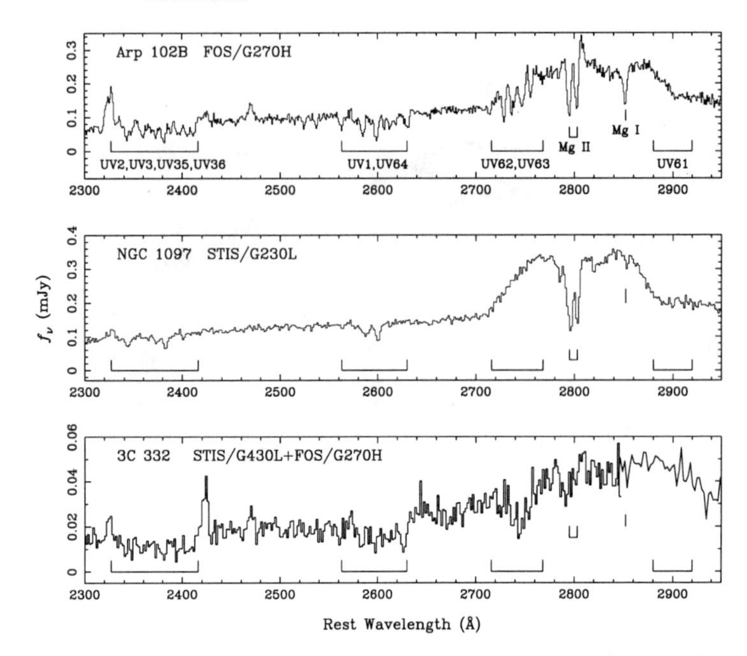

Figure 4. Comparison of the near-UV spectra of Arp 102B, the BLRG 3C 332, and the LINER NGC 1097. All three objects have Fe II and Mg II absorption lines, although they seem to vary in strength from object to object. In particular, the UV62 and UV63 multiplets are noticeably absent in NGC 1097.

the continuum, the absorbing column is still high, $N_{\rm H} > 1 \times 10^{21}$ cm^{-2}, as shown in Figure 3. The large X-ray column does not agree with the optical/UV reddening! The measured X-ray column would imply 3 mag of extinction at the wavelength of the Mg II line, i.e. and attenuation by a factor of 14. Yet the Mg II line is observed to be quite strong, with no sign of any attenuation. *Therefore, the absorber must be dust-free.*

3. Discussion and Interpretation

3.1. The absorber in context.

What is this absorber that absorbs X-ray and emission-line photons and is free of dust? We have one additional clue: the profiles of the Balmer and Mg II emission lines of Arp 102B are double-peaked, indicating that the lines come from an accretion disk (Chen & Halpern 1989; Halpern et al. 1996). So, the absorber must be located above/around the outer parts of the disk ($R > 10^3 R_{\rm g}$), or further out.

Arp 102B is now in good company. In the past few years we have found more examples of such absorption-line systems in NGC 1097, a LINER, and 3C 332, a BLRG (see spectrum in Halpern 1997 and in Figure 4). What all these objects have in common is double-peaked Balmer lines and LINER-like

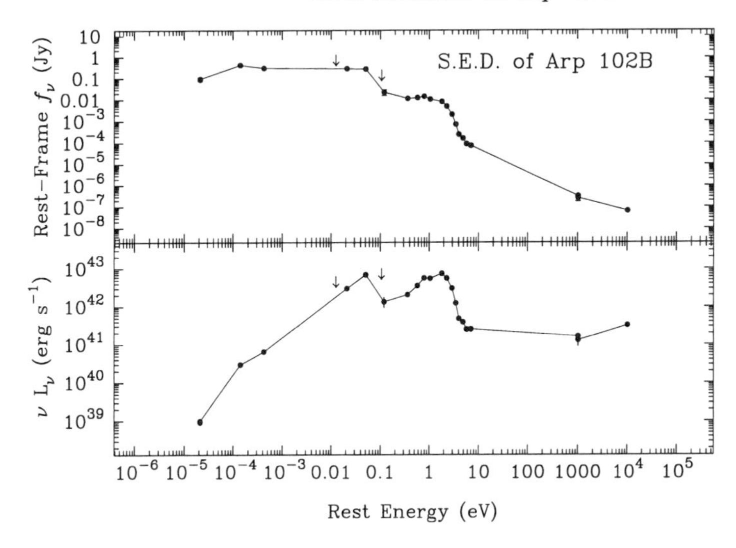

Figure 5. The spectral energy distribution of Arp 102B. It was used as input to CLOUDY calculations. Note that the bump at 1 eV is due to starlight in the host galaxy. Note also that there is no obvious UV bump. Instead, the SED peaks in the IR, at 30 μm, and is reminiscent of emission from an ADAF.

narrow-line ratios (see also Shields, these proceedings). In addition, a few radio-loud quasars found recently in the FIRST survey show similar complexes of Fe II absorption lines (e.g., de Kool et al. 2001, and these proceedings). Moreover, some of the well-known Seyfert galaxies (e.g., NGC 3227; Crenshaw et al. 2001) also show similar Fe II absorption lines when observed with the *HST*/STIS.

3.2. Modeling of the absorber.

We are in the process of making models for the absorber using CLOUDY (Ferland 1996), in an effort to interpret the absorption-line spectrum. We use the spectral energy distribution (SED) shown in Figure 5, which consists of a relatively hard power-law with no UV bump (drawn mostly from Chen & Halpern 1989), It resembles the SEDs of LINERs (Ho 1999) and weak-line radio galaxies (Sambruna, Eracleous, & Mushotzky 1999). We also assume that the column density measured in the X-rays is the same as the column density of the UV absorber, so we keep it fixed.

Under these assumptions we have computed a grid of models using the latest version of CLOUDY, which includes the Verner model for the Fe II ion (371 levels, including the metastable levels responsible in the transitions that we observe). The grid is defined by the density of the gas and the ionization parameter, with the following values: $n = 10^2 - 10^8$ cm^{-3} and $U = 10^{-3.0} - 10^{-0.5}$.

Our preliminary conclusions are (a) the ionization parameter can be constrained to $U < 10^{-2.0}$, since at higher values the abundance of Fe II is too low for any significant absorption to occur, and (b) the density is likely to be in the range $n < 10^4$ cm^{-3}, *probably*, since at higher densities the absorber is extremely optically thick. The latter conclusion is by no means firm, however, until we carry out a detailed comparison of the model predictions with the observed UV spectrum.

3.3. Speculations

From the observed luminosity and SED of Arp 102B, we can infer the output rate of ionizing photons from the central engine. We can combine this rate with the density and ionization parameter inferred from models to constrain the distance of the absorber from the ionizing source. If we take the model implications at face value, we get an absorber distance of $d > 10$ pc. *But*, in view of the current uncertainty in the density (the observed Fe II lines are unresolved and could well be saturated), we can adopt the value inferred for Q 0059–2735 ($n = 10^6$ cm^{-3}), which gives $d > 1$ pc. A particularly interesting case would be that of an even higher density, which would allow the distance to be of order 0.1 pc. This distance is close to the size of the outer accretion disk ($d \sim 10^4\ R_g$). If this last possibility turns out to be true, then the absorber could be a failed accretion disk wind (e.g., Proga, Stone, & Kallman 2000). This would be appealing for another reason: it would explain why the absorber is dust free.

References

Boroson, T. A., Meyers, K. A., & Morris, S. L. 1991, ApJ, 370, L19

Chen, K., & Halpern, J. P. 1989, ApJ, 344, 115

Crenshaw, D. M. Kraemer, S. B., Bruhweiler, F. C., Ruiz, J. R. 2001, ApJ, 555, 633

Cowie, L. L. et al. 1994, ApJ, 432, L83

de Kool, M. et al. 2001, ApJ, 548, 609

Ferland, G. J. 1996, Hazy, University of Kentucky Internal Report

Halpern, J. P. 1997, in Mass Ejection from Active galactic Nuclei, eds. N. Arav, I. Shlosman, & R. J. Weymann (San Francisco: ASP), 41

Halpern, J. P., Eracleous, M., Filippenko, A. V., & Chen, K. 1996, ApJ, 464, 704

Hazard, C., McMahon, R. G., Webb, J. K., & Morton, D. C. 1987 ApJ, 323, 263

Ho, L. C. 1999, ApJ, 516, 672

Proga, D., Stone, J. M., & Kallman, T. R. 2000, ApJ, 543, 686

Sambruna, R. M., Eracleous, M., & Mushotzky, R. F. 1999, ApJ, 526, 60

Wampler, E. J., Chugai, N. N., & Petitjean, P. 1995, ApJ, 443, 586

Mass Outflow in Active Galactic Nuclei: New Perspectives
ASP Conference Series, Vol. 255, 2002
D.M. Crenshaw, S.B. Kraemer, and I.M. George

High Resolution Spectra of Quasar AALs: 3C 191

Fred Hamann

Department of Astronomy, University of Florida, 211 Bryant Space Science Center, Gainesville, FL 32611-2055

T.A. Barlow

Infrared Processing and Analysis Center, California Institute of Technology, MS 100-22, 770 South Wilson Ave., Pasadena, CA 91125

F.C. Chaffee

California Association for Research in Astronomy, W.H. Keck Observatory, 65-1120 Mamalahoa Highway, Kamuela, HI 96734

C.B. Foltz

MMT Observatory, University of Arizona, 933 North Cherry Ave., Tucson, AZ 85721-0065

R.J. Weymann

Observatories of the Carnegie Institution of Washington, 813 Santa Barbara Street, Pasadena, CA 91101-1292

Abstract. We discuss new high-resolution (6.7 km/s) spectra of the associated absorption lines (AALs) in the radio-loud quasar 3C 191. The measured AALs have ionizations ranging from Mg I to N V, and multi-component profiles that are blueshifted by ~400 to ~1400 km/s relative to the quasar's broad emission lines. Excited-state absorption lines of Si II* and C II* imply volume densities of $\sim$300 cm^{-3} and a nominal distance from the quasar of 28 kpc (assuming photoionization). The total column density is $N_H \sim 2 \times 10^{20}$ cm^{-2}. Surprisingly, the absorber only partially covers the quasar emission source along our line of sight. We propose a model for the absorber in which pockets of dense neutral gas are surrounded by bigger clouds of generally lower density and higher ionization. This outflowing material might be left over from a blowout associated with a nuclear starburst, the onset of quasar activity, or a past broad absorption line (BAL) wind phase.

1. Introduction

Associated absorption lines (AALs) are important diagnostics of the gaseous environments of quasars and active galactic nuclei (AGNs). In particular, the lines can trace a variety of phenomena — from energetic outflows like the BALs

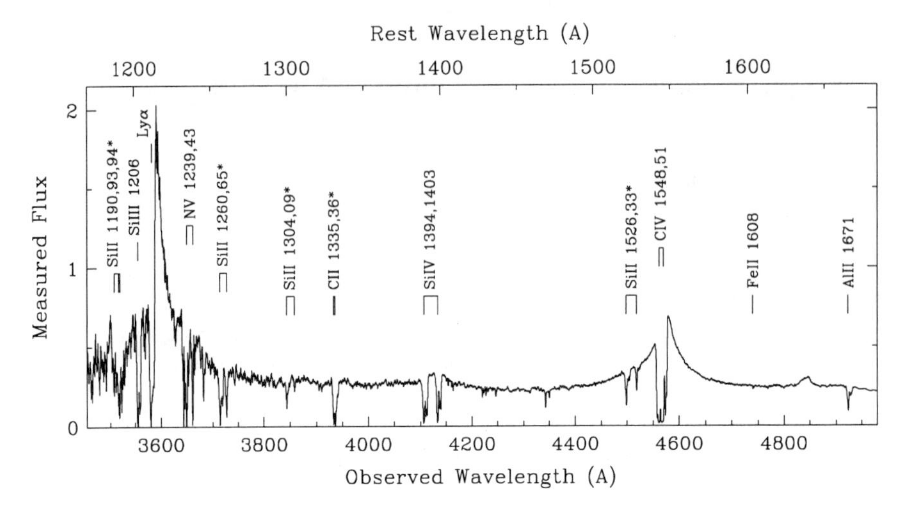

Figure 1. Part of the 3C 191 spectrum showing its strong associated absorption lines (labeled above). The flux has units 10^{-15} ergs cm^{-2} s^{-1} Å^{-1}

to relatively quiescent gas at large galactic or inter-galactic distances (Weymann et al. 1979, Hamann & Brandt 2001). We are involved in a multi-wavelength program to locate individual AAL absorbers, determine their elemental abundances, quantify their kinematic and physical properties, and understand the role of the AGNs and/or host galaxies in providing the source of material and kinetic energy to the absorbing gas.

One interesting property is that, among radio-loud quasars, AALs appear more frequently and with greater strength in sources with "steep" radio spectra and/or lobe-dominated radio morphologies (Wills et al. 1995, Richards 2001, Brotherton et al. 1998). 3C 191 (Q0802+103, emission-line redshift $z_e = 1.956$) is a radio-loud quasar having both strong AALs and a bipolar, lobe-dominated radio structure. It provides a rare opportunity to define the distance between the quasar and the absorbing gas because its AALs include excited-state lines, e.g. C II* λ 1336 and Si II* $\lambda\lambda$ 1265,1533, which constrain the volume density and therefore the quasar–absorber distance (assuming photoionization, Bahcall et al. 1967).

2. Observations and Results

We observed 3C 191 on three occasions between 1997 and 1998 using the High Resolution Echelle Spectrograph (HIRES) on the Keck I telescope on Mauna Kea, Hawaii. On each occasion, a $0.86''$ slit provided spectral resolution $\lambda/\Delta\lambda \approx 45,000$ or 6.7 km s^{-1}, corresponding to 3 pixels on the 2048^2 Tektronix CCD. We reduced the data with standard techniques using the MAKEE software package. The spectra cover the observed wavelengths 3850 – 5975 Å and 6474 – 8927 Å.

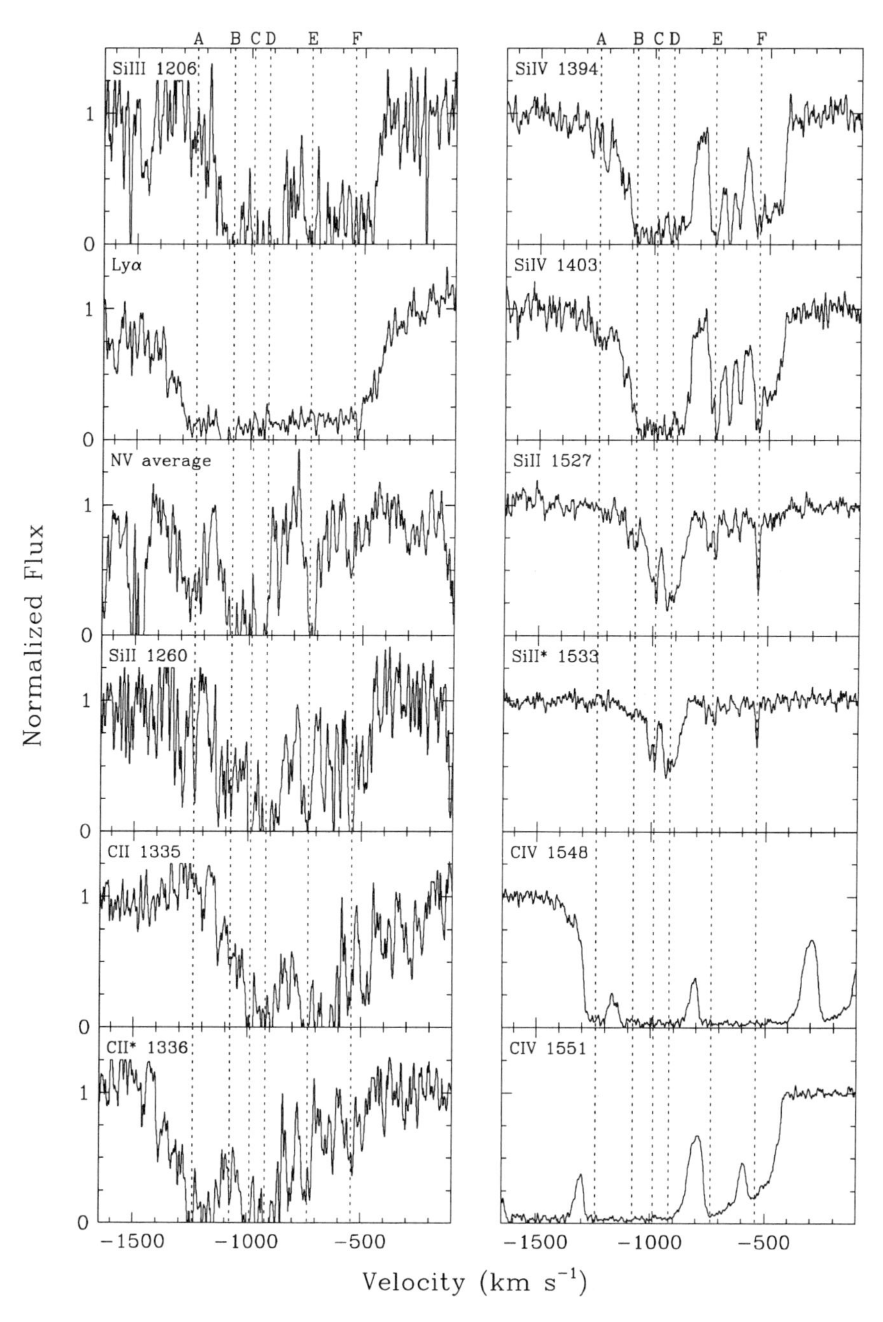

Figure 2. Velocity profiles of the AALs in 3C 191. Zero velocity corresponds to the nominal emission-line redshift, $z_e = 1.956$. Dotted vertical lines mark the positions of strong features.

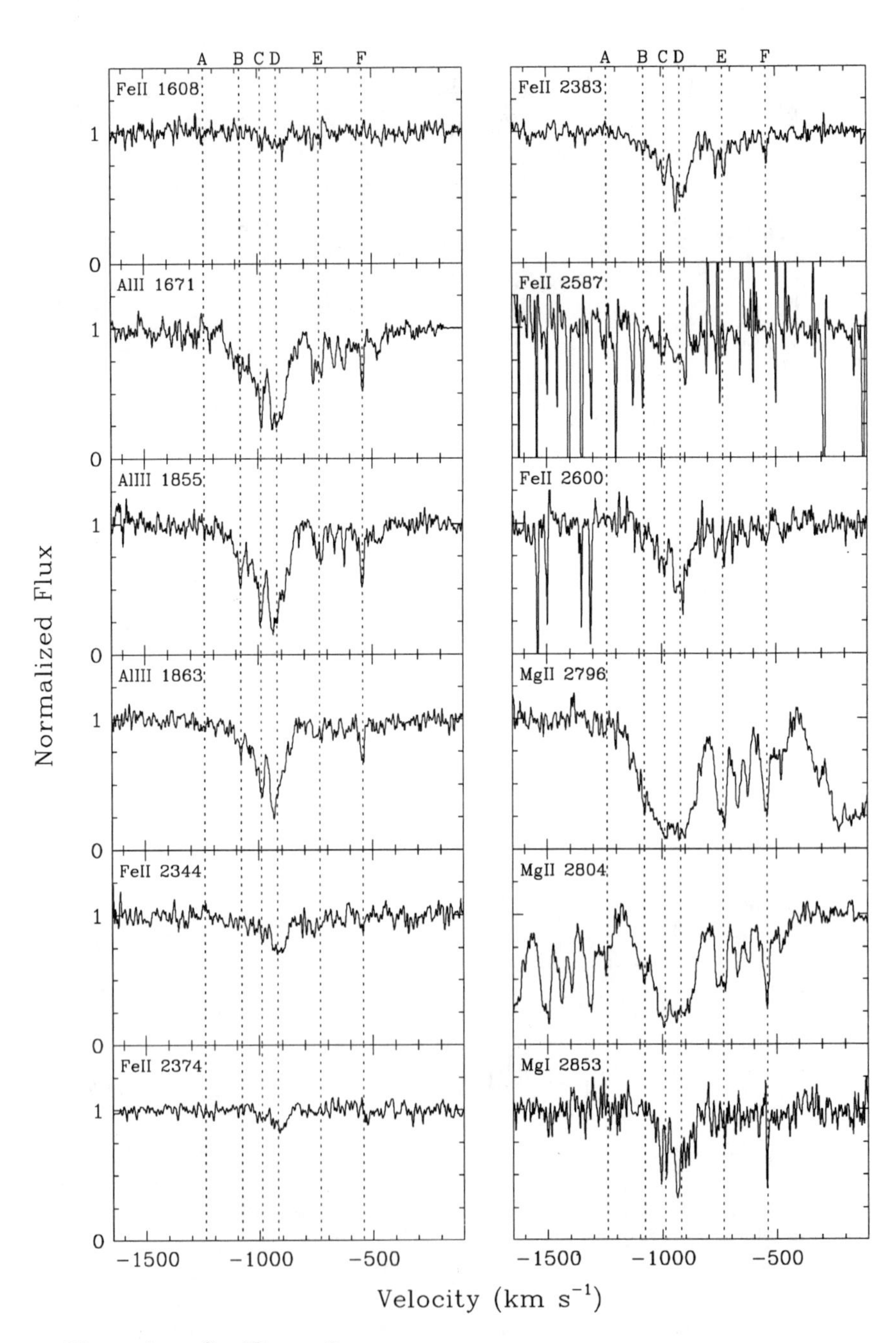

Figure 3. See Figure 2.

The measured AALs have ionizations ranging from Mg I to N V, and multi-component profiles that are blueshifted by ~400 to ~1400 km/s relative to the quasar's broad emission lines (see Figures 1–3). These data yield the following new results. **(1)** The strengths of excited-state Si II* AALs indicate a density of ~300 cm^{-3} in the Si^+ gas. **(2)** If the gas is photoionized, this density implies a distance of ~28 kpc from the quasar. Several arguments suggest that all of the lines form at approximately this distance, with a range of densities determining the range of ionizations. **(3)** Strong Mg I AALs identify neutral gas with very low ionization parameter and high density. We estimate $n_H > 5 \times 10^4$ cm^{-3} in this region, compared to only ~15 cm^{-3} where the N V lines form. **(4)** The total column density is $N_H < 4 \times 10^{18}$ cm^{-2} in the neutral gas and $N_H \sim 2 \times 10^{20}$ cm^{-2} in the moderately ionized regions (up to Al III, Si IV, etc.). There may be larger column densities in more highly ionized gas, however, the total column of $N_H \sim 2 \times 10^{20}$ cm^{-2} is consistent with 3C 191's strong soft X-ray flux and the implied absence of soft X-ray absorption. **(5)** The total mass in the AAL outflow is $M \sim 2 \times 10^9$ $M_\odot$, assuming a global covering factor (as viewed from the quasar) of ~10% . **(6)** The absorbing gas only partially covers the background light source(s) along our line(s) of sight, requiring absorption in small clouds or filaments <0.01 pc across. The ratio N_H/n_H implies that the clouds have radial (line-of-sight) thicknesses <0.2 pc. **(7)** The characteristic flow time of the absorbing gas away from the quasar is $\sim 3 \times 10^7$ yr.

3. Discussion

The physical connection between the quasar 3C 191 and its AALs is established by the excited-state lines (Si II*). However, the absorber–quasar distance is very large, ~28 kpc. Other quasar AALs are known to form much closer to the central engines, possibly within a few pc in outflows similar to the BALs (Hamann et al. 1997, Barlow & Sargent 1997, Barlow, Hamann & Sargent 1997). 3C 191 might contain a different class of absorber (e.g. much farther from the active nucleus) than the majority of AGNs discussed at this meeting. In particular, 3C 191 does not follow the trend identified by Brandt et al (2000) for small X-ray to UV continuum flux ratios (α_{ox}) accompanying large C IV absorption equivalent widths. The outflowing material in 3C 191 might be left over from a blowout associated with a nuclear starburst, the onset of quasar activity, or a past broad absorption line (BAL) wind phase. The flow time of $\sim 3 \times 10^7$ yr might therefore represent the time elapsed since the formation of the quasar and/or an accompanying starburst episode.

We propose a model for the 3C 191 absorber (Figure 4) in which pockets of dense neutral gas (represented by Mg I λ2853) are surrounded by a diffuse, spatially extended medium of generally higher ionization (e.g. C IV and N V). The diffuse clouds contain most of the total column density; their greater size and/or greater numbers lead to more complete coverage in both velocity and projected area. More work is needed to characterize fully this class of distant AAL absorbers (see also Hamann & Brandt 2001, Barlow et al. 1997, Tripp et al. 1996, Morris et al. 1986), and understand its relationship (if any) to other absorption phenomena in quasars/AGNs.

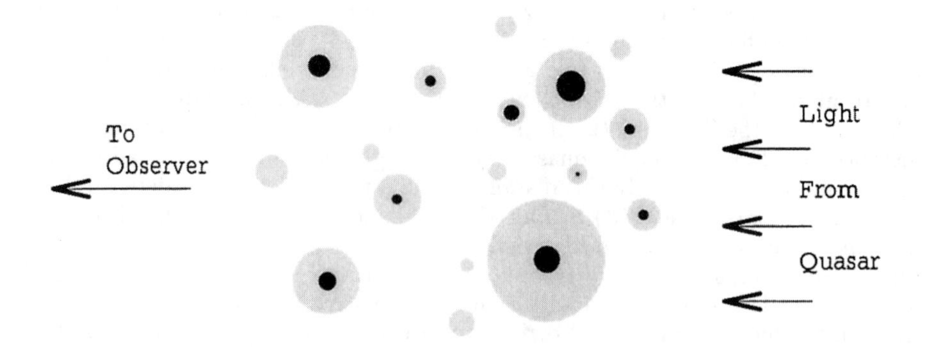

Figure 4. Schematic representation of the AAL environment, showing pockets of dense neutral gas (filled black circles) surrounded by a less dense and more highly ionized medium (grey circles). The more extended regions lead to smoother AAL profiles and more complete line-of-sight coverage of the background light source.

We are grateful to the staff of the Keck Observatory for their willing assistance. FH thanks Bassem Sabra for useful discussions and acknowledges support from NSF grant AST 99-84040. CBF acknowldges NSF grant AST 98-03072.

References

Akujor, C.E., et al. 1994, A&A Supp., 105, 247

Bahcall, J.N., et al. 1967, ApJ, 149, L11

Barlow, T. A., Hamann, F., & Sargent, W. L. W. 1997, in Mass Ejection From AGN, ed. R. Weymann, I. Shlosman, & N. Arav (San Francisco: ASP), 13

Barlow, T. A., & Sargent, W. L. W. 1997, AJ, 113, 136

Brandt, W.N., Laor, A., & Wills, B.J. 2000, ApJ, 528, 637

Brotherton, M.S., et al. 1998, ApJ, 501, 110

Burbidge, E.M., et al. 1966, ApJ, 144, 447

Hamann, F., & Brandt, W.N. 2001, PASP review, in prep.

Hamann, F., et al. 1997, ApJ, 478, 80

Morris, S.L., et al. 1986, ApJ, 310, 40

Richards, G.T. 2000, ApJS, 133, 53

Tripp, T.M., Lu, L., & Savage, B.D. 1996, ApJS, 102, 239

Weymann, R.J., Williams, R.E., Peterson, B.M., & Turnshek, D.A. 1979, ApJ, 234, 33

Williams, R.E., Strittmatter, P.A., Carswell, R.F., & Craine, E.R. 1975, ApJ, 202, 296

Wills, B.J., et al. 1995, ApJ, 447, 139

Mass Outflow in Active Galactic Nuclei: New Perspectives
ASP Conference Series, Vol. 255, 2002
D.M. Crenshaw, S.B. Kraemer, and I.M. George

Narrow C IV λ1549Å Absorption Lines in Moderate-Redshift Quasars

M. Vestergaard

Department of Astronomy, The Ohio State University, 140 West 18th Avenue, Columbus, OH 43210

Abstract. A large, high-quality spectral data base of well-selected, moderate-redshift quasars is used to characterize the incidence of narrow associated C IV λ1549 absorption, and how this may depend on some quasar properties. Preliminary results of this study are presented.

1. Introduction

Associated narrow absorption lines (NALs) in the spectra of active galaxies have widths less than a few hundred km s^{-1} and are located within 5000 km s^{-1} of the emission redshift (Weymann et al. 1979). They are likely physically connected to the active galactic nucleus. The working hypothesis in this study is that NALs are possibly the low-velocity equivalents to the more dramatic broad absorption features (BALs), with line widths reaching tens of thousands of km s^{-1}. The current, commonly adopted physical interpretation is that the line widths of both NALs and BALs trace the outflow velocity of the absorbing gas and that these outflows are somewhat equatorial. The exact solid angle extension of the NAL and BAL matter above the disk is unknown. BALs are predominantly found in the high-luminosity, radio-quiet quasars with a frequency of 10–12%. NALs appear present in 50–70% of the low-luminosity Seyfert galaxies (Hamann 2000), yet the frequency in quasars and how it may depend on source radio power and source axis inclination are not accurately known. The aim of this study is to address this issue with a large, high-quality, UV spectral data base of $z \approx 2$ radio-loud and radio-quiet quasars (hereafter RLQs and RQQs, respectively) for which the data are uniformly processed and analyzed (Vestergaard 2000; Vestergaard et al. 2001, in preparation). The frequency of associated C IV NALs is studied and possible trends with quasar properties, such as luminosity, UV spectral slope, radio loudness, and source inclination are tested for. In particular, if there is a wind evaporating off the accretion disk, then one might naively expect this wind to be stronger in brighter, bluer objects, as the stronger radiation field will blow off more disk matter. If the associated NALs are somehow related to such a wind one would then expect a relationship between the strength of the NALs and the continuum characteristics: the continuum luminosity, L_{cont}, the UV continuum slope, α_{UV}, and/or the absolute magnitude, M_V, of the object.

The first results of this study are presented here. An extended analysis, including a detailed comparison of the NAL properties of the RLQs and the RQQs, will be presented by Vestergaard (2001, in preparation). First, the conclusions are summarized; then the data, measurements, and results are presented and briefly discussed.

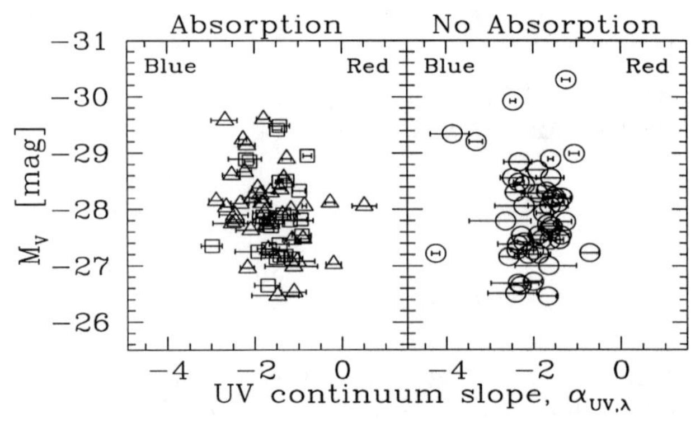

Figure 1. The distribution of M_V and $\alpha_{UV,\lambda}$ for the quasars with (triangles: RLQs; squares: RQQs) and without C IV NALs (circles).

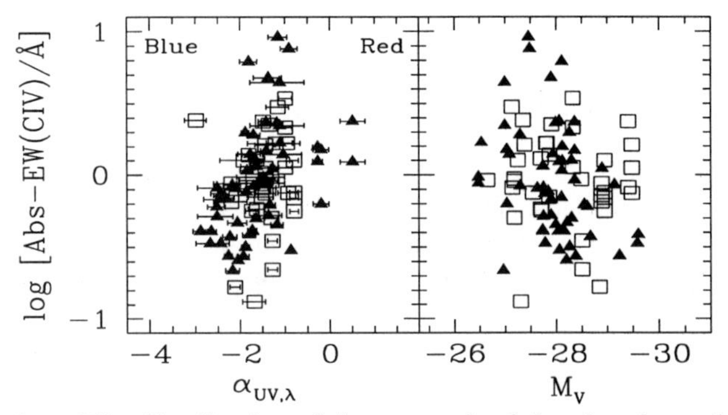

Figure 2. The distribution of the strength of the absorber relative to the UV continuum slope and the quasar luminosity.

2. Summary and Conclusions

The main results of this study are as follows:

- The frequency of high-velocity NALs in RQQs and RLQs is similar.
- RLQs show a small excess of associated NALs with respect to RQQs.
- Relatively fewer *core-dominated* RLQs show associated NALs than lobe-dominated RLQs do.
- The strongest associated NALs ($\gtrsim$ 3Å) are mostly in RLQs.
- The data are consistent with the associated absorption being stronger in more inclined objects where reddening effects may play a larger role. Do these effects dominate radiation pressure effects?
- About 8% (5/66) of the RLQs show strong (EW >3Å) absorption at relatively low velocities, which is not seen in the RQQs here. While RQQs can accelerate the central outflow to high velocities (in BALs), the RLQs are perhaps not capable thereof to the same extent. This is consistent with predictions of disk outflow models. Are RLQ NALs the low-velocity equivalents to the BAL phenomenon seen mostly in RQQs?

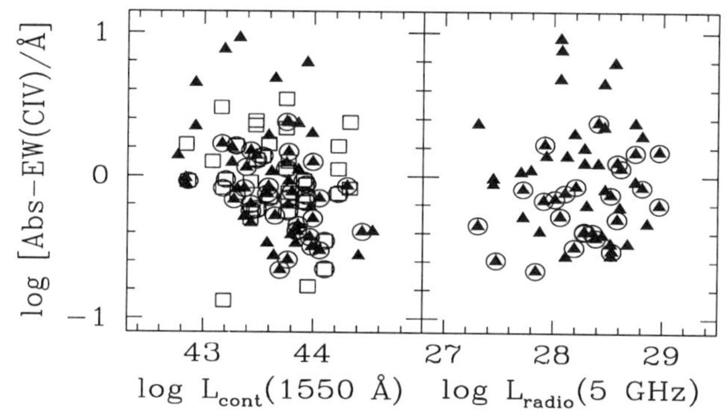

Figure 3. The absorption strength versus continuum luminosities. Encircled symbols are high velocity (>5000 km s^{-1}) absorbers.

Table 1. **Frequency of Absorbed Objects and of C IV NALs**

Sample	**N**	**Absorbed QSOs (All Abs. Vel's)**	**Velocity ≤ 5000 km s^{-1}**	**5000 < Velocity 15000 km s^{-1}**
All QSOs	114	66 = 58%	41 = 36%	36 = 32%
RQQs	48	25 = 52%	14 = 29%	14 = 29%
RLQs	66	41 = 62%	27 = 41%	22 = 33%
CDQs	20	12 = 60%	7 = 35%	7 = 35%
LDQs	46	29 = 63%	20 = 43%	15 = 33%
Absorber Frequency Among Absorbed Quasars Only				
LDQs	–	29/66 = 44%	20/41 = 49%	15/36 = 42%
CDQs	–	12/66 = 18%	7/41 = 17%	7/36 = 19%

3. Sample, Data, and Measurements

The sample of 114 quasars (66 radio-loud and 48 radio-quiet) was selected for a study of the emission lines (Vestergaard 2000; Vestergaard, Wilkes, & Barthel 2000; Vestergaard et al. 2001). The spectra cover a minimum range from ~1000 to ~2100 Å with a spectral resolution of ~5 Å or better. High quality VLA radio maps at 1.4, 5, and 15 GHz are available for most of the RLQs (Barthel et al. 1988; Lonsdale, Barthel, & Miley 1993; Barthel, Vestergaard, & Lonsdale 2000). Vestergaard et al. (2001) provide details on all the data.

The RLQs are further subclassified as lobe-dominated quasars, LDQs (i.e., $R_{5GHz} = S_{5GHz,core}/S_{5GHz,total} < 0.5$), or as core-dominated quasars, CDQs ($R_{5GHz} \geq 0.5$). The quasars were selected such that the RLQs and RQQs have similar z and luminosity, M_V, distributions (e.g., Figure 1).

A cosmology of $H_0 = 50$ km s^{-1} Mpc^{-1} and $q_0 = 0$ is used throughout.

3.1. The Measurements

The C IV emission line profile was reproduced with a smooth fit [see Vestergaard et al. (2001) for details]. This smooth profile fit acts as the local "continuum" for the absorption lines. No Galactic extinction correction is performed on the

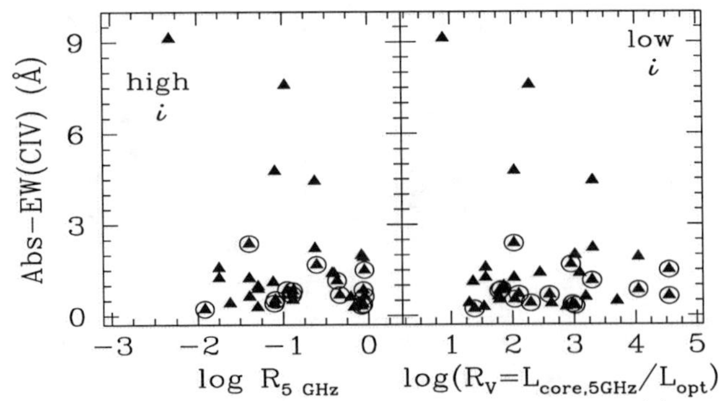

Figure 4. The absorber EW versus source inclination (i) estimators (high inclination to the left, low inclination to the right in each panel).

spectra. The measured absorption lines were defined as C IV doublet absorption based on the following criteria ($\Delta\lambda_{\rm doublet} = 1550.77 - 1548.20$Å $= 2.57$Å):

- The observed equivalent width, $\rm EW_{obs}$, ≥ 0.5Å of each system/blend
- The measured $\Delta\lambda_{\rm separation,rest} = \Delta\lambda_{\rm doublet} \pm \frac{1}{2}$ resolution element
- Rest EW doublet ratio: 0.8 – 2.2 (allows for blending/resolution effects)
- The doublet FWHMs match within the resolution (~200–300 km $\rm s^{-1}$)
- Rest EW of each transition $\geq 3\sigma$ detection limit

4. Results and Brief Discussion

The results are presented in Table 1 and in the figures. RLQs are shown as triangles, while RQQs are shown as open squares. Encircled symbols are high velocity (>5000 km $\rm s^{-1}$) absorbers.

The quasars with NALs (Figure 1) have an average UV slope, $< \alpha_{\rm UV} > = -1.54$ (median $= -1.48$), while unabsorbed quasars have $< \alpha_{\rm UV} > = -1.99$ (median $= -1.89$). However, a K-S test shows no statistically significant differences at the 99.95% confidence level. Both groups have $<M_V> \approx -27.9$ mag. The strength (equivalent width, EW) of the C IV NALs correlates strongly with $\alpha_{\rm UV}$ (Spearman's rank, r = 0.43) with a P <0.1% probability of occuring by chance (Figure 2, left). A slightly weaker correlation (r = 0.21, P = 3.95%) exists with quasar luminosity (Figure 2, right) such that the EW tends to decrease in brighter objects. A stronger trend is seen with UV continuum luminosity (r = −0.32, P = 0.19%; Figure 3, left). This is contrary to the naive expectation (§ 1) that brighter, bluer objects more easily generate stronger disk-winds. As will be clear later, what is encountered is a complication due most likely to inclination and/or reddening effects. No relation is seen with radio luminosity (Figure 3, right).

Table 1 lists the frequency of narrow C IV absorbers among various quasar subgroups. The main results were listed in the summary. Not only do LDQs have a higher NAL frequency than CDQs, but they also tend to be more strongly absorbed (Figure 4; also e.g., Foltz et al. 1988; Barthel, Tytler, & Vestergaard 1997, who used almost the same RLQ sample studied here). LDQs are believed to be intrinsically similar to CDQs, just viewed at a higher inclination, i, of the

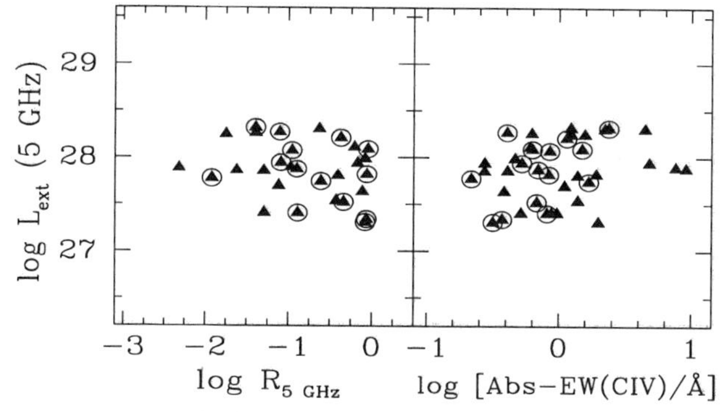

Figure 5. The EW inclination dependence is *not* due to L_{ext} biases.

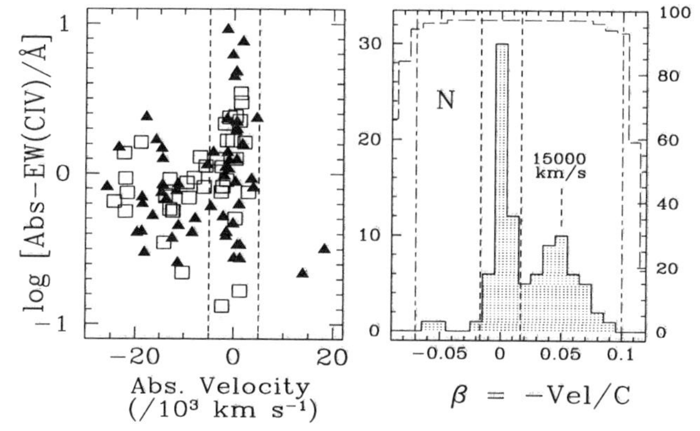

Figure 6. The distribution of absorber strength on the absorber velocity relative to the emission redshift (left). *Right:* The number distribution of the absorbers (shaded; left axis) compared with the spectra measured (dashed curve; right axis) show both peaks are real; see text.

radio axis relative to our line of sight. In this study the i-dependence of the NAL strengths is tested on a more well-suited sub-sample of the RLQs, which is important if the NALs are associated with the quasars. To avoid possible selection biases the LDQs and CDQs were selected to cover the same range in L_{ext} (Figure 5; Vestergaard et al. 2000). Also, no L_{ext} dependence is seen for the EWs. Both $\log R_{5GHz}$ and $\log R_V$ estimate i (e.g., Wills & Brotherton 1995). Also, for this subsample the strongest NALs are seen in LDQs and the strength increases with i (Figure 4). If indeed LDQs are highly inclined, then the correlations of NAL EW with α_{UV}, M_V, L_{cont}, $\log R_{5GHz}$, and $\log R_V$ (Figures 2, 3, 4) can be explained as a combination of inclination and reddening effects, which thus seem to dominate possible radiation pressure effects on disk outflows. It is worth noting that Ganguly et al. (2001) do *not* find an enhanced frequency of C IV NALs for the few LDQs among 59, $z \lesssim 1.2$ quasars from the *HST* Quasar Absorption Line Key Project. Furthermore, their measured NAL strengths are significantly lower indicating a clear redshift evolution of the NALs. Richards et al. (2001) find a small excess of high-velocity NALs in CDQs relative to LDQs. These issues will be addressed in forthcoming work.

The strongest NALs are furthermore *associated* with the quasars. Figure 6 shows the distribution of EW with absorber velocity relative to the quasar restframe. With exception of the usual five strongest RLQ NALs, there is a similar velocity distribution for RLQs and RQQs (Figure 6, left). Figure 6 (right) shows the distribution of the β parameter [$\beta = (r^2 - 1)/(r^2 + 1)$, $r = (1 + z_{\rm em})/(1 + z_{\rm abs})$; e.g., Peterson 1997]. The enhancement of NALs within 5000 km s^{-1} (between the vertical, short-dashed lines) is clear. The long-dashed curve shows the number of spectra available for measurement at a given β bin. The vertical, dot-dashed lines denote the range where at least 95% (92/96) of the spectra are present. The high fraction of spectra across the entire second bump shows that the 15,000 km s^{-1} peak is real. The increased EW around 18,000 km s^{-1} (Figure 6, left) is also quite possibly real. These two peaks are intringuingly close to the terminal velocities typically seen in BALs ($\sim$20,000 km s^{-1}); this may potentially be important.

The fact that RQQs tend to show modest NAL EWs, while RLQs have quite strong NALs is consistent with disk-wind models (e.g., Murray & Chiang 1995) which predict that RLQs are not capable of accelerating the high density outflows to relativistic velocities because the stronger X-ray flux strips the electrons off the atoms, thereby decreasing the radiation pressure on the outflowing gas. The moderately strong, high-velocity NALs seen in the RLQs ($\sim$18,000 km s^{-1} peak, Figure 6, left) adds an interesting twist to this scenario. Perhaps the NALs are a separate subset of the equatorial absorbers as they are seen independently of radio-type. This and the two peaks of high-velocity absorbers will be further addressed by Vestergaard (2001).

References

Barthel, P. D., et al. 1988, A&AS, 73, 515

Barthel, P. D., Tytler, D. R., & Vestergaard, M. 1997, in Mass Ejection from Active Galactic Nuclei, eds. N. Arav, I. Shlosman, & R. J. Weymann, (San Franscisco: ASP), 48

Barthel, P. D., Vestergaard, M., & Lonsdale, C. J. 2000, A&A, 354, 7

Foltz, C. B., et al. 1988, in QSO Absorption Lines: Probing the Universe, ed. C. Blades, D. Turnshek, & C. Norman (Cambridge: CUP), 53

Ganguly, R., et al. 2001, ApJ, 549, 133

Hamann, F. 2000, "Intrinsic AGN Absorption Lines", Encyclopedia of Astronomy and Astrophysics (MacMillan and the Institute of Physics Publishing)

Lonsdale, C. J., Barthel, P. D., & Miley, G. 1993, ApJS, 87, 63

Murray, N., & Chiang, J. 1995, ApJ, 454, L105

Peterson, B. 1997, An Introduction to Active Galactic Nuclei (Cambridge: CUP)

Richards, G. T., et al. 2001, ApJ, 547, 635

Vestergaard, M. 2000, PASP, 112, 1504

Vestergaard, M., Wilkes, B. J., & Barthel, P. D. 2000, ApJ, 538, L103

Weymann, R., Williams, R., Peterson, B., Turnshek, D. 1979, ApJ, 234, 33

Wills, B. J., & Brotherton, M. S., 1995, ApJ, 448, L81

ANYONE WISHING TO
COMPUTER PROJETOR
TALK:
PLEASE SEE MATT
BEFORE
EXIT

DESCRIBVNT
QVATUORIST
TISON

Mass Outflow in Active Galactic Nuclei: New Perspectives
ASP Conference Series, Vol. 255, 2002
D.M. Crenshaw, S.B. Kraemer, and I.M. George

A Deep ASCA Spectrum of Broad Absorption Line Quasar PHL 5200: Clues to Quasar Evolution?

Smita Mathur

The Ohio State University

Giorgio Matt

Universitá Roma Tre, Italy

Paul J. Green, Martin Elvis

Harvard Smithsonian Center for Astrophysics

K. P. Singh

Tata Institute of Fundamental Research, Mumbai, India

Abstract. We discuss results from a deep ASCA observation of the prototype broad absorption line quasar (BALQSO) PHL 5200. This is the best X-ray spectrum of a BALQSO yet. We find that (1) the source is not intrinsically X-ray weak, (2) the line of sight absorption is very strong (3) the covering fraction of the absorber is $\sim$90%. This is consistent with the large optical polarization observed in this source, implying multiple lines of sight. The most surprising result of this observation is that (4) the spectrum of this BALQSO is *not* exactly similar to other radio-quiet quasars. The hard X-ray spectrum of PHL 5200 is steep with the power-law spectral index $\alpha \approx 1.5$. This is similar to the steepest hard X-ray slopes observed so far. At low redshifts, such steep slopes are observed in narrow line Seyfert 1 galaxies, believed to be accreting at a high Eddington rate. This observation strengthens the analogy between BALQSOs and NLS1 galaxies and supports the hypothesis that BALQSOs represent an early evolutionary state of quasars (Mathur 2000). It is well accepted that the orientation to the line of sight determines the appearance of a quasar; age seems to play a significant role as well.

1. Introduction

Broad absorption line quasars, in which the kinetic energy carried out in the absorbing outflow is a significant fraction of the bolometric luminosity of the quasar, offer a challenge to our understanding of the quasar energy budget, as suggested in Mathur, Elvis & Wilkes (1995; see also Krolik 1999). At the same time, they also offer new insights into the nuclear structure of quasars (Ogle 1998, Elvis 2000). X-ray observations are important in this investigation as they offer precise measurements of absorbing column densities. However, BALQSOS are elusive X-ray sources, with most X-ray observations resulting in non-detections.

For many years now we have been arguing that BALQSOs are not intrinsically X-ray faint, but it must be the heavy absorption that makes them appear faint in X-rays (Green et al. 2002). However, until last year, we could not rule out the alternative that the BALQSOs are intrinsically X-ray weak. This situation changed with our observations of a carefully selected BALQSO PG 0946+301 (this multiwavelength project was conceived in the Pasadena meeting; see also Arav [2002] for HST observations of PG 0946+301). In a deep ASCA observation, the source was clearly detected in one of the gas imaging spectrometers, but not in any other detectors. This led to an unambiguous conclusion that the observed X-ray weakness of this BALQSO is due to strong absorption, and not due to intrinsic weakness (Mathur et al. 2000). We also concluded that the absorber must be at least partially ionized and may be responsible for attenuation in the optical and UV.

While most X-ray observations of BALQSOs resulted in non-detections, some BALQSOs were detected. Comparing the properties of those detected with the ones which were not, we had proposed that the highly polarized BALQSOs may be the X-ray brighter members of the class (Gallagher et al. 1999). In this picture, there would be two separate lines of sight to the nuclear continuum source, one direct and absorbed, and one scattered and unabsorbed.

All these results to date were based on the assumption that the X-ray spectra of BALQSOs are similar to those of non-BALQSOs. However, none of the parameters could be determined with certainty for lack of a good X-ray spectrum. A 1994 ASCA observation of PHL 5200 yielded the first X-ray spectrum of a BALQSO (Mathur, Elvis & Singh 1995, MES95 here after), and remained the only one until this year. However, the quality of the ASCA spectrum was poor and the parameters of the fit were not well determined in the 17.7 ksec observation. The photon index of the best-fit power-law was uncertain to ± 0.9 (all the errors are quoted to 90% confidence, unless noted otherwise). While excess absorption with $N_H = 1.3^{+2.3}_{-1.1} \times 10^{23}$ cm^{-2} at the source provided a better fit, a model with only Galactic absorption was acceptable (see also Gallagher et al. 1999). A good signal to noise spectrum of a BALQSO was very much needed and PHL 5200 remained the best target. So we re-observed PHL 5200 with ASCA, to obtain better quality data and better constrain the parameters of an X-ray spectrum of a BALQSO. We find that the spectrum contains more surprises that affect the interpretation of the BAL phenomenon.

2. Observations & Analysis

We observed PHL 5200 with ASCA for ~ 100 ksec (see Mathur et al. 2001 for the details of observations and data analysis). The source was clearly detected in all the four instruments, two solid-state imaging spectrometers (SIS) and two gas imaging spectrometers (GIS). With a total of 2610 "good" counts, we were able to obtain a spectrum of the source. This is the best X-ray spectrum of a BALQSO yet.

Our observations were taken towards the end of ASCA mission, and the CCDs on the satellite were significantly degraded over time. So we took into account all the calibration uncertainties carefully in our spectral extraction and analysis. The extracted spectra were analyzed using **XSPEC** (Arnaud 1996). As a first step we fitted the spectra with a simple power-law and Galactic ab-

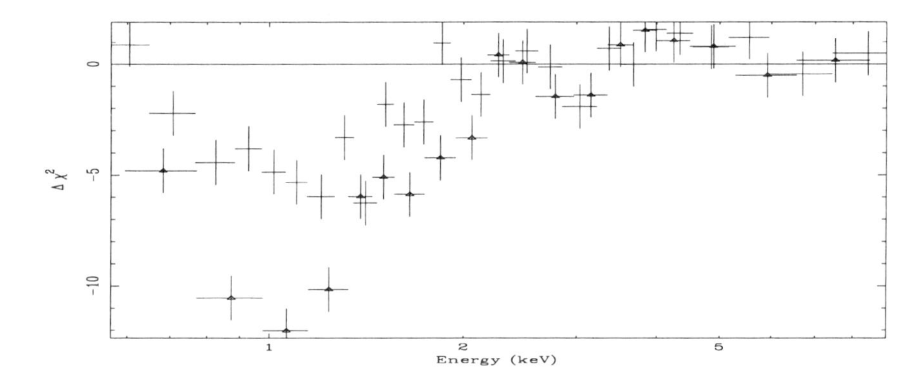

Figure 1. Residuals ($\Delta\chi^2$) to a power-law fit showing clear signature of a partially covering absorber. Here the power-law was fitted only to the data above 2 keV and extrapolated to lower energies. Note the strong turn-over below 2 keV and recovery below about 1 keV. Only SIS data are shown for clarity; dots: SIS0, triangles: SIS1.

sorption. The fit was not good, with $\chi^2 = 226.5$ for 137 degrees of freedom. As excess absorption was indicated in the earlier spectrum of PHL 5200 (MES95) and in other BALQSOs (Mathur *et al.* 2000 and references there in), absorption at the source was added as a next step. The fit was better, but clearly showed negative residuals around one keV, and positive residuals at lower energies (Figure 1). The implied recovery of the underlying power-law is a typical signature of partial covering by the absorber. So we fitted the spectrum with a partial covering model and the fit was significantly better ($\chi^2 = 182.5$, 135 degrees of freedom). The resulting $N_H = 5 \pm 1 \times 10^{23}$ cm^{-2} and covering factor is $0.9^{+0.05}_{-0.06}$. The confidence contours of the interesting parameters are plotted in Figure 2. The scale over which Figure 2a is plotted shows the range of 3σ uncertainty in the earlier observation (MES95). The column density at the source is greater than 2×10^{23} cm^{-2} and the power-law slope α is greater than 1.2 at 99% confidence (flux $f(E) \propto E^{-\alpha}$ where E is the energy).

3. Results & Interpretation

With this deep ASCA observation of PHL 5200, the parameters of the spectral fit are well constrained; the X-ray flux is highly absorbed with $N_H = 5 \pm 1 \times 10^{23}$ cm^{-2}. While consistent with the conclusions of MES95, the earlier low signal to noise spectrum was also consistent with a model with no excess absorption (see also Gallagher *et al.* 1999). We unambiguously confirm large absorbing column density in PHL 5200. Thus, the X-ray weakness of BALQSOs is surely a result of absorption.

The absorber does not cover the continuum fully, with a covering fraction 90±5% (Figure 2b). This might be the reason for the apparent lack of excess absorption and flatter spectrum in the low signal to noise spectrum (see also Green *et al.* 2001). Partial covering of the absorber is consistent with the large optical polarization observed in this source (Figure 3, Goodrich & Miller 1995) implying that there are at least two lines of sight to the nucleus: one direct

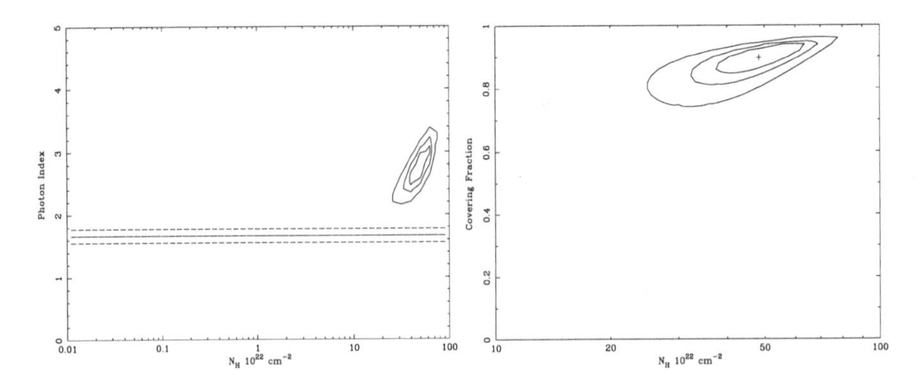

Figure 2. Confidence contours of: (a) Power-law photon index against the absorbing column density (the photon index is $= 1 + \alpha$). Note the large best fit column density and the steep power-law spectrum. The horizontal lines represent the average spectral slope and range in z$\gtrsim$2 radio-quiet quasars. The scale over which the plot is made shows 3σ uncertainty in the previous observation. (b) Absorber covering fraction against the column density. It is clear that the absorber doe not cover the source completely.

and highly absorbed, and another scattered but unabsorbed. The CIV BAL in PHL 5200 does not reach zero intensity (fig. 3). The covering fraction, deduced from the trough intensity, is 85%, consistent with our X-ray result. Gallagher *et al.* (1999) have discussed the possibility that partial covering by the absorber might be responsible for highly polarized BALQSOs to be relatively X-ray bright. Here we find that indeed, that is the best fit model to the PHL 5200 spectrum.

The most surprising result of this observation is the steep spectrum, with X-ray power-law slope $\alpha = 1.7 \pm 0.4$. This might be an additional reason behind the unexpected non-detections of BALQSOs, even in sensitive hard X-ray observations. The mean ASCA slope for radio-quiet quasars at high redshifts is $\alpha = 0.67 \pm 0.11$, with a dispersion of $\sigma = 0.07$ (Vignali *et al.* 1999, in the redshift interval between z=1.9 and 2.3). Our observations imply that the spectra of BALQSOs may *not* be exactly like other radio-quiet quasars, but are steeper. The observed, absorption corrected, 2–10 keV flux is 3.7×10^{-13} ergs $\mathrm{cm}^{-2}\ \mathrm{s}^{-1}$ (SIS0), consistent with the 1994 ASCA observation (MES95). So, the observed steep spectrum is unlikely to be a result of a short lived high-state. The only other BALQSO, PG 2112+059, for which a spectrum is available, has $\alpha = 0.98^{+0.4}_{-0.27}$ (Gallagher *et al.* 2001). As noted by Gallagher et al. this is consistent with the mean quasar slope. However, it is also consistent with as α steep as 1.38 at 90% confidence.

A Note of Caution: as noted earlier, CCDs on ASCA have degraded over time. While we have been very careful in our analysis, there might be some unknown calibration uncertainties which have affected the results presented here. Future observations with Chandra and XMM-Newton will be useful in this respect. The discussion below assumes that these ASCA results are correct.

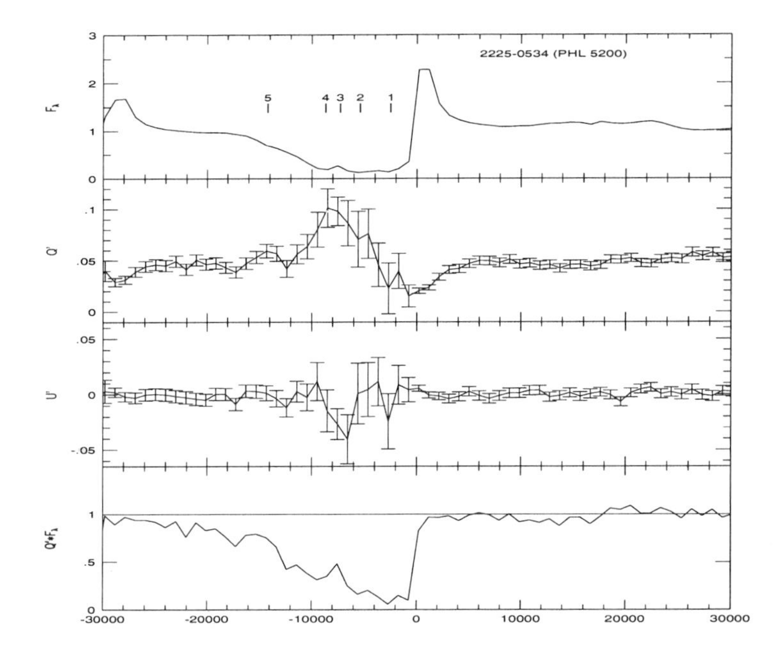

Figure 3. PHL 5200 CIV BAL region. Panels top to bottom show the normalized total flux, rotated Stokes Q′, rotated Stokes U′, and normalized polarized flux (from Ogle 1998).

At low redshifts, Brandt, Mathur & Elvis (1997) found that the hard X-ray spectra of Seyfert galaxies are typically flatter than about $\alpha = 1.0$. The Seyfert galaxies with steeper spectra are the narrow line Seyfert 1 galaxies (NLS1s). Brandt & Gallagher (2000) have discussed the analogy between low ionization BALQSOs and NLS1 galaxies. Mathur (2000) has discussed the analogy between BALQSOs and NLS1s further and has argued that NLS1 may be active galactic nuclei in the making. In this scenario, young radio-quiet AGNs are accreting at a high Eddington rate and have steep spectra. Over time, the accretion rate drops and the X-ray spectrum flattens. BALQSOs may be in that early evolutionary phase when the shroud surrounding the nuclear black-hole is being blown away and a quasar emerges (e.g. Fabian 1999). If the observed steep spectrum of PHL 5200 is a general property of BALQSOs, then it supports the evolutionary hypothesis of Mathur (2000) and further supports their analogy with NLS1 galaxies.

The unified models (Antonucci & Miller 1985) explain various subclasses of AGNs as orientation effects. This has been an accepted AGN paradigm for years. The evolutionary hypothesis supplements rather than contradicts unified models. While the unified models describe the quasar structure in three dimensions, the evolutionary hypothesis deals with the fourth dimension of time.

After all, quasars are not static phenomena and must evolve with time. It is of great interest to determine how evolution changes the appearance of a quasar. The X-ray observations presented here might be providing us with a clue.

References

Antonucci, R., & Miller, J.S. 1985, ApJ, 297, 621

Arav, N. 2002, in Mass Outflow in Active Galactic Nuclei: New Perspectives, eds. D.M. Crenshaw, S.B. Kraemer, & I.M. George (San Francisco: ASP), p. 179

Arnaud, K. A. 1996 in ASP Conf. Ser. 101 Astronomical Data Analysis Software and Systems V, ed. G. Jacoby & J. Barnes (San Fracisco: ASP), 17

Brandt, W. N., & Gallagher, S. 2000, in New Astronomy Reviews, 4, 461

Brandt, W. N., Mathur, S., & Elvis, M., 1997, MNRAS, 285, L25

Elvis, M. 2000, ApJ, 545, 63

Fabian, A.C. 1999, 308, 39

Gallagher, S., Brandt, W. N., Sambruna, R., Mathur, S., & Yamasaki , N. 1999, ApJ, 519, 549

Gallagher, S., Brandt, W. N., Laor, A., Elvis, M., Mathur, S., Will s, B. J. 2001, ApJ, in press.

Goodrich, R. & Miller, 1995, ApJL, 448, 73

Green, P.J., et al. 2002, in Mass Outflow in Active Galactic Nuclei: New Perspectives, eds. D.M. Crenshaw, S.B. Kraemer, & I.M. George (San Francisco: ASP), p. 19

Green, P. J., Aldcroft, T. L., Mathur, S., Wilkes, B. J., & Elvis, M. 2001, ApJ, in press

Krolik, J. 1999, Active Galactic Nuclei (Princeton: Princeton University Press)

Mathur, S. 2000, MNRAS Letters, 314, L17

Mathur, S., Elvis, M., & Singh, K. P. 1995, ApJ, 455, L9 (MES95)

Mathur, S., Elvis, M., & Wilkes, B. 1995, ApJ, 452, 230

Mathur, S. *et al.* 2000, ApJL, 533, 79

Mathur, S., Matt. G., Green, P.J., Elvis, M., & Singh, K.P. 2001, ApJL, 551, 13

Ogle, P. 1998, Ph.D. Thesis, California Institute of Technology.

Vignali, C., Comastri, A., Cappi, M., Palumba, G., Matsuoka, M., & Kubo, H., 1999, ApJ, 516, 582

Mass Outflow in Active Galactic Nuclei: New Perspectives
ASP Conference Series, Vol. 255, 2002
D.M. Crenshaw, S.B. Kraemer, and I.M. George

Radio-Selected Broad Absorption Line Quasars

Michael S. Brotherton
National Optical Astronomy Observatories, Kitt Peak National Observatory,950 N. Cherry Ave., Tucson, AZ 85719

Abstract. While it is apparently still true that the most powerful radio quasars never show broad absorption lines (BALs), there is a significant population of radio-moderate and formally radio-loud BAL quasars. Deep radio surveys are an effective way to find such BAL quasars. The properties of these quasars differ from those of unabsorbed quasars in several clear ways and rule out simple orientation schemes which explain BAL quasars as normal quasars seen edge-on. Such schemes must be modified or give way to alternatives that rely on evolutionary or environmental differences.

1. Introduction

Blueshifted broad absorption lines (BALs) of ionized species are seen in the optical/UV spectra of ~10% of bright, optically selected quasars, indicating high-velocity outflow ($\leq 0.2c$) along the line-of-sight. The column density of these BAL clouds implied by the UV absorption is $N_{\rm H} \sim 10^{20} - 10^{21} cm^{-2}$ (Hamann, Korista, & Morris 1993), with metallicities as high as ten times solar (Hamann 1997), although saturation probably tempers these conclusions. The location of these outflows and their mass fluxes are unknown. Understanding their geometry and relationship to the central quasar engine may also provide clues to the accretion process which appears invariably tied to outflows/jets.

"HiBAL" quasars show absorption from only high-ionization species (e.g., C IV λ1549) while "LoBAL" quasars also show absorption from low-ionization species (e.g., Mg II λ2798) and are seen more rarely (1% of optically selected quasars). Until a few years ago it was widely accepted that BAL quasars are always radio-quiet (e.g., Stocke et al. 1992). This was regarded as an important clue toward unraveling the puzzle of the radio-loud/radio-quiet dichotomy. But Becker et al. (1997) reported the first discovery of a radio-loud BAL quasar, and a population of such BAL quasars has since been found (Brotherton et al. 1998; Becker et al. 2000; Menou et al. 2001).

2. Properties of Radio-Selected BAL Quasars

The largest and brightest sample of radio-selected BAL quasars is that of Becker et al. (2000), sporting ~27 BAL quasars from the FIRST Bright Quasar Survey (White et al. 2000). Their radio properties can tell us some things that cannot be learned from optically selected BAL quasars. First, a wide range of spectral indices are present, including both flat and steep radio spectra; unified radio models (e.g., Orr & Browne 1982) would indicate that therefore a range of orientations are present. Second, the radio sources are almost all compact (90%), whereas a matched parent population from the FBQS consists of only 60% compact sources.

Optically selected BAL quasars possess other properties that distinguish them from unabsorbed quasars, properties that radio-selected BAL quasars share. In collaboration with FIRST Survey team members I am conducting follow-up investigations of the Becker et al. (2000) BAL quasar sample examining polarization, reddening, X-ray emission, and rest-frame optical properties. Many ($\sim 25\%$) BAL quasars are highly polarized in the optical ($P > 3\%$) compared to $< 1\%$ for unabsorbed quasars. BAL quasars are redder on average, as if seen through dust. They have excess Fe II emission and weak emission from extended narrow-line regions ([O III] λ5007 in particular is a diagnostic). They are exceedingly X-ray faint. LoBAL quasars in particular show the more extreme properties.

3. Implications of Radio-Selected BAL Quasars

It has been popular to believe that BAL quasars are simply normal unabsorbed quasars seen at a preferred orientation. The similarity of the emission lines in BAL quasars and unabsorbed quasars (Weymann et al. 1991) suggests that the central engine and surroundings are the same for both classes and that our line of sight just happens to intersect an outflow common to all quasars. If this is true, spectopolarimetry of highly polarized BAL quasars may indicate the geometry: Goodrich & Miller (1995), Hines & Wills (1995), and Cohen et al. (1995) all suggested that BAL quasars are seen along a line of sight skimming the edge of a disk or torus through an equatorial wind, and polarized continuum light scattered above along a less obscured path. LoBAL quasars are those seen at the largest inclinations. This geometry is very similar to that invoked for Seyfert 1/2 unification (e.g., Antonucci 1993).

This popular orientation scheme faces problems in light of a better understanding of BAL quasar properties. Radio spectral indices of BAL quasars show both flat and steep spectrum sources, the same as unabsorbed quasars (Barvainis & Lonsdale 1997; Becker et al. 2000), whereas edge-on systems should show a preference for steep spectra, suggesting that the BAL quasars are randomly oriented. Nearly ubiquitous compact radio emission in radio-loud BAL quasars implies that they are small (frustrated or young like compact steep spectrum sources) and favors gas-rich interacting systems. And there **are** significant emission-line differences whose importance has sometimes been overlooked. BAL quasars, especially LoBAL quasars, show strong Fe II emission and weak narrow-line emission (e.g., [O III] λ5007) (e.g., Turnshek et al. 1997); quasars with these properties are thought to possess high accretion rates (like narrow line Seyfert 1 galaxies with steep soft X-ray spectra) and obscuring material with large covering fractions (e.g., Boroson & Green 1992; Laor et al. 1997). Accretion rate should not depend on orientation.

An alternative to "unification by orientation" is "unification by time," with BAL quasars characterized as young or recently refueled quasars. Voit et al.

(1993) argue that LoBALs are a manifestation of a "quasar's efforts to expel a thick shroud of gas and dust," consistent with the scenario of Sanders et al. (1988) in which quasars emerge from a dusty, gas-rich merger-produced ultraluminous infrared galaxies. BAL quasars are a large fraction of *IRAS*-selected quasars and Markarian 231 (Smith et al. 1995), *IRAS* 07598+6508 (Boyce et al. 1996), PG 1700+518 (Hines et al. 1999; Stockton et al. 1998), and FIRST 1556+3517 (Najita et al. 2000) all show evidence for recent mergers or interactions, including young starbursts.

I thank my collaborators on the FIRST Bright Quasar Survey and follow-up projects related to BAL quasars, especially Bob Becker, Michael Gregg, Sally Laurent-Muehleisen, Nahum Arav, Rick White, Hien Tran, and Mark Lacy.

References

Antonucci, R. 1993, ARA&A
Barvainis, R., & Lonsdale, C. 1997, AJ, 113, 144
Becker, R., White, R., & Helfand, D., 1995, ApJ, 450, 559
Becker, R., et al. 1997, ApJ, 479, L93
Becker, R., et al. 2000, ApJ, 538, 72
Boroson, T. A., & Green, R. F. 1992, ApJS, 80, 109
Boroson, T. A. & Meyers, K. A. 1992, ApJ, 397, 442
Boyce, P. J., et al. 1996, ApJ, 473, 760
Brotherton, M. S., et al. 1998, ApJ, 505, L7
Cohen, M., et al., 1995, 448, L65
Goodrich, R., & Miller, J. 1995, ApJ, 448, L73
Hamann, F. 1997, ApJS, 109, 279
Hamann, F., Korista, K., & Morris, S. 1993, ApJ, 415, 541
Hines, D. & Wills, B. 1995, ApJ, 448, L69
Hines, D., et al. 1999, ApJ, 512, 140
Krolik, J. H., & Voit, G. M. 1998, ApJ, 497, L5
Laor, A., Fiore, F., Elvis, M., Wilkes, B., & McDowell, J. 1997, ApJ, 477, 93
Menou, K., et al. 2001, astro-ph/0102410
Najita, J., Dey, A., & Brotherton, M. 2000, AJ, 120, 2859
Orr, M., & Browne, I. 1982, MNRAS, 200, 1067
Sanders, D., et al. 1988, ApJ, 328, L35
Smith, P. S., Schmidt, G.., Allen, R., & Angel, J. 1995, ApJ, 444, 146
Stocke, J. T., Morris, S. L., Weymann, R. J., & Foltz, C. B. 1992, ApJ, 396, 487
Stockton, A., Canalizo, G., Close, L. 1998, ApJ, 500, L121
Turnshek, D., Monier, E., Sirola, C., & Espey, B. 1997, ApJ, 476, 40
Voit, G. et al., 1993, ApJ, 95, 109
Weymann, R., Morris, S., Foltz, C. & Hewett, P. 1991, ApJ, 373, 23
White, R. et al. 2000, ApJS, 126, 133

Mass Outflow in Active Galactic Nuclei: New Perspectives
ASP Conference Series, Vol. 255, 2002
D.M. Crenshaw, S.B. Kraemer, and I.M. George

Extreme BAL Quasars from the Sloan Digital Sky Survey

Patrick B. Hall, J. E. Gunn, G. R. Knapp, V. K. Narayanan, M. A. Strauss
Princeton University Observatory, Princeton NJ 08544-1001

S. F. Anderson
University of Washington

D. E. Vanden Berk
Fermi National Accelerator Laboratory

T. M. Heckman, J. H. Krolik, Z. I. Tsvetanov, W. Zheng
The Johns Hopkins University

G. T. Richards, D. P. Schneider
The Pennsylvania State University

X. Fan
Institute for Advanced Study

D. G. York
The University of Chicago

T. R. Geballe
Gemini Observatory

M. Davis
University of California at Berkeley

R. H. Becker
Lawrence Livermore National Laboratory

R. J. Brunner
California Institute of Technology

Abstract. The Sloan Digital Sky Survey has discovered a population of broad absorption line quasars with various extreme properties. Many show absorption from metastable states of Fe II with varying excitations; several objects are almost completely absorbed bluewards of Mg II; at least one shows stronger absorption from Fe III than Fe II, indicating temperatures T>35000 K in the absorbing region; and one object even seems to have broad Hβ absorption. Many of these extreme BALs are also heavily reddened, though 'normal' BALs (particularly LoBALs) from SDSS also show evidence for internal reddening.

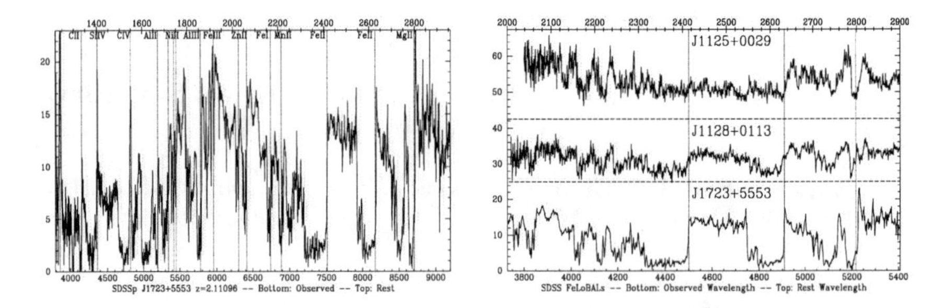

Figure 1. BALs with Fe II* absorption. a) SDSS 1723+5553, z=2.11. b) Comparison of 2000-2900 Å regions in SDSS 1723+5553 (bottom) and two lower-z FeLoBALs, SDSS 1128+0113 (middle, z=0.894) and SDSS 1125+0029 (top, z=0.865); dashed lines show zero flux for each, and vertical dotted lines show Fe II* λ2414 & λ2632 and Mg II λ2798.

1. Introduction

The Sloan Digital Sky Survey (York et al. 2000) is using dedicated instruments on a 2.5m telescope (Gunn et al. 1998) to image 10^4 deg^2 of sky to $\sim 23^m$ in five bands (Fukugita et al. 1996), and obtain spectra of $\sim 10^6$ galaxies and $\sim 10^5$ quasars selected primarily as outliers from the stellar locus. Its area, depth, and selection criteria make SDSS effective at finding unusual quasars. The first data release (Stoughton et al. 2001, in prep.). contains ~4500 spectroscopically confirmed quasars, including ~200 BALs, a few percent of which have extreme properties of one sort or another. All these extreme BALs are LoBALs, which show absorption from both low- and high-ionization transitions, instead of the more common HiBALs with only high-ionization absorption. Full analysis is underway (Hall et al. 2001, in prep.), but already these objects confirm the existence of a population of extreme BALs, as suspected from previous discoveries of individual extreme BALs (Becker et al. 1997, Djorgovski et al. 2001).

2. BAL Quasars With Fe II* Absorption

The rare LoBAL quasars with absorption from metastable excited states of Fe II (Fe II*) have been dubbed FeLoBALs (Becker et al. 2000; Hazard et al. 1987; Menou et al. 2001). They are valuable because photoionization modelling of them can constrain n_e in the BAL clouds (e.g., de Kool et al. 2001). Fig. 1a shows a spectacular example, SDSS 1723+5553, with absorption from over twenty transitions in at least a dozen elements. Fig. 1b compares SDSS 1723+5553 to two lower-z FeLoBALs with [O II] emission line redshifts. Both low-z objects show Fe II* absorption blueward of 2414 Å and 2632 Å from states up to ~1 eV above ground, but SDSS 1125+0029 also shows absorption near 2500 Å from even more excited levels. In both low-z objects, the Mg II BAL absorption apparently extends 2000 km s^{-1} *redward* of the systemic z.

Some SDSS FeLoBALs show very abrupt drops in flux near Mg II λλ2796,2803 (e.g., SDSS 0300+0048 in Fig. 2a). SDSS 0300+0048 has associated Mg II

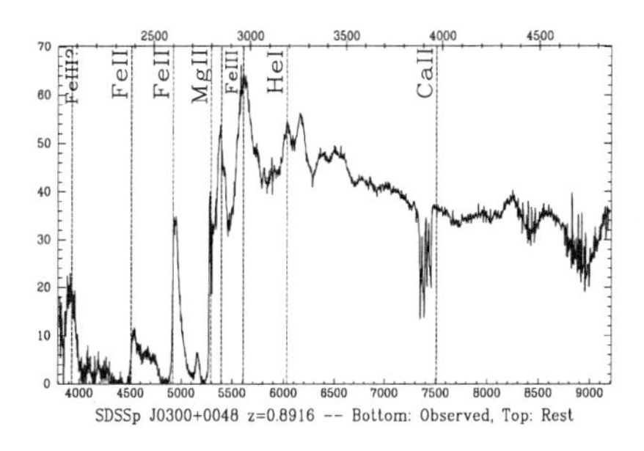

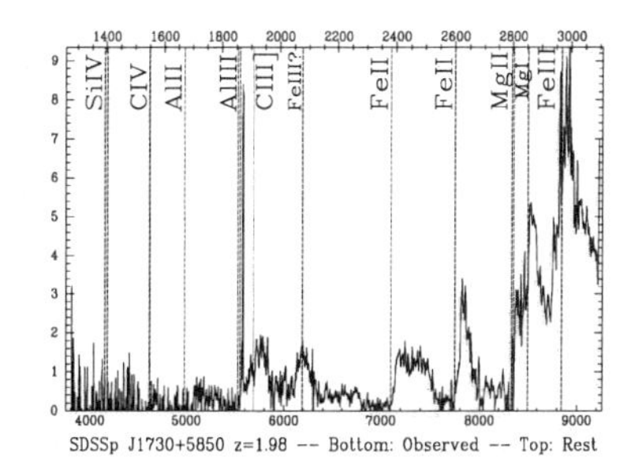

Figure 2. SDSS LoBALs with strong absorption blueward of Mg II. a) SDSS 0300+0048 at z=0.8916. b) SDSS 1730+5850 at z=1.98.

absorption at z=0.8916, at least 4 narrow Ca II H&K absorption systems located 2350 to 3900 km s^{-1} blueward of the Mg II system, and broad Ca II absorption extending a further 2000 km s^{-1} blueward. Broad, near-total Mg II absorption is associated with the highest-z Ca II system, but broad Fe II* absorption is associated instead with the *strongest* Ca II system, at slightly lower z. Fig. 2b shows SDSS 1730+5850, which is clearly a higher redshift analogue of SDSS 0300+0048. Our spectrum extends farther into the UV for this z~2 object, and shows a weak recovery at C III] λ1908 but essentially zero flux below Al II λ1670. Quasars such as these at $z \geq 2$ will obviously be greatly underrepresented in optical surveys.

3. BAL Quasars With Fe III Absorption

Fig. 3 shows SDSS 2215−0045, a LoBAL at z=1.47548 (measured from associated Mg II absorption, as with SDSS 0300+0048). Its absorption troughs are unusual for a LoBAL: they are very broad, detached, and strongest near the high velocity end rather than at low velocity. By comparison to SDSS 1723+5553 (Fig. 1), we initially identified the strong trough at λ_{obs}~4900 Å as Cr II. However, the implied abundance of Cr relative to Mg is implausible, and the expected corresponding Zn II is missing. We now believe this absorption is due to Fe III (multiplet UV 48), with additional Fe III (UV 34) absorption at λ_{obs}~4500 Å, redward of Al III. Since Fe II absorption is weak or absent, the large Fe III/Fe II ratio suggests that the BAL clouds in this object have T>35000 K, sufficient to collisionally ionize Fe II to Fe III. Fe III absorption is seen in several other SDSS LoBALs (e.g., Fig. 4) and in a few previously known LoBALs, but nowhere as strongly (alone or relative to Fe II) as in SDSS 2215−0045. Note the different spectral slopes blueward & redward of ~2400 Å, indicating reddening which must occur outside the BAL region since dust cannot survive long at T>35000 K.

4. A LoBAL With Broad Hβ Absorption

Fig. 4 shows an optical (Keck) plus NIR (UKIRT) spectrum of SDSS 0437-0045 which reveals a strongly absorbed quasar with z=2.74389 from [O III]. The absorption extends 2900 km s^{-1} redward of this z (cf. Fig. 2b). Even more

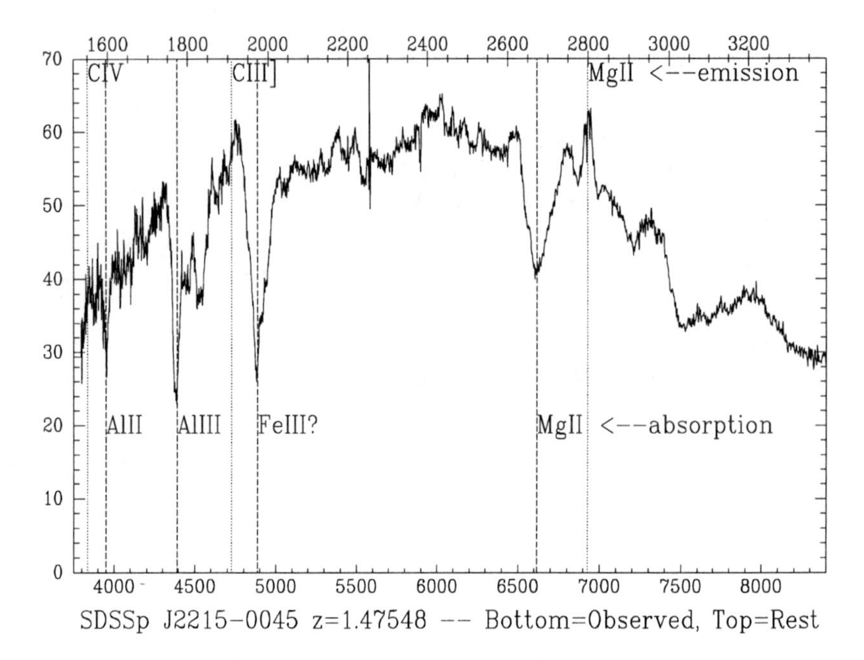

Figure 3. SDSS 2215−0045 at z=1.47548. Note the absence of strong Fe II absorption at 2200–2600 Å in the rest frame.

remarkable is the probable presence of Hβ absorption nearly 10^4 km s^{-1} wide and of REW$\sim$100 Å. Hβ absorption in AGN has previously been seen only in NGC 4151 (Anderson & Kraft 1969; Sergeev et al. 1999), but with $\leq$1000 km s^{-1} width and $\leq$3 Å REW. This object is also unusual because the FeIII trough at $\sim$2070 Å has been seen to vary with nearly unprecendented amplitude and speed.

5. Heavily Reddened BAL Quasars

SDSS has found evidence for a population of red quasars (Richards et al. 2001), and BALs in general are redder than the typical quasar (Menou et al. 2001), but for the most extreme objects the reddening is unambiguous. Fig. 5 shows SDSS 1456+0114, an extremely reddened LoBAL (and FIRST source) at z=2.367 (measured from weak C III], the only broad line visible). Several other similar objects have been found by SDSS, but with even weaker broad emission. Since reddening does not affect equivalent widths, this may indicate that in these objects the broad line region is even more heavily reddened than the continuum. This would be quite plausible if most of the continuum light in those objects is scattered light, a hypothesis which can easily be tested with polarization data.

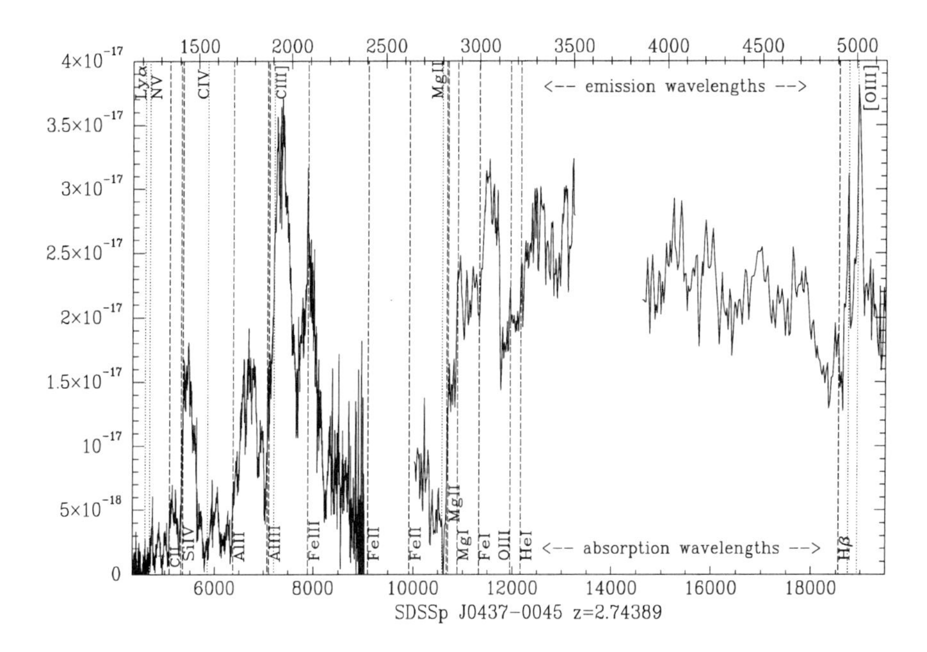

Figure 4. Optical-NIR spectra of SDSS 0437−0045 at z=2.74389. Strong lines are labelled at the expected wavelengths for emission (top) and absorption (bottom). Note the nearly complete absorption near the expected wavelength of C IV, and the probable broad Hβ absorption.

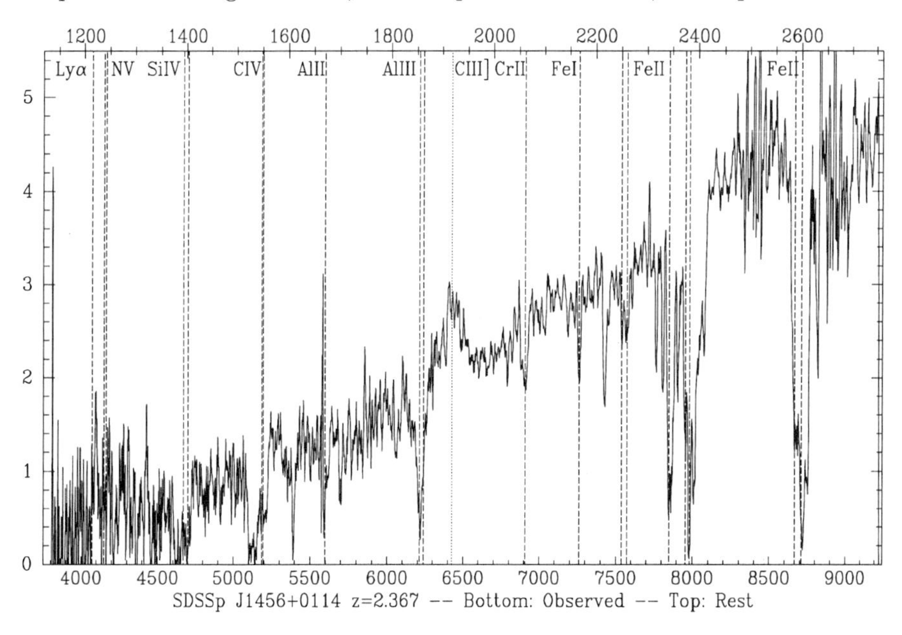

Figure 5. SDSS 1456+0114, an extremely reddened LoBAL at z=2.367.

6. Discussion

The area and depth of SDSS, plus its simple selection of quasar candidates as outliers from the stellar locus, makes it efficient at finding quasars with unusual properties. Moreover, the discovery of typically several examples of each type of extreme LoBAL quasar presented here means that a *population* of extreme LoBAL quasars exists and that only now, with SDSS, are we beginning to sample the full range of properties that exist in BAL outflows, and thus around quasars on the whole.

Acknowledgments. The Sloan Digital Sky Survey (SDSS) is a joint project of The University of Chicago, Fermilab, the Institute for Advanced Study, the Japan Participation Group, The Johns Hopkins University, the Max-Planck-Institute for Astronomy (MPIA), the Max-Planck-Institute for Astrophysics (MPA), New Mexico State University, Princeton University, the United States Naval Observatory, and the University of Washington. Apache Point Observatory, site of the SDSS telescopes, is operated by the Astrophysical Research Consortium (ARC). Funding for the project has been provided by the Alfred P. Sloan Foundation, the SDSS member institutions, the National Aeronautics and Space Administration, the National Science Foundation, the U.S. Department of Energy, the Japanese Monbukagakusho, and the Max Planck Society. The SDSS Web site is http://www.sdss.org/.

References

Anderson, K., & Kraft, R. 1969, ApJ, 158, 859

Becker, R., Gregg, M., Hook, I., McMahon, R., White, R., & Helfand, D. 1997, ApJ, 479, L93

Becker, R., et al. 2000, ApJ, 538, 72

de Kool, M., et al. 2001, ApJ, 548, 609

Djorgovski, S. G., et al. 2001, to appear in Mining the Sky, eds. A. Bandel et al. (Berlin: Springer-Verlag) (astro-ph/0012489)

Fukugita, M., Ichikawa, T., Gunn, J., Doi, M., Shimasaku, K., & Schneider, D. 1996, AJ, 111, 1748

Gunn, J. E., & The SDSS Collaboration 1998, AJ, 116, 3040

Hazard, C., McMahon, R., Webb, J., and Morton, D. 1987, ApJ, 323, 263

Menou, K., & The SDSS Collaboration 2001, ApJ, in press (astro-ph/0102410)

Richards, G. T., & The SDSS Collaboration 2001, AJ, 121, 2308

Sergeev, S., Pronik, V., Sergeeva, E., & Malkov, Y. 1999, A&A 341, 740

York, D. G., & The SDSS Collaboration 2000, AJ, 120, 1579

Mass Outflow in Active Galactic Nuclei: New Perspectives
ASP Conference Series, Vol. 255, 2002
D.M. Crenshaw, S.B. Kraemer, and I.M. George

HST UV and Keck HIRES Spectra of BALQSOs

V. Junkkarinen, R. D. Cohen

University of California, San Diego, CASS 0424, La Jolla, CA 92093-0424

T. A. Barlow

IPAC, California Institute of Technology, MS 100-22, Pasadena, CA 91125

F. Hamann

University of Florida, Dept. of Astronomy, Gainesville, FL 32611-2055

Abstract. In the analysis of broad absorption line (BAL) quasar spectra, Keck HIRES spectra are a useful complement to lower resolution HST and ground based spectra. The HIRES spectra provide accurate parameters for narrow, intervening type absorption systems including Lyman limit systems and a direct measurement of the smoothness of BAL features. The smoothness of the troughs is related to the number of "clouds" in the BAL region if the BAL region consists of clouds. The HIRES spectra, especially for $z \sim 2$ BAL quasars, are also a source of high quality absorption templates. All of the BAL features are apparently resolved in the high resolution spectra (R $\sim$ 45 000), while low resolution (R $\sim$ 1000) spectra do not always resolve features belonging to the outflow.

1. Introduction

BALs are present in about 10% of all optically selected quasars (Weymann et al. 1991). The absorption features result from mass outflows with velocity widths of order 2000 to 20,000 km s^{-1} and maximum velocities to 0.1c and possibly higher in a few cases. The chemical abundances in the BAL regions are of interest as possible indicators of chemical enrichment in galactic nuclei (e.g. Korista et al. 1996, Hamann and Ferland 1999). The absorption features in BAL quasars are not easy to interpret because the BAL region may cover only a part of the background source(s) (Hamann 1998, Arav 1997). Partial covering along the line of sight is common in narrower intrinsic absorbers (e.g. Barlow, Hamann, and Sargent 1997), but more difficult to determine in BAL quasars where the strong doublets of Si IV and C IV are typically blended into wide features. In narrower intrinsic absorption systems, the partial covering is both ion and velocity dependent (Barlow and Sargent 1997). Partial covering definitely influences some narrow BAL components as determined from HIRES observations of CSO 755 (Barlow and Junkkarinen 1994) and Q0226−1024.

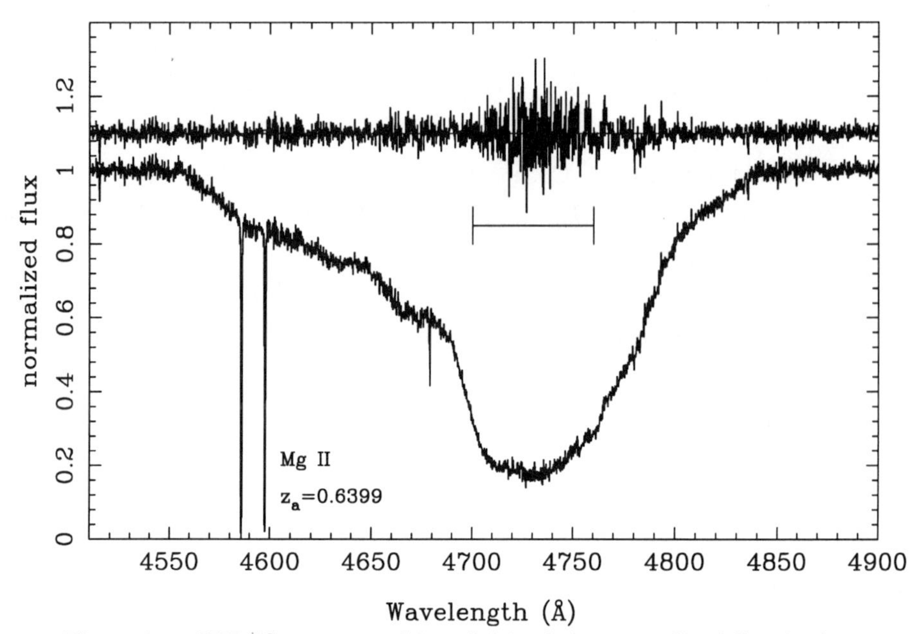

Figure 1. HIRES spectrum, binned 4:1, of the normalized flux in the C IV BAL in Q1246-057. The residual in the region indicated by the bar (4700 – 4760Å) is discussed in the text.

In this conference proceedings paper, we report on a program that uses Keck HIRES spectra in the analysis of the BALs. BAL features in the UV (observed frame) are analyzed using lower resolution spectra obtained from the HST archive. The HIRES spectra contribute in a number of areas to the analysis. Here we report on: 1) BAL smoothness, and 2) absorption templates. As a by–product of the template generation, the HIRES spectra provide some information on partial covering.

2. Smoothness of BAL Features

The observed smoothness of the BAL features in HIRES spectra rules out a "picket fence" model for the BALs. A "picket fence" model (partial covering by turbulent components in velocity space) was proposed by Kwan (1990) in order to explain the relative weakness of the H I Ly α BAL without resorting to non–solar chemical abundances or partial covering along the line of sight. The smooth BAL features also imply a large number of "clouds" in the BAL region if the BAL region is composed of clouds rather than a smooth outflow.

In order to quantify the smoothness of the BAL features, the C IV BAL in Q1246−057 has been divided by a piecewise polynomial. A small number of polynomials of low order were used to preserve residual fluctuations on scales less than $\sim$ 200 km s^{-1}. The result of that division is shown in Figure 1. The entire C IV BAL feature spans absorption velocities from 10 000 km s^{-1} to 27 000 km s^{-1}.

In the C IV BAL, from 4700 Å to 4760 Å (chosen because the optical depth is roughly constant), there are 375 8.8 km s^{-1} bins. Each bin is four pixels, about the FWHM resolution for this HIRES spectrum. If we ignore the fact that the bins are not completely independent and and also ignore the doublet nature of the C IV transition, an estimate of the number of independent "clouds" can be made from the noise in the divided C IV spectrum. The residual intensity is assumed to be given by $r = e^{-\tau}$ with $\tau = N\tau_c$ where N is the number of clouds contributing to the absorption in each bin and τ_c is the optical depth per cloud. If the clouds obey a simple gaussian distribution with its $N^{1/2}$ rms variation in number, then the rms fluctuations in the divided spectrum, $\sigma^2 = < (\delta r/r_s)^2 >$, will be given by: $\sigma \sim \tau \delta N/N = \tau N^{-1/2}$. Here $\delta r = r - r_s$ and r_s is the smooth fit to the data. Putting in the observed $\sigma = 0.059$ and $\tau \sim 1$ in this wavelength range, gives N $\sim$ 300 clouds and multiplying by 375 bins gives $N_{total} \sim 10^5$ clouds. This is a very crude estimate, but it is clear that a large number of clouds are needed to make a smooth trough. The residuals observed are at a level slightly higher than expected from the noise estimate in the spectrum (${\chi_\nu}^2 = 1.8$, while 1.0 is expected if purely from noise). Given the process of piecing together the HIRES spectral orders, fitting a continuum, and then fitting the trough with a piecewise polynomial, it is possible that all of the observed variation is noise plus some fitting errors.

This analysis is most sensitive to a large number of clouds at the resolution of the observation. Above a few hundred km s^{-1}, for the C IV trough in Q1246-057, some structure is removed by the polynomial fit used to model the smooth trough. The bin size can easily be varied and repeating this exercise shows that for "cloud" (or component) widths from 5 to 200 km s^{-1} a large number ($\geq 3 \times 10^4$) of clouds is needed. The assignment of a typical velocity width for a typical cloud is uncertain because the physical mechanism behind BAL cloud confinement is not known. Photoionization at parsec distances or greater leads to a much larger number of very small individual clouds and small filling factors (e.g. Junkkarinen, Burbidge, and Smith 1987). Velocity widths $\sim$ 8 km s^{-1} (FWHM) are produced by thermal broadening for carbon at T $\sim$ 15 000 K, near the temperature expected from photoionization equilibrium.

The smoothness of the troughs in BAL quasars like Q1246−057 may instead indicate a smooth outflow. Models that avoid the need for small clouds have been calculated by Murray et al. (1995). In these models the emission and absorption regions are both formed very close to the central engine (starting point as near as $\sim$0.003 pc), and the absorption is produced by a smooth outflow that crosses the line of sight at an angle.

3. Generation of Absorption Templates

The conventional analysis of BAL spectra involves the extraction of absorption templates from the data. Doublets like C IV $\lambda\lambda$ 1548.2,1550.8 and Si IV $\lambda\lambda$ 1393.8,1402.8, can be iteratively corrected for the weaker line (Junkkarinen, Burbidge, and Smith 1983) to produce optical depth versus velocity. These templates are shifted to the wavelengths of other transitions and scaled in a χ^2

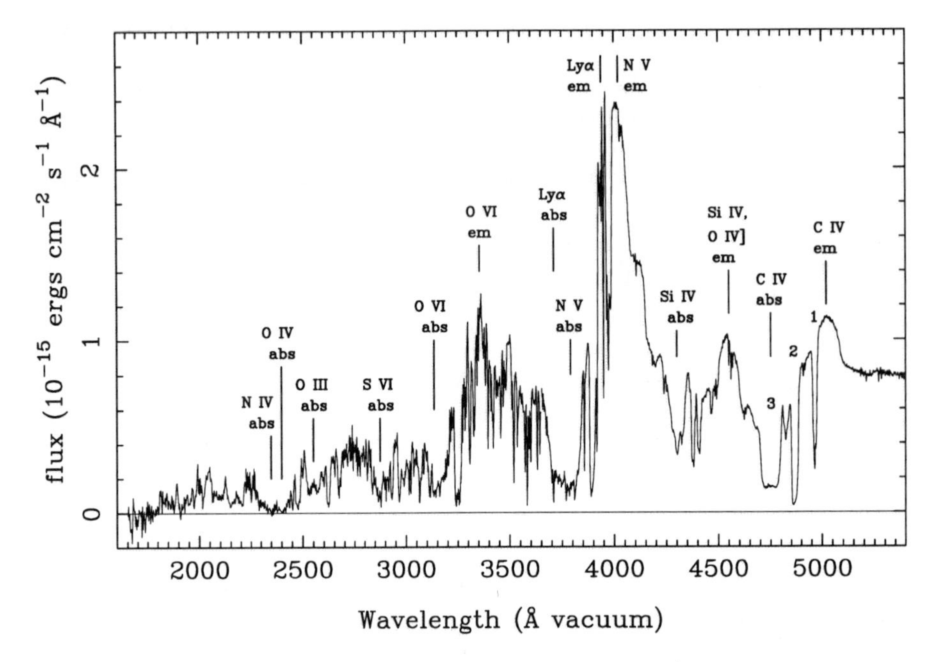

Figure 2. Spectrum of Q0226-1024. The UV HST/FOS spectra were obtained on 1991 June 13 with the G270H grating and 1992 July 25 for the G190H grating. The HST spectra have been smoothed using 8:1 binning for the G190H and 4:1 binning for the G270H. The optical spectrum was obtained on 1996 October 9 with the Lick 3m.

fitting procedure to match the data and give approximate (lower limit) column densities (e.g. Korista et al. 1992). Because BAL quasars like Q0226−1024 (Fig. 2) and Q1246−057 have smooth troughs with no apparent features at scales around 30 km s^{-1}, spectra obtained at the HIRES resolution (typically 7 - 9 km s^{-1} FWHM) can be smoothed to produce high quality templates.

Figure 3 shows the C IV and Si IV absorption templates extracted from our Keck HIRES spectrum of Q0226−1024. The spectra were smoothed to produce these templates and could be further smoothed without losing any of the apparent structure. At the HIRES resolution of 7 km s^{-1} FWHM for Q0226−1024 and at a resolution of 9 km s^{-1} FWHM for Q1246−057, the troughs are completely resolved. The troughs are not without structure as is illustrated in the case of Q0226−1024. The lowest velocity BAL feature (marked "1" in Figs. 2 & 3) breaks up into three "spikes" of absorption repeated in both C IV and Si IV. These features are not resolved with $R \sim 1000$ spectra.

Analyzing BAL quasar spectra using multiple templates that cover only a part of the velocity space may be possible when the BALs occur in clumps well separated in velocity. Troughs 1 and 2 in Q0226−1024 are narrower than the Si IV λλ 1393.8,1402.8 doublet. Trough 2 shows partial covering in Si IV with $C_f \approx 0.70$ based on measurements of the doublet using: $C_f = \frac{I_1{}^2 - 2I_1 + 1}{I_2 - 2I_1 + 1}$, where

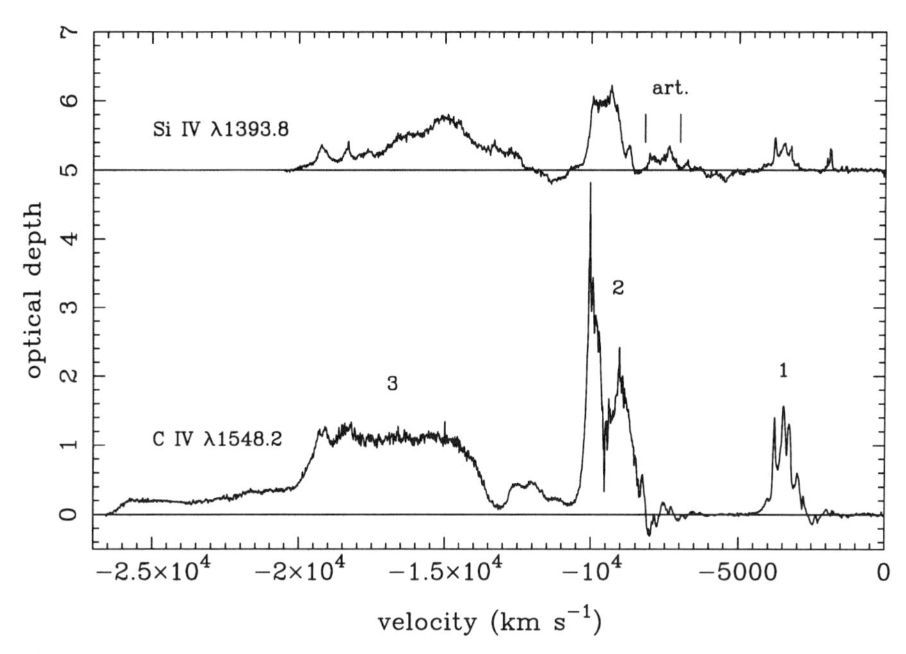

Figure 3. Q0226-1024 absorption templates for C IV and Si IV extracted from a Keck HIRES spectrum smoothed to 18 km s^{-1} FWHM.

I_1 and I_2 are the residual intensities in the stronger and weaker lines in the doublet (e.g. Hamann et al. 1997). The template extraction procedure assumes full covering and produces a false absorption feature around 7500 km s^{-1} in the extracted template (marked "art."Fig. 3). Such weak, echo–like features and regions where the template goes negative are indications that the assumption of complete coverage of the background source has broken down. Artifacts can also be produced by a poor choice of continuum, so the templates must be evaluated allowing for some uncertainty in the continuum. For very wide features with smooth edges compared to the doublet separation, like trough 3 in C IV in Q0226−1024, it is easy to construct partial covering models that mimic the data. The data are consistent with either 100% covering over those velocities or with high optical depths and a covering function that is chosen to produce the shape of the observed feature. The important issue of partial covering dominating the shape of the broad BAL troughs is not one that is easily resolved.

4. Summary

The BALs in Q1246−057 and Q0226−1024 are found to be smooth and completely resolved in the Keck HIRES spectra at a resolution of 7 – 9 km s^{-1} FWHM. The smoothness of the C IV trough in Q1246−057 leads to an estimate of at least 3×10^4 clouds (or components) of width 5 to 200 km s^{-1} comprising the deepest part of the trough. Keck HIRES spectra are a good source for generating absorption templates for analyzing lower resolution ground based and HST UV spectra of BAL quasars. The spectrum of Q0226−1024 shows BAL features with structure that is not resolved in low resolution spectra. The Q0226−1024 Si IV template also shows artifacts produced by a breakdown of the assumption of complete covering of the background source. In broader BAL features (broad relative to the doublet separation), the effects of optical depth and partial covering are not easy to separate. Spectropolarimetry (e.g. Cohen et al. 1997) or photoionization equilibrium analysis could yield estimates of the partial covering and the true column densities in the broad features.

Acknowledgments. This work has been supported by an archival research grant (AR-08355.01-97A) from the Space Telescope Science Institute which is operated by AURA, Inc., under NASA contract NAS 5-26555.

References

Arav, N. 1997 in Mass Ejection from AGN, eds. N. Arav, I. Shlosman, & R. Weymann (San Francisco: ASP), p. 208

Barlow, T. A., & Sargent, W. L. W. 1997, AJ, 113, 136

Barlow, T. A., & Junkkarinen, V. T. 1994, BAAS, 26, 1339

Barlow, T. A., Hamann, F., & Sargent, W. L. W. 1997 in Mass Ejection from AGN, N. Arav, I. Shlosman, & R. Weymann (San Francisco: ASP), p. 13

Boksenberg, A., Carswell, R. F., Smith, M. G., & Whelan, J. A. J. 1978, MNRAS, 184, 773

Cohen, M. H., Ogle, P. M., Tran, H. D., Vermeulen, R. C., Miller, J. S., Goodrich, R. W., & Martel, A. R. 1997, ApJ, 448, L77

Hamann, F. 1997, ApJS, 109, 279

Hamann, F. 1998, ApJ, 500, 798

Hamann, F., Barlow, T. A., Cohen, R. D., Junkkarinen, V., & Burbidge, E. M. 1997 in Mass Ejection from AGN, eds N. Arav, I. Shlosman, & R. Weymann (San Francisco: ASP), p.19

Hamann, F., and Ferland, G. J. 1999, ARA&A, 37, 487

Junkkarinen, V. T., Burbidge, E. M., & Smith, H. E. 1983, ApJ, 265, 51

Junkkarinen, V. T., Burbidge, E. M., & Smith, H. E. 1987, ApJ, 317, 460

Korista, K., Hamann, F., Ferguson, J., & Ferland, G. 1996, ApJ, 461, 641

Korista, K., et al. 1992, ApJ, 401, 529

Kwan, J. 1990, ApJ, 353, 123

Murray, N., Chiang, J., Grossman, S. A., & Voit, G. M. 1995, ApJ, 451, 498

Weymann, R.J., Morris, S.L., Foltz, C.B., & Hewett, P.C. 1991, ApJ, 373, 23

Young, P., Sargent, W. L. W., & Boksenberg, A. 1982, ApJS, 48, 455

Mass Outflow in Active Galactic Nuclei: New Perspectives
ASP Conference Series, Vol. 255, 2002
D.M. Crenshaw, S.B. Kraemer, and I.M. George

Polarimetric & Infrared Constraints on Absorbing Material in (BAL) QSOs

Dean C. Hines

Steward Observatory, The University of Arizona, 933 N. Cherry, Ave., Tucson, AZ 85715

Abstract.
I briefly discuss the polarization and (sparse) mid-to-far Infrared properties of BALQSOs, and place these properties in context with other classes of QSO. A simple axisymmetric model is then used to qualitatively reproduce the relationship between polarization and infrared properties.

1. The Polarization of BALQSOS

Moore & Stockman (1984) and Stockman, Moore & Angel (1984) showed that high linear optical polarization in radio quiet QSOs was rare. They found only two out of 53 radio quiet QSOs with high polarization ($p > 3\%$), both of which showed strong broad absorption lines (BALQSOs). In a large broad-band optical polarimetry survey of BALQSOs conducted at Steward and McDonald observatories, Schmidt & Hines (1999) found that 22/53 BALQSOs showed $p \geq 1\%$ and eight which had $p \geq 2\%$. This is in sharp contrast the 12/114 objects with $p \geq 1\%$, and only two with $p \geq 2\%$ in the PG QSOs (Berriman et al. 1991). Objects exhibiting absorption from low-ionization species such as Na I, Mg II and Al III (lo-BALQSOs), tend to be more highly polarized than those objects having high-ionization absorption only. Schmidt & Hines (1999) also identified a possible trend with the balnicity index (Weymann et al. 1991).

Similar broad-band polarimetry results were obtained by Hutsemekers, Lamy & Remy (1998), but they also identified a strong trend with the detachment index (Weymann et al. 1991) such that the polarization decreases with increased blue offset of the initial absorption edge compared with the emission line wavelength. Since both surveys used broad-band, aperture polarimetry, the observed high polarization shows that BALQSOs as a class must contain a structure(s) that imparts an overall asymmetry to light escaping the system. However, the location of this structure relative to the emission regions, and the nature of the polarizing mechanism requires spectropolarimetry. Several investigations have been performed on individual objects and two comprehensive surveys have been conducted (Schmidt & Hines 1999; Ogle 1999). The following overall trends are revealed:

- The continuum polarization increases towards shorter wavelengths, but this dependence is not caused by dilution from host galaxy starlight. Wavelength variations in the continuum polarization position angle are small ($< 10°$), suggesting a single dominant polarizing mechanism.

- The polarization generally increases within the BAL troughs implying multiple light paths to the continnum source, and ruling out polarization induced by transmission through aligned dust grains like the interstellar polarization (ISP) in our galaxy (ISP is a line of sight effect only). This leaves scattering by small particles as the most likely polarizing mechanism.
- The polarization decreases across the emission lines with little or no rotation in position angle implying that the scattering region is just beyond or mixed with the broad emission line region (but is within the broad absorption line region).

With few exceptions, these trends persist over a large range of observed polarizations (Ogle 1999; Hines, Schmidt & Smith 1999). This strongly suggests a common geometry for these objects. Their appearance would then be a manifestation of orientation, detailed scattering geometry or possibly evolutionary state.

2. Placing BALQSOS in Context

The similarity of optical emission line properties between non-BALQSOs and BALQSOs (Hartig & Baldwin 1986; Weymann et al. 1991) suggests that all QSOs contain BAL regions (BALR) but the absorbing material is not always along our line of sight. Therefore, it is important to compare BALQSOs with known types of (radio-quiet) QSO, not just those selected optically. In particular, if the BALRs cover only a fraction of the sky as seen from the continuum source, then estimating covering factor by comparing numbers of BALQSOs to the number of non-BALQSOs drawn from optically-selected sample only, can produce erroneous results, especially if a significant fraction of all QSOs are obscured by dust along our line of sight (e.g. Wills & Hines 1997; Hines, Schmidt & Smith 1999).

2.1. The Hyperluminous Infrared Galaxies (Type II QSOs)

IRAS showed clearly that the mid-IR ($10 - 60\mu$m) spectral energy distributions (SEDs) of (radio-quiet) QSOs were often dominated by thermal emission from dust heated by a QSO-like continuum. The lack of obvious optical extinction features in QSO spectra means that the dust does not lie along our direct line of sight. Therefore, there must be a population of objects where our sight line intercepts the dust. Searches based on "warm" mid-to-far IR colors identified many examples of reddened QSOs (*IRAS*-QSOs: Beichman et al. 1986; Low et al. 1988, 1989), as well as Hyperluminous Infrared Galaxies (HIGs) with extreme (QSO-like) infrared luminosities, but dominated optically by strong narrow emission lines and host galaxy starlight (i.e., a Type II AGN spectrum: Kleinmann et al. 1988; Cutri et al. 1994).

Imaging and polarimetry often reveals highly polarized (biconical) extended emission and polarized broad lines (Type I spectrum). The high luminosity, high polarization and (often) large narrow line equivalent widths require a strong source of UV continuum that is seen by the gas and dust, but not by us directly. Thus many if not all HIGs contain powerful QSOs hidden from our direct view by an axisymmetric distribution of dust (probably a torus), but revealed in light polarized by scattering (Hines et al. 1995, 1999; Young et al. 1996; Tran et al. 2000). Viewed from the vantage point of the scattering material and narrow emission line gas, these Type II QSOs would appear as typical, optically-selected luminous Type I QSOs.

2.2. The 2MASS QSOs

There must also be intermediate objects that are optically reddened but otherwise indistiguishable from PG-type QSOs in the infrared. Based on J,H,K_s color selection, Cutri et al. (in preparation) have identified several hundred objects with QSO (K-band) luminosities in the Two Micron All Sky Survey (2MASS). The objects exhibit Type I or Type II optical spectra, and are intermediate between the PG-QSOs and HIGs in the ratio of infrared-to-optical luminosities [$\log(L_{IR}/L_{opt}) \sim 1 - 2$]. They also generally exhibit highly polarized Type I spectra regardless of their classification in total optical light (Smith et al. 2000; Smith et al., in preparation). Like HIGs, the 2MASS QSOs would be indistinguishable from PG-type QSOs if viewed from the vantage point of the polarizing (scattering) material.

2.3. The Optical to far-IR Spectral Energy Distribution of BALQSOs

Mid-to-far Infrared observations of BALQSOs are rare since *IRAS* was only able to detect lensed high-ionization BALQSOs, and few low-ionization BALQSOs (C IV BALs preferentially select for $z \sim 2$). Figure 1 shows the SEDs of the six BALQSOs with *IRAS* detections compared with template SEDs for radio quiet PG QSOs, HIGs (Type II QSOs) and, for completeness, an Ultraluminous Infrared Galaxy powered by star-formation. The BALQSOs have spectral slopes intermediate between the PG-type QSOs and the HIGs, and there is a hint that the lo-BALQSOs are redder than hi-BALQSOs (see also, Sparyberry & Foltz 1992). Unfortunately, the unexpectedly low sensitivity of the *Infrared Space Observatory* did not enable any new insights into the mid-to-far infrared properties of BALQSOs. We anticipate new results from the *Space Infrared Telescope Facility (SIRTF)*.

2.4. Polarization of PG QSOs, BALQSOs, 2MASS QSOs & HIGs

Scattering from an ensemble of particles and dilution from direct, unpolarized light from the QSO and host galaxy will reduce measured polarization. Therefore, high polarization from an unresolved object indicates strong asymmetry. High polarization also requires large scattering angles. Figure 2 shows the cumulative distributions of observed broad-band optical (linear) polarization for the PG QSOs, BALQSOs, 2MASS QSOs and HIGs. The distributions differ dramatically above $p = 1\%$, with a progression of increasing fraction of highly polarized objects for each class. The redshifts of the PG QSOs, 2MASS QSOs and HIGs overlap well, so wavelength-dependent dilution from the QSO host galaxies is not responsible for the differences in the polarization properties between classes. Redshift dependence for the BALQSO polarization is also small (Schmidt & Hines 1999). The results in Figure 2 suggest an increase in projected asymmetry or a decrease in dilution by direct unpolarized QSO light, or both.

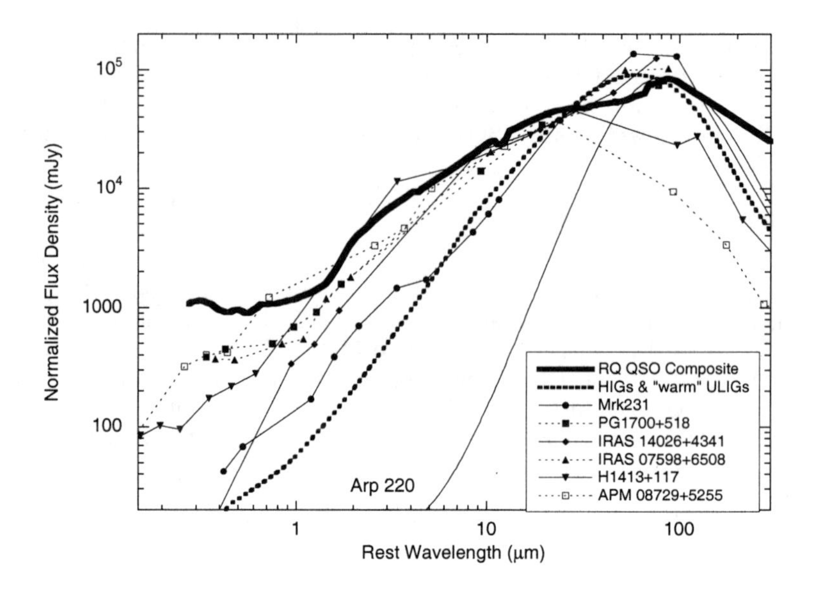

Figure 1. SEDs of the six BALQSOs with *IRAS* detections compared with template SEDs for radio quiet PG QSOs, HIGs (Type II QSOs) and an Ultraluminous Infrared Galaxy powered by star-formation. Except the two lensed (high-z, high-ionization) BALQSOs, the SEDs are normalized near their peak $\sim 80\mu$m in the rest frame. Increased amplification due to differential lensing of the compact short wavelength component in the lensed hi-BALQSOs overemphasizes the shorter wavelength flux. Therefore, these two objects were normalized at 12μm to compare the spectral slopes with the PG-type QSOs.

In addition, the fraction of objects where the Type I spectrum is revealed in polarized light increases as we move from PG QSOs to the HIGs.

3. An Axisymmetric Model for QSOs

Both the polarization and the ratio of infrared to optical luminosity increase from the PG QSOs to BALQSOs to 2MASS QSOs and finally the *IRAS*-QSOs/HIGs. The absorption/obscuration of the QSO also increases, with the PG QSOs apparently completely unobscured, the BALQSOs exhibiting absorption from gas and some dust extinction, the 2MASS QSOs being highly extincted and the HIGs showing QSOs that are totally obscured along the direct line of sight, revealed in scattered light.

The spectral and polarization properties of PG-QSOs, 2MASS QSOs and HIGs have been successfully explained by a simple unification model analogous to that for the Seyfert galaxies (Hines et al. 1995; 1999; Smith et al. 2000; Smith et al., in preparation). BALQSOs have also been tied to this scheme whereby the absorbing material is ablated from the surface of a torus of gas and dust surrounding the QSO (Weymann et al. 1991; Voit, Weymann & Korista

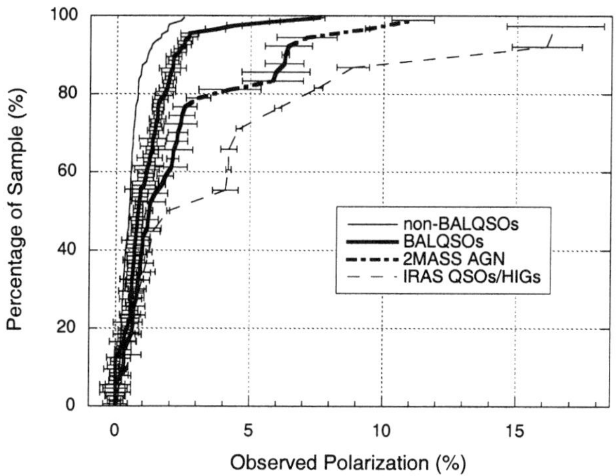

Figure 2. Cumulative distributions of observed broad-band optical (linear) polarization for the PG QSOs, BALQSOs, 2MASS QSOs and HIGs (see Smith et al., in preparation).

1993). Therefore it is tempting to model the overall population of (radio-quiet) QSOs with a luminous QSO (characterized by a strong UV continuum and broad emission lines: Type I spectrum) surrounded by an axisymmetric dusty torus. BAL material is ablated from the skin of the torus, providing a natural stratification. Low-ionization material lies closest to the torus, is mixed with dust, and has a lower velocitiy spread. Higher ionization material occupies a much larger height above the torus where decreased dust shielding allows strong ionization and higher velocities from radiation pressure. For a fixed torus half-opening angle θ, the properties of the classes of (radio-quiet) QSO depend primarily on the torus inclination (i). For $i << \theta$ (pole-on), we see a PG-type QSO with low polarization and unreddened Type I spectrum. When $i \sim \theta$, the probability of our line of sight passing through obscuring and absorbing material increases as does the polarization from scattering. For $i > \theta$ both the Type I nuclear spectrum and the BAL region are blocked from direct view, and we only see (extended) narrow line emission and perhaps a scattering region. The polarization is high, revealing a scattered Type I spectrum. The ratio of infrared to optical luminosity also increases with inclination, because extinction affects the UV/optical flux but not the mid-to-far IR flux. This geometry is also consistent with the trends seen in polarization versus BAL detachment and balnicity. Deeper, broader absorption occurs when $i \approx \theta$, while shallower or more detached BALs would be viewed for $i \leq \theta$. The model of the BAL region presented by Elvis (2002) fits nicely with this scheme.

4. Concluding Remarks

This simple model does not easily distinguish highly inclined objects with large half-opening angles from those objects with lower inclination and smaller half-opening angles (i.e., the model has a degeneracy between orientation and dust cover). However, the mid-to-far infrared SEDs may provide an important clue since the ratio of infrared to optical luminosity is sensitive to both orientation and dust cover, but the infrared luminosity itself should depend primarily on dust cover (for a given QSO luminosity). If the QSO luminosity can be determined independently, from the narrow emission-line luminosity for example (Hines et al. 1995, 1999), then comprison between the infrared luminosity and the ratio of infrared to directly observed optical flux (luminosity) should help distinguish orientation from dust cover. Planned GTO observations with SIRTF, and in particular with the Multiband Imaging Photometer for *SIRTF* (MIPS), should provide the data needed to address this issue.

Acknowledgments. DCH acknowledges from MIPS Science Development, supported by the National Aeronautics and Space Administration (NASA) and the Jet Propulsion Laboratory (JPL) under Contract 960785 to the University of Arizona. Special thanks to B. Wills, G. Schmidt, F. Low, P. Smith and R. Cutri for many years of fruitful collaboration.

References

Cutri, R.M. et al. 1994, ApJ, 424, L65

Elvis, M. 2002, in Mass Outflow in Active Galactic Nuclei: New Perspectives, eds. D.M. Crenshaw, S.B. Kraemer, & I.M. George (San Francisco: ASP), p. 303

Hartig, G.F. & Baldwin, J.A. 1986, ApJ, 302, 64

Hines, D.C. et al. 1995, ApJ, 450, 1

Hines, D.C. et al. 1999, ApJ, 512, 145

Hines, D.C., Schmidt, G.D. & Smith, P.S. 1999, ApJ, 514, L91

Hutsemekers, D., Lamy, H. & Remy, M. 1998, A&A, 340, 371

Kleinmann, S.G. et al. 1988, ApJ, 328, 161

Low, F.J. et al. 1988, ApJ, 327, L41

Low, F.J. et al. 1989, ApJ, 340, L1

Moore, R.L. & Stockman, H.S. 1984, ApJ, 279, 465

Ogle, P. et al. 1999, ApJS, 125, 1

Schmidt, G.D. & Hines, D.C. 1999, ApJ, 512, 125

Smith, P.S. et al. 2000, ApJ, 545, L19

Sprayberry,D., Foltz,C.B. 1992, ApJ, 390, 39

Stockman, H.S., Moore, R.L. & Angel, J.R.P. 1984, ApJ, 279, 485

Tran, H.D. et al. 2000, AJ, 120, 562

Voit, G.M., Weymann, R.J., & Korista, K.T. 1993, ApJ, 413, 95

Weymann, R.J. et al. 1991, ApJ, 373, 23

Wills, B.J. & Hines, D.C. 1997, in Mass Ejection from AGN, eds. N. Arav, I. Shlosman & R. Weymann (San Francisco: ASP), p. 99

Young, S., et al. 1996, MNRAS, 280, 291

Mass Outflow in Active Galactic Nuclei: New Perspectives
ASP Conference Series, Vol. 255, 2002
D.M. Crenshaw, S.B. Kraemer, and I.M. George

Ionization Equilibrium and Chemical Abundances in BALQSO PG 0946+301

Nahum Arav

Astronomy Department, UC Berkeley, Berkeley, CA 94720, University of California, Davis, CA 95616

Martijn de Kool

Research School of Astronomy and Astrophysics, ANU ACT, Australia

Kirk T. Korista

Western Michigan Univ.,Dept. of Physics,1120 Everett Tower, Kalamazoo, MI 49008

D. Michael Crenshaw

Department of Physics and Astronomy, Georgia State University, Atlanta, GA 30303

Abstract. We present preliminary results of 40 orbits of *HST*/STIS observations of the quasar PG 0946+301. These observations are the major part of a multi-wavelength campaign on this object aimed at determining the ionization equilibrium and abundances in broad absorption line (BAL) QSOs. We find that the outflow's metalicity is a few times solar, while the phosphorus abundance relative to hydrogen is enhanced by about a factor of ten. These findings are based on diagnostics that are not sensitive to saturation and partial covering effects in the BALs, which considerably weakened previous claims for enhanced metalicity. Ample evidence for these effects is seen in the spectrum.

1. Introduction

Establishing the ionization equilibrium and abundances (IEA) of the Broad Absorption Line (BAL) material is a fundamental issue in quasar studies (Weymann, Turnshek, & Christiansen 1985; Hamann 1996; Korista et al. 1996; Arav et al. 2001a; de Kool et al. 2001). Furthermore, determining the IEA is crucial for understanding the dynamics of the flows, especially in radiative acceleration scenarios that are strongly coupled to the ionization equilibrium (Arav, Li & Begelman 1994; Murray et al 1995; Proga, Stone & Kallman 2000). Finally, the mass flux and kinetic luminosity associated with the flow cannot be constrained without a reliable IEA determination.

Careful spectroscopic analysis has made it apparent that the BALs are often saturated while not black (Arav 1997; Telfer et al. 1998; Arav et al. 1999;

de Kool et al. 2001). Support for this picture comes from spectropolarimetry studies (Cohen et al. 1995). As a result, BAL ionic column densities (N_{ion}) determined in the traditional way (from equivalent width or direct conversion of the residual intensity into optical depth) can only serve as lower limits to the real column densities. Since BAL column densities are the foundation of any attempt to determine the IEA in the outflows, it became clear that more sophisticated analyses are essential for any progress in understanding the BAL phenomenon. To derive real N_{ion} we have to account for saturation and partial covering factor in the BALs. In the UV, this approach is currently feasible only for exceptionally bright BALQSOs and still requires long integration times. To achieve these goals we observed the best available target, PG 0946+301, using 40 orbits of *HST*/STIS time. The *HST*/STIS observations and template fitting results are described in the Korista et al. (2002), and the full analysis is found in Arav et al. (2001b).

2. Constraints on the IEA in PG 0946+301

In this paper we concentrate on constraining simple-slab photoionization models (using the photoionization code CLOUDY; Ferland 1996). Such models are commonly used in the study of quasar outflows (e.g., Weymann, Turnshek, & Christiansen 1985; Arav, Li & Begelman 1994; Hamann 1996; de Kool et al. 2001) and assume that the absorber consists of a constant density slab irradiated by an ionizing continuum. The two main parameters in these models are the thickness of the slab as measured by the total hydrogen column density (N_H) and the ionization parameter U (defined as the ratio of number densities between hydrogen ionizing photons and hydrogen in all forms). The spectral shape of the ionizing continuum can also be varied, but its influence on the model's result is weaker. We therefore use a single continuum for the current discussion. We discuss the quantitative effects of other continua in Arav et al. (2001b).

A particularly useful way to constrain such models is to plot curves of constant N_{ion} on the plane of N_H vs. U. On such a plot any type of information regarding the ionic column densities can be used to constrain the allowed parameter space. Lower limits (derived from fits and/or apparent optical depth measurements) exclude the area below the constant N_{ion} curve, N_{ion} upper-limits (from non-detections) exclude the area above their associated constant-N_{ion} curve. In Figure 1 we show the most significant N_{ion} constraints extracted from the STIS data of PG 0946+301, where we use the standard Mathews–Ferland AGN spectrum (Mathews and Ferland 1987) as the incident ionizing continuum and assume solar metalicity.

Several important conclusions can be drawn from Figure 1. The combined constraints from H I, S IV, C II, Ar VII as well as from other available CNO ions (which for clarity's sake are not shown here), suggest that the absorber is characterized by $19.8 < \log(N_H) < 20.3$ and $-2 < \log(U) < -1.5$ and a small metalicity enhancement compared to solar. The metalicity enhancement is necessary since the strips allowed by the S IV and H I constraints do not overlap, and most of the metals' lower limit curves lie somewhat above the allowed H I zone. Assuming a higher metalicity lowers the metals' ionic curve and thus produces overlapping regions. However, the enrichment cannot be larger than a factor of

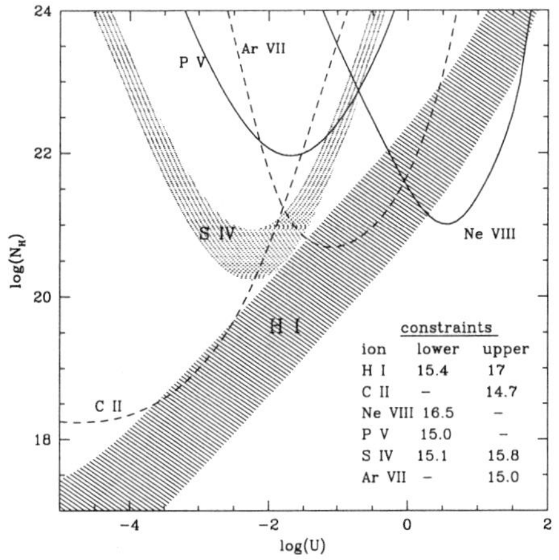

Figure 1. Curves of constant ionic column density plotted on the N_H/U plane, using solar abundances. Lower limits are shown as solid lines, upper limits as dashed lines, and shaded regions designate cases where we have both limits for a given ion. The inserted table gives the $\log(N_{ion})$ for the various constraints.

~ 3 since then we would expect to detect C II, Si II, a strong Si III, or Ar VII. An exception to this inferred IEA is phosphorus. The firm detection of P V yields the associated lower limit on Figure 1, which is incompatible with the allowed parameter space we derived above. The simplest way to resolve this apparent contradiction is to assume that phosphorus is over-abundant by a factor of 10–30 compared to its solar abundance relative to hydrogen. All else equal, this will lower the P V curve to the point where it intersects with the allowed parameter space. The infered phosphorus over-abundance value is similar in optically thick photoionization models, in particularly in models that include a He II ionization front in the absorber (full discussion of this issue and the inconsistency evident from the Ne VIII curve are found in Arav et al. [2001b]).

References

Arav, N., Li, Z.Y., & Begelman, M. C. 1994, ApJ, 432, 62

Arav, N., 1997 in Mass Ejection from AGN, ed. N. Arav, L. Shlossman, & R. Weymann (San Francisco: ASP), p. 208

Arav, N., et al., 1999, ApJ, 524, 566

Arav, N., et al., 2001a, ApJ, 546, 140

Arav, N., et al., 2001b, ApJ, in press

Cohen, M.H., et al., 1995, ApJ, 448, L77

de Kool, M., et al., 2001, ApJ, in press

Ferland, G.J., 1996, Univ. of Kentucky Dept. of Phy. and Ast. Internal Report

Hamann, F. 1997, ApJS, 109, 279

Korista, K.T. 2002, in Mass Outflow in Active Galactic Nuclei: New Perspectives, eds. D.M. Crenshaw, S.B. Kraemer, & I.M. George (San Francisco: ASP), p. 201

Korista, K.T., Hamann, F., Ferguson, J., & Ferland, G. J. 1996, ApJ, 461, 645

Mathews, W.G., & Ferland, G.J., 1987, ApJ, 323, 456

Murray, N., Chiang, J., Grossman, S.A., & Voit, G.M. 1995, ApJ, 451, 498

Proga, D., Stone, J.M., Kallman, T.R., 2000, ApJ, 543, 686

Telfer, R.C., et al., 1998, ApJ, 509, 132

Weymann, R.J., Turnshek, D.A., & Christiansen, W.A. 1985, in Astrophysics of Active Galaxies and Quasi-stellar Objects, 333

Mass Outflow in Active Galactic Nuclei: New Perspectives
ASP Conference Series, Vol. 255, 2002
D.M. Crenshaw, S.B. Kraemer, and I.M. George

Absorption from Excited States as a Density Diagnostic for AGN Outflows

Martijn de Kool

Research School of Astronomy and Astrophysics, Mt. Stromlo Observatory, Cotter Road, Weston Creek, ACT 2611, Australia

Robert H. Becker, Michael D. Gregg

University of California Davis and IGPP, Lawrence Livermore National Laboratories

Nahum Arav

University of California Davis

Richard L. White

Space Telescope Science Institute

Kirk T. Korista

Physics Department, Western Michigan University

Abstract. We present some examples of how absorption lines from excited levels can be used to determine the density in AGN outflows, and lead to an estimate of their size. The best example is the BALQSO FBQS 1044+3656, in which a detailed analysis of the population of several excited states of Fe II has been possible. These levels are populated but not in LTE, leading to an electron density of $\sim 4 \times 10^3 \mathrm{cm}^{-3}$, and an outflow size estimate of a few hundred parsec. We also show two other examples (FBQS 0840+3633 and FBQS 1214+2803) of QSOs with Fe II absorbers at large outflow velocities of a few thousand km s^{-1}. In these systems the lines are much more saturated, but it still appears to be possible to derive electron density limits of $\sim 10^5 \mathrm{cm}^{-3}$ and $> 10^6 \mathrm{cm}^{-3}$ respectively. We also present a first look at the absorption lines from excited states of S IV and O III in the high ionization BALQSO PG 0946+301. Surprisingly, the relative strengths of the lines in the S IV multiplets are best fitted with an electron density as low as $10^5 \mathrm{cm}^{-3}$.

1. Introduction

This paper consists of two parts. In the first we will present a brief summary of an analysis of the low-ionization intrinsic absorption lines in KECK HIRES spectra of the QSOs FBQS1044+3656, FBQS 0840+3633 and FBQS 1214+2803. These objects were discovered in the FIRST Bright Quasar Survey (White et al. 2000) which consists of 636 radio-selected quasars. More details on these sources and low-resolution spectra can be found in Becker et al. 2000. In the second part, we briefly discuss the possible implications of the absorption lines from excited states of S IV and O III that are present in the spectrum of the high-ionization BALQSO PG0946+301 (see Arav et al. 2002)

2. FBQS 1044+3656

The spectrum of FBQS 1044+3656 contains blue-shifted absorption lines of Mg I, Mg II and Fe II. The characteristic velocity and width of the Mg I and Fe II lines are intermediate between BALs and associated absorption lines (AALs). The Mg II absorption is very broad, and fulfils the strict criterion for BALQSOs of Weymann et al. 1991. Although the Mg II lines are much broader, the excellent correspondence in velocity between the strongest Mg II components and the Mg I and Fe II lines shows that they are formed in the same outflow (de Kool et al. 2001). The source is unique in that the Fe II lines are not very wide nor heavily saturated, behaving roughly as expected from their relative oscillator strengths. This makes quantitative estimates of the column density in several low-lying excited states possible. We found that some excited levels are populated, but considerably less than expected for LTE. Based on the predictions of the Fe II level populations as a function of density from Verner et al. (1999), we estimated an electron density in the outflow of a few times $10^3 \mathrm{cm}^{-3}$.

We constructed simple homogeneous slab photo-ionization models using the photo-ionization code CLOUDY, trying to fulfil the constraints on the electron density and the column densities of Fe II, Fe I and Mg I simultaneously. These constraints are best fulfilled for a slab containing dust, with an ionization parameter $\log U \sim -3$, a density $n_H \sim 10^5 \mathrm{cm}^{-3}$ and a moderate amount of depletion of Fe. For a QSO with a luminosity of 10^{46}ergs s^{-1}, this implies a distance between the slab and the source of radiation of a few hundred parsec. This distance is about two orders of magnitude larger than the size usually associated with outflows in BALQSOs. Details can be found in de Kool et al. (2001).

3. FBQS 0840+3633 and FBQS 1214+2803

The intrinsic absorption in these two sources indicates a much higher column density than in FBQS 1044+3656 , and the Mg I line is not detected. The higher column densities imply that many lines from the ground term of Fe II are saturated, and that much weaker lines, both from more excited Fe II levels and other iron group elements are becoming visible. This increases the number of lines that have to be taken into account to several thousand. Because of the large velocity spread of each line, blending problems become so strong that very few lines can be analyzed individually. In addition, the analysis is complicated by the fact that the Fe II broad emission lines are very strong in these QSOs so that the level of the continuum is poorly defined.

We have tried to address all these problems by trying to fit the spectra globally with a "physical" model taking all these effects into account. The model contains the following parameters:

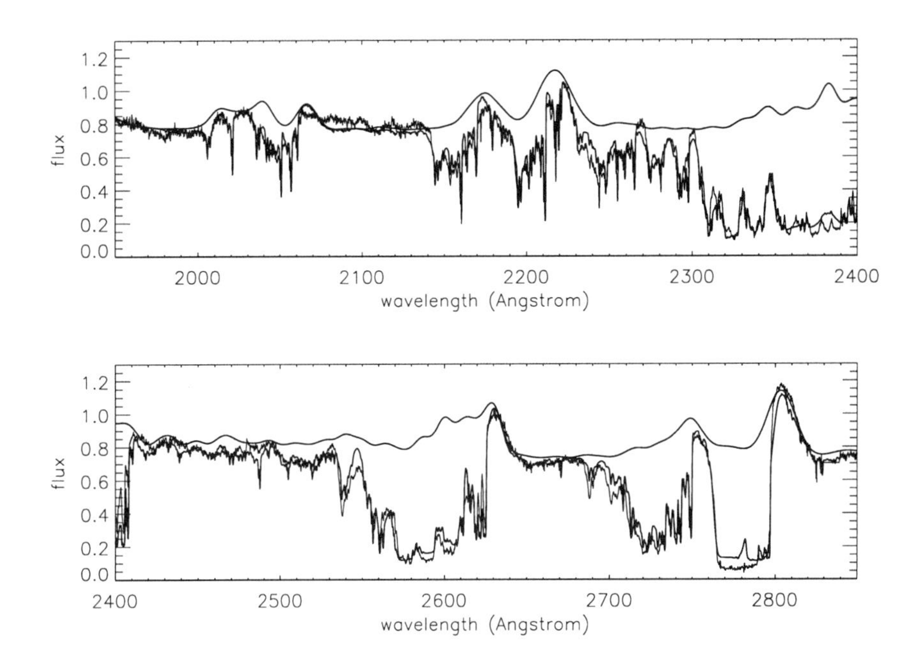

Figure 1. The spectrum of FBQS 0840+3633, overplotted with the result of the spectral fit and the effective continuum.

1. column densities (Fe II,Ni II,Mn II, Co II, Zn II, Ti II, Cr II, Mg II and Fe I)
2. the excitation temperature
3. a power law continuum
4. an Fe II emission model. Detailed modelling of the Fe II BELs was beyond the scope of this work. We have used 5 different Fe II BEL models as calculated by Verner et al. 1999, and allow the fit to use a linear superposition of these.
5. covering factors, different for power law, Fe II BELs

The resulting fit of the spectrum of FBQS 0840+3633 is shown in Figure 1. To obtain a reasonable fit, it was necessary to introduce several non-Fe II broad emission lines, most notably two lines at 2175 and 2220 Å. The wavelengths of these two lines correspond to strong Ni II lines. The group of weaker emission lines from 2000 - 2070 Å is not identified. Systematic differences between model and observations over a longer wavelength range are likely to be due to the poor model of the effective continuum, but discrepancies in the form of individual narrow features are likely to be due to the model of the absorber. The most obvious difference is a narrow line at 2490 Å, which appears to be at least five times weaker than predicted by the best fit model. This is the strongest line from an excited state of Fe II at 21251.61 cm^{-1}. Using the figure showing the deviation from LTE for this level as a function of density from Verner et al. 1999, the weakness of this line would imply an electron density of the order of 10^5 cm^{-3}. The other fit parameters are given in Table 1. The column densities found are in reasonable agreement with standard abundance ratios, which leads us to believe that they are not wrong by more than a factor of a few, and that the effects of saturation do not dominate the results. The covering factor of the power law continuum source is close to one, and the covering factor for the Fe II BEL region is significantly smaller at 0.44, in agreement with expectations for simple geometrical models. Note that the covering factor of Mg II must be larger than that of Fe II. The excitation temperature is not very reliable, since it is likely that the level population is also determined by density effects as mentioned above. If we perform a fit describing the level population by two parameters, an excitation temperature and a low-density break energy (the energy above which the level populations are strongly reduced due to low-density effects), we obtain a temperature of 7800 K and a break energy of 1.2 10^4 cm^{-1}. The column

Table 1. PARAMETERS OF GLOBAL FITS

Parameter	FBQS 0840+3633	FBQS 1214+2803
f_{cov}(powerlaw)	0.89	0.73
f_{cov}(Fe IIBELs)	0.44	0.60
T_{exc} (K)	5800	7600
n_e (cm^{-3})	$\sim 10^5$	$> 10^6$
log $N_{Fe\,II}$	16.47	17.20
log $N_{Ni\,II}$	15.77	16.22
log $N_{Mn\,II}$	14.21	15.25
log $N_{Zn\,II}$	14.12	?
log $N_{Cr\,II}$	14.81	15.41
log $N_{Fe\,I}$	<13.5	<13.5

density of the absorber in FBQS 1214+2803 is larger than in FBQS 0840+3633 (Figure 2). Our model is not able to reproduce the Fe II BELs very well, with the longer wavelength BELs being too weak by a factor of ~ 2 in the model. Most of the information is obtained from unsaturated lines from relatively highly excited states of Fe II in the wavelength range 2850-3200 Å. There is no indication of an under-population of the more highly excited states due to low density, leading to a lower limit on the electron density of 10^6 cm^{-3}. In this source the covering factor of the power law continuum source and the Fe II BEL region are found to be more similar. Note from the fit in the region just longward of the Mg II BEL how many overlapping lines can make it very difficult to determine the level of the continuum in sources like this. This region appears to be a well-defined section of continuum, but in fact contains significant absorption from many overlapping Cr II and Fe II lines.

4. Absorption from excited states in PG 0946+301

It is clear that the analysis of absorption lines from excited states is a very powerful technique to determine the density in AGN outflows, thus resolving

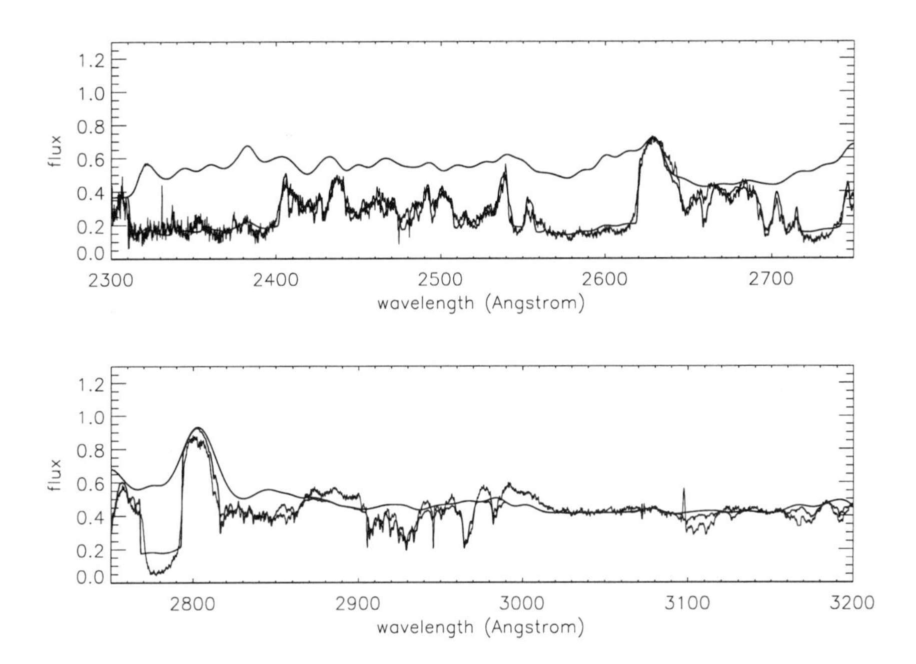

Figure 2. The spectrum of FBQS 1214+2803, overplotted with the result of the spectral fit and the effective continuum.

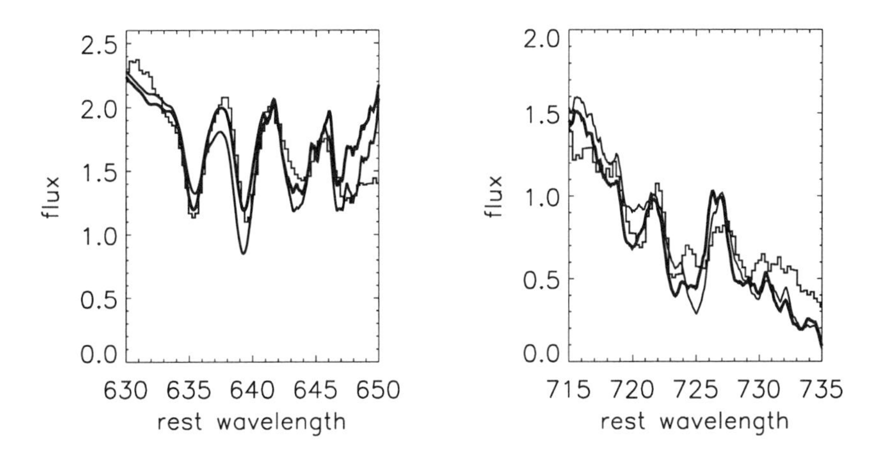

Figure 3. A comparison of the relative strengths of S IV lines in two multiplets for an LTE model (thin line) and a model in which the population of the excited level is 1/3 of the LTE value (thick line). The histogram represents the data.

the long-standing problem of the scale of the outflow. However, so far this technique has only been applied to absorption lines from very low-ionization species with outflow velocities ranging from zero to a few thousand km s^{-1}. It is not clear whether the findings obtained for these lines also apply to classical high-ionization BALQSOs. We have tried to look for absorption from excited states of more highly ionized species in the high-quality STIS spectrum that we obtained of the high-ionization BALQSO PG 0946+301. Possible candidates are lines from O III, S IV and the isoelectronic sequence C III, N IV and O V. There has been a claim of detecting some of the latter species in the literature (Pettini & Boksenberg 1986) on the basis of IUE spectra, but inspection of a much higher quality FOS spectrum (Arav et al. 1999) did not confirm their presence. Perhaps this is not surprising, since not only do the levels from which these lines originate have very high critical densities ($> 10^{10}$ cm^{-3}), but also the level energies are very high and temperatures around 10^5 K would be required to populate them. The metastable levels of O III and S IV however have much lower critical densities and excitation energies. All lines from these metastable levels are detected in the spectrum of PG 0946+301. The relative strengths of the lines within the O III multiplets at 835 and 703 Å appear to be consistent with LTE, from which we derive a lower limit on the electron density of 10^4 cm^{-3}. Absorption from a more highly excited state (20273 cm^{-1}) of O III at 600 Å is present in the data but is hard to quantify because of the very badly defined continuum at this position. In Figure 3 we show the expected relative strengths of lines within the two strongest S IV multiplets at 660 and 750 Å for two models, one which assumes that the two levels within the ground state are in LTE, and one in which the population of the upper level at 951 cm^{-1} is only 1/3 of the LTE value. The latter model reproduces the relative strengths of the lines much better, and would imply an electron density of $\sim 10^5$ cm^{-3}. This result needs further investigation, especially to determine to what extent partial covering / saturation effects could mimic a non-LTE level population.

References

Arav, N. 2002, in Mass Outflow in Active Galactic Nuclei: New Perspectives, eds. D.M. Crenshaw, S.B. Kraemer, & I.M. George (San Francisco: ASP), p. 179

Arav, N., et al. 1999, ApJ, 516, 27

Becker, R.H., et al. 2000 ApJ, 538, 72

de Kool, M., et al. 2001, ApJ, 548, 609

Pettini, M. & Boksenberg, A. 1986, in " ESA Proceedings of an International Symposium on New Insights in Astrophysics. Eight Years of UV Astronomy with IUE" p 627-631

Verner, E.M., et al. 1999, ApJS, 120, 101

Weymann, R. J., et al. 1991, ApJ, 373, 23

White, R.L. et al. 2000, ApJS, 126, 133

Mass Outflow in Active Galactic Nuclei: New Perspectives
ASP Conference Series, Vol. 255, 2002
D.M. Crenshaw, S.B. Kraemer, and I.M. George

Observational Evidence for a Multiphase Outflow in QSO FIRST 1044+3656

John Everett, Arieh Königl

Department of Astronomy and Astrophysics, University of Chicago, 5640 S. Ellis Avenue, Chicago, IL 60637

Nahum Arav

Astronomy Department, University of California, Berkeley, 601 Campbell Hall, Berkeley, CA 94720

Abstract. Recent Keck spectra of QSO FIRST J104459.6+365605, analyzed by de Kool et al. (2001) in terms of a single-phase outflow, imply that outflowing absorbing gas in the system lies approximately 700 parsecs from the central source. This is about two orders of magnitude greater than the commonly accepted distance for the Broad Absorption Line Region (BALR), and makes it difficult to understand the partial covering factor inferred for the absorbing gas. We show that this problem is avoided if the gas is not homogeneous (e.g., a continuous wind with embedded dense clouds). Our models place the BALR at approximately four parsecs from the central source, and are observationally constrained to have a dust-free continuous wind shielding dense, dusty clouds from the central continuum.

1. Introduction

Researchers have commonly believed that the Broad Absorption Line Region (BALR) in Active Galactic Nuclei (AGNs) lies just outside of the Broad Emission Line Region (BELR), but measurements of its distance remain uncertain. Recently, de Kool et al. (2001) analyzed a Keck spectrum of QSO FIRST J104459.6+365605, using different excited Fe II lines to determine the level populations of several low-lying Fe states. These level populations show that Fe II is not in local thermodynamic equilibrium, since those levels are less populated than expected in LTE. This observation implies $n_e \sim 4 \times 10^3$ cm^{-3}, and this constraint, together with the observed broad absorption in Mg I, compels single gas phase photoionization simulations to place this BALR at approximately 700 parsecs from the central source. Not only does this surprising result contradict the idea of a close-in BALR, but, in addition, de Kool et al.'s in-depth absorption analysis shows that the gas only partially covers the continuum source. It is not easy to understand how the absorbing gas could be approximately 1 kpc from the central source and yet only partially cover the AGN core. Finally, the Keck spectra indicate distinct groupings of gas in velocity space; it is also difficult to explain how these apparent clumps could move to such a great distance from the source and yet remain bunched up in this manner, retaining individual identities.

The distance estimate of 700 parsecs comes from a photoionization analysis where the gas is assumed to lie in thin, constant-density slabs. In this contribution we argue that, if this assumption is relaxed, then the absorbing gas can be placed at traditional BALR distances and the above difficulties disappear. Instead of a homogeneous outflow, the outflowing gas could be multiphased, consisting, for example, of a continuous wind with embedded dense clouds. Such a situation may arise naturally in the context of centrifugally driven disk outflows, where the clouds remain confined by the magnetic pressure of the surrounding MHD wind (e.g., Emmering et al. 1992; Königl & Kartje 1994 [KK94]; Kartje et al. 1999; Everett, Königl, & Kartje 2001). Alternatively, the 'clouds' may represent transient density enhancements that arise in a turbulent medium (e.g., Bottorff & Ferland 2001) or through the formation of shocks in a radiatively driven wind (e.g., Arav et al. 1994). For the purpose of illustration, we adopt here the 'clouds uplifted and confined by an MHD wind' picture.

2. Simulation Setup

To model the inner regions of this AGN, we use the photoionization code Cloudy (Ferland 2000), and with it, simulate a much different gas model than those attempted by de Kool et al. The dependence of number density on radius, $n(r)$, is given by the self-similar magnetocentrifugal wind model, which for our choice of parameters for AGN winds (following KK94) has $n(r) \propto r^{-1}$.

In order to simulate the full range of gas between the BALR and the central continuum source, we have to break the gas up into different zones. We simply model the innermost region of the wind as a "Thomson shield": a region where we decrease the initial incident spectrum by $e^{-\tau}$, with the optical depth τ determined by the column of gas in this (fully ionized) region. The rest of the continous, dust-free wind is simulated as another zone by Cloudy, and contains the Fe II column that is observed by de Kool et al. After that continuous wind, we place the contant-pressure zone that simulates confined, higher density clouds.

For the Cloudy simulations, we adopted a total luminosity of 10^{46} ergs s^{-1} coming from a central source of mass $5 \times 10^8 M_\odot$, the same values assumed by de Kool et al. in their analysis. The inner edge of the wind model starts at $r_{\rm in} = 10\ GM/c^2$, and we stop the continuous wind model when n_e drops below 4×10^3 cm^{-3}. We then simulate how the resultant, transmitted continuum will affect clouds at the end of the wind. We vary the gas number density at the inner face of the wind as well as the gas number density within the clouds embedded in the wind.

2.1. Simulation Results

We find that we can satisfy all of the observational constraints with an initial hydrogen number density of approximately $10^{8.75}$ cm^{-3}. This yields the observed electron number density in the region where the Fe II column density attains the observed value of 3×10^{15} cm^{-2}, which is significant since the n_e measurement comes from the Fe II absorption lines. However, we do have to cut off the outflow very close to the end of the hydrogen recombination front so as not to exceed the observed Fe II column (see Fig 1). A very abrupt end to the absorbing gas at approximately 4 parsecs from the central source could be explained by a combination of two factors: a lower ionization fraction that reduces the efficiency of gas–magnetic field coupling, and the possible transition from a gaseous to a 'clumpy' disk on that scale (see Shlosman & Begelman 1987).

We also point out that for this model, the Mg II front occurs about a factor of 2.5 closer to the central source than the Fe II front. In self-similar magnetohydrodynamic winds, the final wind velocity varies as $r_{injection}^{-1/2}$, so the smaller initial radius of the Mg II front in this model could help explain the higher observed Mg II velocities observed in FBQS 1044. Approximately the same radial dependence occurs for the final velocity of radiatively accelerated winds (exactly the same dependence is found for winds of constant ionization parameter).

Having simulated the entire continuous wind, we then model higher-density clouds that would be illuminated by the spectrum transmitted through such an outflow. Our models imply a typical cloud density of $10^{8.5}$ cm^{-3}. Using a string of such constant-density clouds gives the column density as a function of distance as shown in Figure 2. It must be noted, however, that we could accept cloud densities in the range of $10^{7.75}$ cm^{-3} up to approximately 10^{9} cm^{-3}: a range allowed by the uncertainty in observed extinction at 2500 Å from approximately 1.0 down to 0.2 magnitudes (de Kool et al. inferred these values by comparing 1044's spectrum to an averaged QSO spectrum from Weymann et al. 1991). If a column of lower-density clouds is present with a significant fraction of the observed Mg I column, the dust within those clouds would have $A_{2500} > 1$. If we have densities much greater than 10^{9} cm^{-3}, the models predict $A_{2500} < 0.25$. However, with a modeled cloud density of $10^{8.5}$ cm^{-3}, including ISM dust (dust must be present in order to deplete iron and ensure more Mg I in the clouds than Fe I), we satisfy all of the observational constraints, and produce an extinction at 2500 Å of approximately 0.5. This fits well within the limits on the observed extinction.

In Table 1 we compare our model to the observational constraints for QSO FIRST 1044. This model solves the "distance problem" by implying that the observed gas lies at approximately 4 parsecs from the central engine, over two orders of magnitude closer than the earlier estimate. The radius where the observed gas leaves the disk surface is, in general, smaller yet.

Table 1. A Comparison of the Two-Phase Outflow Model to Observational Results for FIRST QSO 1044+3656.

Quantity	Observational Result	Model Result
N(Fe II)	$\sim 3 \times 10^{15}$cm^{-2}	3×10^{15}cm^{-2}
N(Fe I)	$< 10^{13}$cm^{-2}	2.5×10^{11}cm^{-2}
N(Mg II)/N(Mg I)	$\gtrsim 30$	500
N(Mg I)	$\sim 2 \times 10^{13}$cm^{-2}	2×10^{13}cm^{-2}
n_e	~ 4000	4000
A_{2500}	$\lesssim 1$	~ 0.5

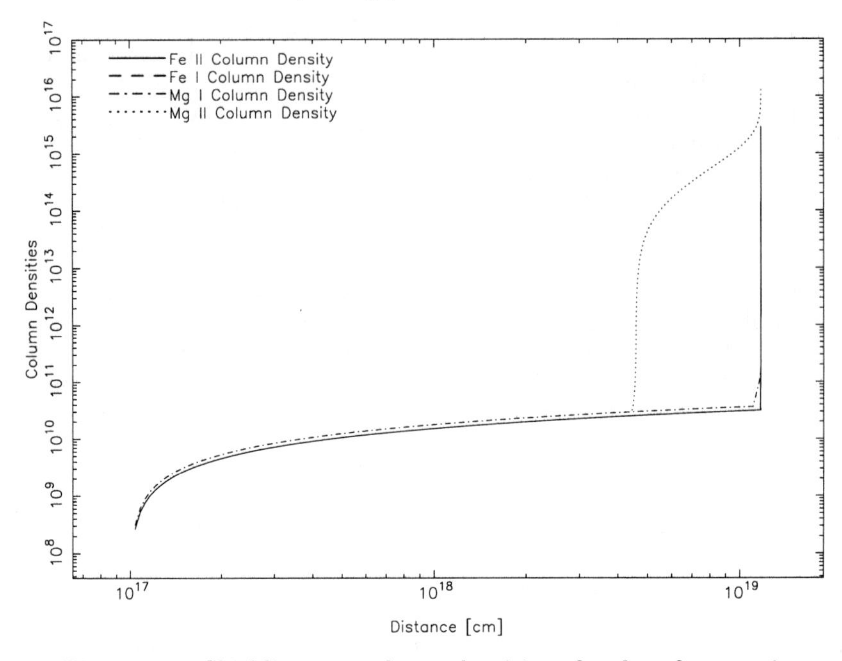

Figure 1. Shielding gas column densities of a dust-free continuous wind with an initial density of $10^{8.75}$ cm^{-3}.

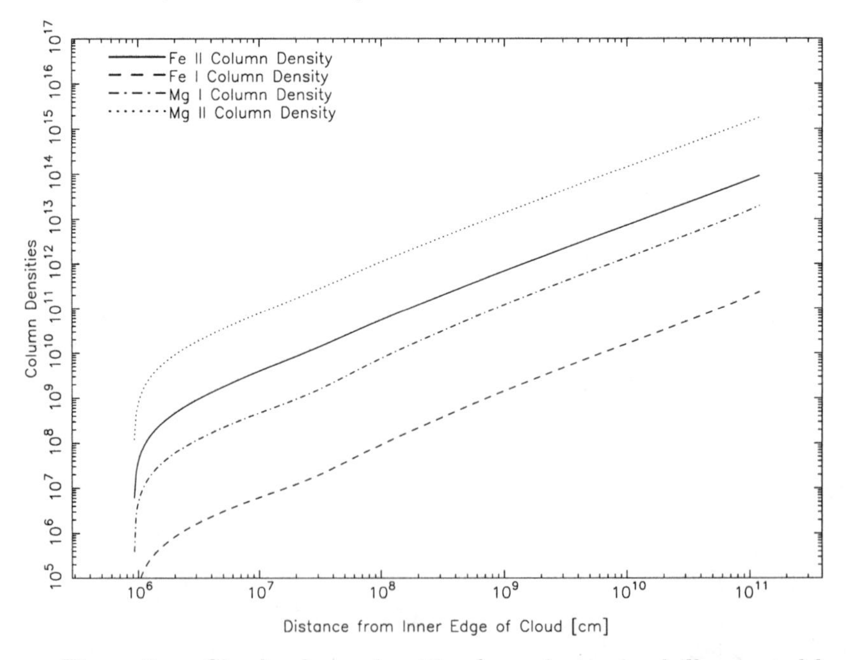

Figure 2. Cloud column densities for a dusty cloud illuminated by an AGN spectrum attenuated by a dust-free shield with a cloud density of $10^{8.5}$ cm^{-3}.

We can directly estimate the mass outflow rate associated with the absorbing gas through the relation $\dot{M}_{wind} \approx 2\pi f r_{max} N_H m_p v$, where f is the fraction of 4π steradians into which the wind flows, N_H is the total hydrogen column density of *only* the inferred Mg II and Fe II absorbing region ($\approx 3.9 \times 10^{23}$ cm^{-2} for our best model), v is the observed outflow speed ($\sim 10^8$ cm s^{-1}), r ($= 1.2 \times 10^{19}$ cm) is the inferred distance, and where we assume a vertically and azimuthally continuous wind. We take $f \sim 0.1$ as is typical for BALQSO sources. Then, for the above parameters, we find $\dot{M}_{wind} \approx 8\ M_\odot$ yr^{-1}.

By equating the thermal pressure in the clouds ($\approx 2.7 \times 10^{-4}$ dynes cm^{-2}) to the confining wind magnetic pressure, $B^2_{wind}/8\pi$, we deduce $B_{wind} \approx 8.2 \times 10^{-2}$ Gauss. It is encouraging that, when this value of B is used in the $n(r) \propto r^{-1}$ self-similar MHD wind model (which results in an approximately constant ionization parameter for the confined clouds), it implies a local mass outflow rate that is comparable to the above estimate of $\dot{M}_{wind}$, yielding $\approx 4 M_\odot$ yr^{-1}. Going back to other cloud models, if we had chosen $n_{cloud} = 10^9$ cm^{-3}, the magnetic pressure necessary to confine those clouds would yield an outflow with $\dot{M}_{wind} \approx 9 M_\odot$ yr^{-1}, in even better agreement with observational estimates. These higher-density clouds give $A_{2500} = 0.26$, near the bottom of the inferred range of plausible extinction values.

We have found the extinction limit to be the most difficult constraint to satisfy in our simulations. In order to reproduce the inferred upper bound of one magnitude of extinction at 2500 Å, our models require a continuous wind that is dust free from the base of the wind. However, a dust-free wind must have a larger gas column than a dusty wind (in order to shield the clouds), so the lack of dust in the wind implies prohibitively large mass outflow rates (much larger than could be launched by the magnetic fields that confine clouds in our model). Thus, only part of the shield can be outflowing at the observed velocities. Perhaps the inner outflow has a lower velocity than in a self-similar MHD wind (e.g. a Thomson scattering-launched wind; Blandford 2001), or may not even be outflowing, such as a disk corona (Emmering et al. 1992) or a "failed wind" from a radiatively driven disk outflow (see Murray et al. 1995 and Proga 2000).

3. Conclusions

We have demonstrated that it is not necessary to invoke absorbing gas that is on the order of 700 parsecs from the central source and that the observations of FBQS 1044 can be accounted for with a more traditional BALR scale ($\sim$ 4 pc) if one allows the absorption region to be multiphased. We have illustrated this possibility with the help of a "clouds embedded in a continuous MHD disk wind" model, but our conclusions also apply to other plausible multiphase outflow scenarios.

In addition to explaining the observations of this particular source, this scenario may be important for interpreting absorption features in similar objects where a large distance to the central source has been inferred (e.g., 3C 191; Hamann et al. 2001). Furthermore, the multiphase picture might help explain other observations where inconsistent UV/X-ray columns are implied. For example, this model may be relevant to the interpretation of distinct columns of UV and X-ray absorbing gas in Seyfert galaxies and radio-quiet QSOs such as PG 1114+445 (Mathur et al. 1998) if one attributes the X-ray absorption to a hot intercloud medium associated with a continuous disk wind and the UV features to embedded clouds. This model could also help explain, through its multiple phases, features in the warm absorber (e.g., Morales et al. 2000) and the BELR spectrum (e.g., Goad et al. 1999) in Seyfert galaxies. Overall, multiple-phase scenarios of AGN outflows might be germane to a great variety of observations, and may indeed be a key element in future AGN models.

Acknowledgments. J.E. and A.K. thank NASA for support through grant NAG5-9063. We also thank the conference participants for their comments, which improved this work.

References

Arav, N., Li, Z.-Y., & Begelman, M.C. 1994, ApJ, 432, 62

Blandford, R.D. 2001, Phil. Trans. R. Soc. Lond. A, 358, 811

Blandford, R.D., & Payne, D.G. 1982, MNRAS, 199, 883

Bottorff, M., Ferland, G. 2001, ApJ, 549, 118

de Kool, M., Arav, N., Becker, R.H., Gregg, M.D., White, R.L., Laurent-Muehleisen, S.A., Price, T., Korista, K.T. 2001, ApJ, 548, 609

Emmering, R.T., Blandford, R.D., Shlosman, I. 1992, ApJ, 385, 460

Everett, J.E., Königl, A., & Kartje, J.F. 2001, in Probing the Physics of AGN by Multiwavelength Monitoring, eds. B.M. Peterson, R.S. Polidan, & R.W. Pogge (San Francisco: ASP), p. 441

Ferland, G.J. 2000, Hazy, A Brief Introduction to Cloudy 94.00

Goad, M.R., Koratkar, A.P., Axon, D.J., Korista, K.T., & O'Brien, P.T. 1999, ApJ, 512, L95

Hamann, F.W., Barlow,T.A., Chaffee, F.C., Foltz, C.B., Weymann, R.J. 2001, ApJ, 550, 142

Kartje J.F., Königl, A., & Elitzur, M. 1999, ApJ, 513, 180

Königl, A., & Kartje, J.F. 1994, ApJ, 434, 446 (KK94)

Mathur, S., Wilkes, B., & Elvis, M. 1998, ApJ, 503, L23

Morales, R., Fabian, A.C., Reynolds, C.S. 2000, MNRAS, 315, 149

Murray, N., Chiang, J., Grossman, S. A., Voit, G. M. 1995, ApJ, 451, 498

Proga, D. 2000, ApJ, 538, 684

Shlosman, I. & Begelman, M.C. 1987, Nature, 329, 29

Weymann, R.J., Morris, S.L., Foltz, C.B., & Hewett, P.C. 1991, ApJ, 373, 23

Mass Outflow in Active Galactic Nuclei: New Perspectives
ASP Conference Series, Vol. 255, 2002
D.M. Crenshaw, S.B. Kraemer, and I.M. George

Low-Ionization BAL QSOs in Ultraluminous Infrared Systems

Gabriela Canalizo
Institute of Geophysics and Planetary Sciences, Lawrence Livermore National Laboratory, 7000 East Avenue, L413, Livermore, CA 94550

Alan Stockton
Institute for Astronomy, University of Hawaii, 2680 Woodlawn Drive, Honolulu, HI 96822

Abstract. Low-ionization broad absorption line (BAL) QSOs present properties that cannot generally be explained by simple orientation effects. We have conducted a deep spectroscopic and imaging study of the host galaxies of the only four BAL QSOs that are currently known at $z < 0.4$, and found that all four objects reside in dusty, starburst or post-starburst, merging systems. The starburst ages derived from modeling the stellar populations are in every case a few hundred million years or younger. There is strong evidence that the ongoing mergers triggered both the starbursts and the nuclear activity, thus indicating that the QSOs have been recently triggered or rejuvenated. The low-ionization BAL phenomenon then appears to be directly related to young systems, and it may represent a short-lived stage in the early life of a large fraction of QSOs.

1. Introduction

Broad absorption line (BAL) QSOs comprise $\sim 12\%$ of the QSO population in current magnitude-limited samples. The standard view is that the BAL clouds have a small covering factor as seen from the QSO nucleus, implying that essentially all radio-quiet QSOs would be classified as BAL QSOs if observed from the proper angle.

An even rarer class, comprising only $\sim 1.5\%$ of all radio-quiet QSOs in optically-selected samples, are the low-ionization BAL (hereafter lo-BAL) QSOs, which pose serious problems to orientation-based models. Lo-BAL QSOs have considerably different broad emission line properties, are substantially redder than non-BAL QSOs, and are intrinsically X-ray quiet (Weymann et al. 1991; Sprayberry & Foltz 1992; Green 2001). In addition, the radio properties of lo-BAL QSOs may indicate that no preferred viewing orientation is necessary to observe BAL systems in the spectra of quasars since BALs are present in both flat and steep spectrum quasars, *i.e.*, objects that are presumably viewed along the jet axis and at high inclination respectively (Brotherton 2002; Becker et al. 2000; Gregg et al. 2000). Thus lo-BAL QSOs are thought to constitute a different class of QSOs, having more absorbing material and more dust (Voit et al. 1993; Hutsémekers et al. 1998).

2. A Sign of Youth?

An intriguing possibility is that the lo-BAL phenomenon may represent a stage in the early life of QSOs, either in the form of young QSOs "in the act of casting off their cocoons of gas and dust" (Voit et al. 1993; see also Egami et al. 1996, and Hazard et al. 1984), or as the result of outflows driven by supermassive starbursts (Lípari 1994; Shields 1996).

We have carried out deep Keck spectroscopic observations of the host galaxies of the four lo-BAL QSOs that are currently known at $z < 0.4$: PG 1700+518 (Canalizo & Stockton 1997), Mrk 231 and IRAS 07598+6508 (Canalizo & Stockton 2000), and IRAS 14026+4341 (Canalizo & Stockton 2001, in preparation). These objects have nuclear properties that are relatively rare in the classical QSO population (*e.g.*, Boroson & Meyers 1992; Turnshek et al. 1997): (1) strong Fe II emission, with the flux ratio Fe II $\lambda4570/\mathrm{H}\beta \geq 1$, and (2) very weak or no [O III] emission (see Table 1). In addition, we have found the following properties in the low redshift sample: (3) every lo-BAL QSO resides in an ultraluminous infrared galaxy (ULIG; *i.e.*, log $L_{ir}/L_{\odot} \geq 12$); (4) they have a small range in far infrared (FIR) colors, intermediate between those characteristic of ULIGs and QSOs; (5) the host galaxies show signs of strong tidal interaction, and they appear to be major mergers (Fig. 1); (6) spectra of their host galaxies (Fig. 2) show unambiguous interaction-induced star formation, with post-starburst ages $\lesssim$ 250 Myr.

Table 1. Properties of low-redshift low-ionization BAL QSOs

Object Name	Z	log $L_{ir}/L_{\odot}$	REW[a] [O III]	Fe II/ Hβ	Tail(s) Length[b]	Dyn. Age[c]	Sb. Age[c]
IRAS 07598+6508	0.1483	12.41	0	2.6	50	160	30
Mrk 231	0.0422	12.50	0	2.1	35	110	40
IRAS 14026+4341	0.3233	12.77	0	1.0	37	120	tbd
PG 1700+518	0.2923	12.58	2	1.4	13	40	85

[a] Rest equivalent width (REW) in Å
[b] Tail lengths in kpc, assuming $H_0 = 75$ km s^{-1} Mpc^{-1} and $q_0 = 0.5$
[c] Dynamical and starburst ages in Myr

The spatially resolved spectra of the hosts clearly show a concentration of material towards the central regions in timescales that are consistent with the dynamical age for the tidal interaction (Canalizo & Stockton 2000; Table 1). Thus, it is clear that there has been a recent flow of gas towards the central regions which fueled the centrally concentrated starbursts in each of these objects. Moreover, there is strong evidence that these QSOs have been recently fueled (either for the first time, or simply rejuvenated). In agreement with this scenario, Mathur et al. (2001) find that the X-ray flux of the lo-BAL QSO PHL 5200 is highly absorbed with a very steep power-law slope of $\alpha = 1.7 \pm 0.4$ (compared to the mean slope for non-BAL QSOs of $\alpha = 0.67 \pm 0.11$). Such a steep slope may be the result of a high accretion rate close to the Eddington limit, which may in turn be indicative of a recent fueling of the black hole (*i.e.*, a young QSO).

3. Proposed Model

Our results support those interpretations of the lo-BAL phenomenon which imply young systems. Here we propose a model that accounts for the observed properties in the four low-redshift lo-BAL QSOs. A major merger between galaxies of similar mass triggers intense bursts of star formation. As the gas concentrates in nuclear regions, the QSO activity is ignited. Along with the gas, dust is concentrated in the central 1 or 2 kpc, resulting in a dust-enshrouded QSO. The dust cocoon shields the narrow line region from the ionizing radiation coming from the central continuum source. The resulting low ionization parameter and the dusty environment increases the relative prominence of Fe II emission. A lo-BAL phase comes next, consisting of widespread outflows whereby the QSO expels the shroud of gas and dust (e.g. Voit, Weymann, & Korista 1993). As the ionizing photons are able to escape through holes on the cocoon poked by the outflows, nuclear and extended [O III] appear. A cocoon with holes may explain the different lightpaths in some BAL QSOs inferred from polarimetric studies, where some continuum is seen to escape without passing through dust (e.g. Hines & Wills 1995). Powerful QSOs, especially powerful radio sources, are able to break through the dust cocoon more rapidly, and this is why we do not see many strong radio-loud quasars in cocoon phase (Canalizo & Stockton 2001; Gregg et al. 2000).

4. Biases and Future Work

What keeps the evidence from having more weight is the fact that the sample of lo-BAL QSOs at $z < 0.4$ is incomplete and may be significantly biased. Only a fraction of low-redshift QSOs have been observed in the UV (where the BAL features are evident), and these observations have different biases. For example, Turnshek et al. (1997) conducted an HST FOS survey of low-z BALs in a sample of QSOs with weak [O III] (and strong Fe II). These two properties have an unusually high incidence in young IR-loud QSOs (Canalizo & Stockton 2001). The Turnshek et al. sample may then be biased towards young objects. We are therefore obtaining further spectroscopic and high resolution imaging observations of the host galaxies of BAL QSOs at somewhat higher redshifts (for which the BAL features appear in the optical).

The possibility that the lo-BAL phenomenon represents a short phase in the early life of QSOs is of great interest because it could potentially provide a method to answer one of the fundamental questions regarding QSOs, namely, how long does the QSO activity last? Determining the ages of starbursts that are related to the fueling of QSOs in a sample of objects could place limits on the duration of the BAL phase. This, along with an estimate of the fraction of QSOs go through a lo-BAL phase (as seen from our line of sight) would place upper limits on the mean lifetime of the QSO activity.

Acknowledgments. Part of this work was performed under the auspices of the U.S. Department of Energy, National Nuclear Security Administration by the University of California, Lawrence Livermore National Laboratory under contract No. W-7405-ENG-48, and was also partially supported by NSF under grant AST 95-29078. This study is based in part on observations with the NASA/ESA Hubble Space Telescope, obtained from the data archive at the Space Telescope Science Institute, which is operated by the Association of Universities for Research in Astronomy, Inc. under NASA contract No. NAS5-26555.

References

Becker, R. H., White, R. L., Gregg, M. D., Brotherton, M. S., Laurent-Meuleisen, S. A., Arav, N. 2000, ApJ, 538, 72

Bruzual A., G. & Charlot, S. 1996, unpublished (ftp://gemini.tuc.noao.edu/pub/charlot/bc96)

Brotherton, M. S. 2002, in Mass Outflow in Active Galactic Nuclei: New Perspectives, eds. D.M. Crenshaw, S.B. Kraemer, & I.M. George (San Francisco: ASP), p. 157

Canalizo, G., & Stockton, A. 1997, ApJ, 480, L5

Canalizo, G., & Stockton, A. 2000, AJ, 120, 1750

Canalizo, G., & Stockton, A. 2001, ApJ, 555, in press

Egami, E., Iwamuro, F., Maihara, T., Oya, S., Cowie, L. L. 1996, AJ, 112, 73

Green, P. J. 2001, in Mass Outflow in Active Galactic Nuclei: New Perspectives, eds. D.M. Crenshaw, S.B. Kraemer, & I.M. George (San Francisco: ASP), p. 19

Gregg, M. D., Becker, R. H., Brotherton, M. S., Laurent-Meuleisen, S. A., Lacy, M., White, R. L. 2000, ApJ, 544, 142

Hazard, C., Morton, D. C., Terlevich, R., & McMahon, R. 1984, ApJ, 282, 33

Hines, D. C., & Wills, B. J. 1995, ApJ, 448, L69

Hutsemékers, D., Lamy, H., & Remy, M. 1998, A&A, 340, 371

Lípari, S. 1994 ApJ, 436, 102

Mathur, S., Matt, G., Green, P. J., Elvis, M., & Singh, K. P. 2001, ApJ, 551, L13

Shields, G. A. 1996, ApJ, 461, L9

Sprayberry, D. & Foltz, C. B. 1992, ApJ, 390, 39

Voit, G. M., Weymann, R. J., Korista, K. T. 1993, ApJ, 413, 95

Weymann, R. J., Morris, S.L., Foltz, C.B., & Hewett, P.C. 1991, ApJ, 373, 23

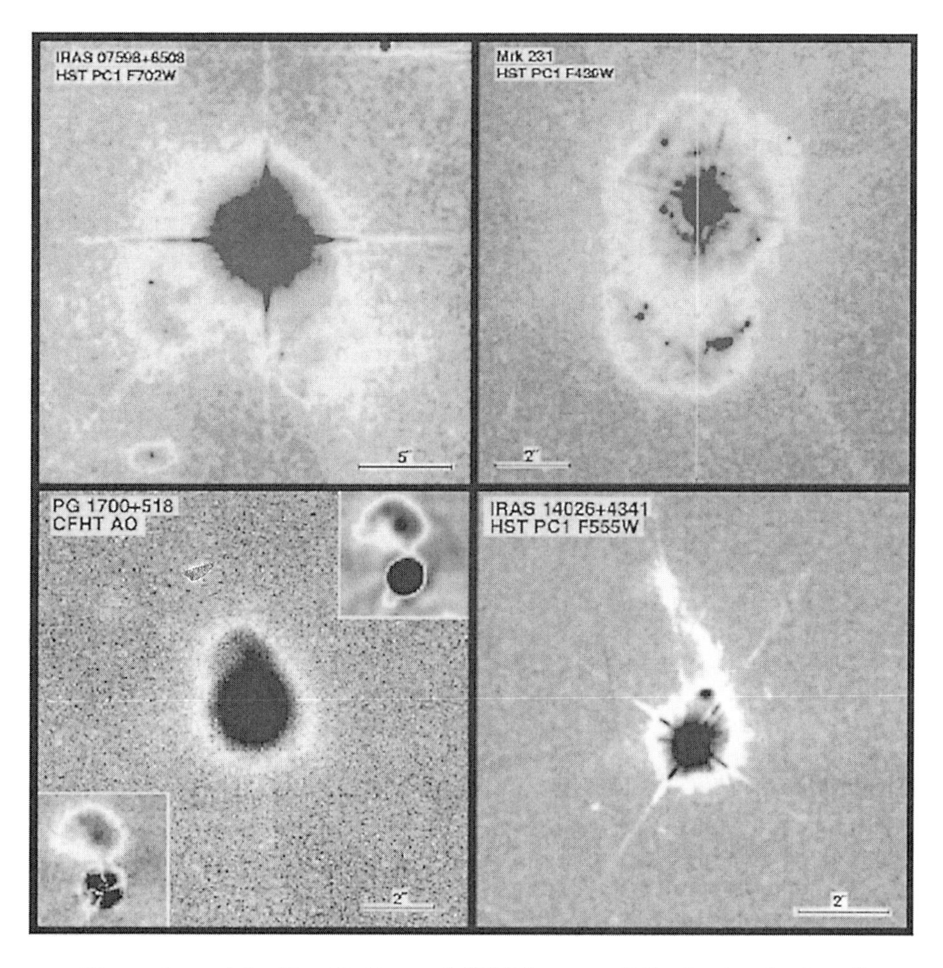

Figure 1. Adaptive Optics and HST images of the host galaxies of low-redshift lo-BAL QSOs. Every host galaxy shows signs of recent strong tidal interaction.

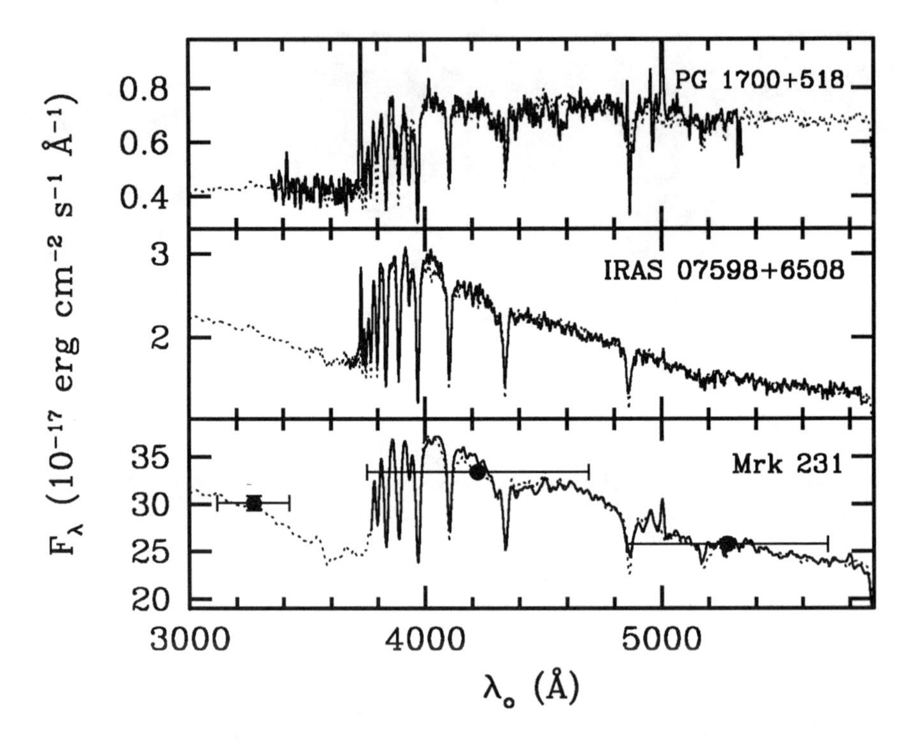

Figure 2. Stellar populations in the host galaxies of low-z lo-BAL QSOs. Panels display Keck LRIS spectra of the host galaxies of 3 of the lo-BAL QSOs in rest frame (solid trace). Superposed are the best fitting models (dotted traces), consisting of the sum of an underlying old stellar population model and a young instantaneous starburst model (Bruzual & Charlot 1996). For details, see Canalizo & Stockton 1997. We have recently obtained Keck ESI spectroscopy of the fourth object, IRAS 14026+4341, which shows a large H II region $\sim 2''.5$ North of the nucleus, as well as several post-starburst regions with populations similar to those of the other three objects (Canalizo & Stockton 2001, in preparation).

Mass Outflow in Active Galactic Nuclei: New Perspectives
ASP Conference Series, Vol. 255, 2002
D.M. Crenshaw, S.B. Kraemer, and I.M. George

HST/STIS Spectra of PG 0946+301: Spanning 1000 Å in the UV Rest Frame of a BALQSO

Kirk T. Korista

Dept. of Physics, Western Michigan University, Kalamazoo, MI 49008-5252

Nahum Arav

Astronomy Dept., UC Berkeley, Berkeley, CA 94720

Martijn de Kool

Research School of Astronomy & Astrophysics, ANU ACT, Australia

D. Michael Crenshaw

Department of Physics and Astronomy, Georgia State University, Atlanta, GA 30303

Abstract. We describe deep *HST*/STIS observations (40 orbits) of the BALQSO PG 0946+301. These observations were the major part of a multi-wavelength campaign on this object aimed at determining the ionization equilibrium and abundances (IEA) in a broad absorption line (BAL) QSO. We present simple template fits to the entire data set, which yield firm identifications of more than two dozen BALs from 18 ions and give lower limits for their column densities. Upper limits to the column densties of 7 other ions (due to their absence in less-confused spectral regions) are also established.

1. Introduction

As described in Arav (1997) and elsewhere in these proceedings (Arav et al. 2002), it has become apparent that the troughs of BALQSOs are almost always saturated with the absorbing gas only partially covering the effective continuum source. This means that the column densities derived from simple optical depth template fits to the spectrum are only lower limits (see Korista et al. 1992). To disentangle the effects of saturation and partial coverage so that reliable limits may be derived for the ionic column densities, we must use doublet ratios or inter-multiplet ratios. Since most doublets have separations significantly smaller than the BAL velocity spread, we must generally find ions with several sets of transitions out of the same ground state with a significant range in their oscillator strengths. Thus wide spectral coverage is necessary not only to pick up as many ions as possible, but also to use measured inter-multiplet optical depth ratios to constrain upper limits to the ionic column densities. After studying the available

BALQSO data in the *HST* archive and hundreds of ground based spectra, we concluded that PG 0946+301 was the best candidate for such analysis (Arav 1997; Arav et al. 1999, hereafter FOS99). It has a substantially higher flux between 1200 Å – 2500 Å (observed frame) than any other BALQSO that is suitable for IEA studies. Thus, we could obtain very high-quality data for a wide rest frame spectral region (500 Å – 1700 Å), and the object suffers only minor contamination by Lyα forest lines due to its low redshift ($z = 1.223$).

It was during the 1997 Pasadena *Mass Ejection from AGN* meeting that many QSO researchers joined together to propose the study of PG 0946+301 with deep, high-quality multi-wavelength observations. The components of the campaign were: *HST* UV spectroscopy, *ASCA* X-ray observations (Mathur et al. 2000), *FUSE* UV spectroscopy, high-resolution optical spectroscopy and optical spectropolarimetry. Here we describe the Space Telescope Imaging Spectrograph (STIS) observations and present simple template fits for the BALs with their resulting lower limits to the ionic column densities.

2. Data Acquisition and Reduction

We observed PG 0946+301 with the Space Telescope Imaging Spectrograph (STIS) on the *Hubble Space Telescope* (*HST*) over a three-month period in early 2000. All of the spectra were obtained through a 52″ × 0.2″ slit to maximize throughput without a significant loss in spectral resolution. We used the MAMA detectors and G140L and G230L gratings to obtain full UV coverage at a resolution of 1.2 Å and 3.2 Å, respectively. We also used the CCD detectors and the G430M grating to obtain limited coverage in adjacent optical regions at a resolution of 0.56 Å. The details of the observations are given in Table 1.

Table 1. *HST*/STIS Observations of PG0946+031

Grating	Range (Å)	Exposure (s)	Date (UT)
G140L	1140 – 1715	12,960	2000 February 26
G140L	1140 – 1715	12,960	2000 March 28
G140L	1140 – 1715	12,960	2000 March 30
G140L	1140 – 1715	12,960	2000 April 4
G140L	1140 – 1715	12,960	2000 May 11
G430M	3022 – 3306	5,274	2000 April 12
G430M	3163 – 3447	7,250	2000 April 12
G230L	1640 – 3148	12,960	2000 April 23
G230L	1640 – 3148	12,960	2000 May 2

Each G140L or G230L entry in Table 1 represents a five-orbit visit, since STIS MAMA observations are limited to five consecutive orbits due to constraints imposed by the South Atlantic Anomaly. To maximize the signal-to-noise of the spectra, we obtained exposures at five different locations separated by 0.3″ along the slit (one location per orbit). We used this strategy to place the spectra at slightly different positions on the detector, to avoid possible fixed pattern noise not removed by the flat-fielding of the spectra (Kaiser et al. 1998). For the G430M observations, we obtained two spectra per orbit (at the standard

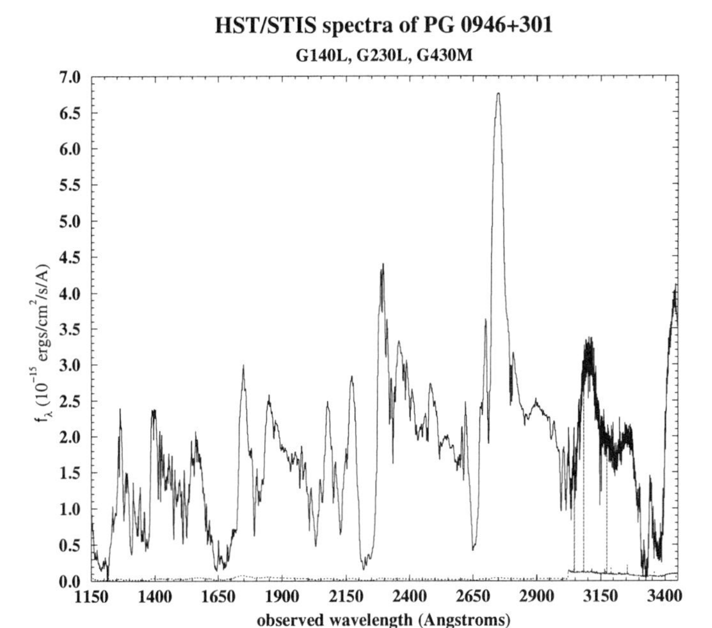

Figure 1. Forty orbits of *HST*/STIS spectra of the BALQSO PG 0946+301. The statistical noise spectrum is plotted at the bottom.

slit locations) to accumulate enough images to allow for accurate cosmic-ray rejection.

We reduced the spectra using the IDL software developed at NASA's Goddard Space Flight Center for the STIS Instrument Definition Team (Lindler 1999). The individual spectra from each orbit were independently calibrated in wavelength and flux, and resampled to a linear wavelength scale (retaining the same average dispersion). Careful examination of the spectra from each orbit reveals no discernible change in flux over time or as a function of position on the detectors. We therefore averaged the spectra (weighted by exposure time) to obtain final versions of the G140L, G230L, and two G430M spectra. These were then combined into a single spectrum using Galactic absorption lines for a final wavelength calibration. The resultant spectrum and its noise are shown in Figure 1, spanning 1000 Å in the rest frame. The STIS spectra in ascii format are available on-line at: http://unix.cc.wmich.edu/korista/ftp/pg0946/spectra/.

3. Spectral Fitting Procedures & Results

In order to make measurements or fit the observed BALs we must use a model for the unabsorbed emission (continuum plus BELs, hereafter "the effective continuum") of the quasar. At wavelengths longer than $\sim$ 1000 Å, the spectrum of PG 0946+301 is very close to the non-BALQSO composite of Weymann et al. (1991), which we therefore used as our long wavelength effective continuum

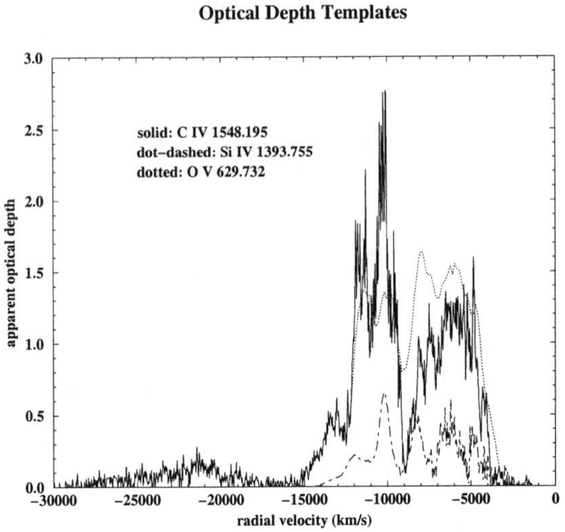

Figure 2. The optical depth templates used in fitting the spectrum.

with only small modifications to some emission line strengths. Shortward of 830 Å we found only two narrow regions (624 Å – 631 Å and 567 Å – 571 Å) free from BAL features. This complicated the task of determining the effective continuum based on the data at hand, and very little is known observationally about intrinsic QSO spectra in the range 500 Å – 900 Å.

With these uncertainties in mind, we used a similar approach to the one we employed in the analysis of the *HST*/FOS data of PG 0946+301 (FOS99). Three effective continua were employed shortward of 1045 Å. The first was a fixed effective continuum, very similar to the one in FOS99. Emission features of roughly the same shape and wavelengths are also seen in a smoothed version of the quasar composite spectrum produced by Zheng et al. (1997). The second and third effective continua were allowed to vary during the fitting process for wavelengths shorter than 1045 Å (rest frame). See Arav et al. (2001) for details.

The column densities resulting from the optical depth template (see Figure 2) fits are summarized in Table 2, and the fit itself for the fixed effective continuum is illustrated in Figure 3. We re-emphasize that column densities obtained from template fitting are in reality only lower limits because of partial covering effects. Further constraints on the ionic column densities and on the IEA are discussed in Arav et al. (2001) and in Arav et al. (2002).

References

Arav, N., 1997 in Mass Ejection from AGN, eds. N. Arav, I. Shlosman, & R. J. Weymann (San Francisco: ASP), p. 208

Table 2. *Apparent* Ionic Column Densities

Ion	$\log N_{ion}$ (cm^{-2})	Ion	$\log N_{ion}$ (cm^{-2})
H I (Lyα)	15.4	Mg X	small
H I (Ly limit)	16.3	Si II	< 14.2
He I	< 15.5	Si III	< 14.2
C II	< 14.7	Si IV	15.1
C III	15.2	P IV	< 15.0
C IV	16.1	P V	15.0
N III	15.6	S III	14.5::
N IV	16.0:	S IV	15.1
N V	16.2	S V	15.4
O III	16.2	S VI	15.8
O IV	16.8	Ar III	small
O V	16.2	Ar IV	small
O VI	16.6	Ar V	small
Ne IV	16.6:	Ar VI	small
Ne V	16.7	Ar VII	< 15.1
Ne VI	16.8::	Ar VIII	small
Ne VIII	16.7:	Fe III	< 15.3
Na IX	small	Fe IV	15.5::

(:) indicates spectral features from this species definitely present, but column density uncertain. (::) indicates spectral features from this species only marginally detected. (<) indicates an upper limit. (small) indicates undetected with no firm limit due to confusion with other species.

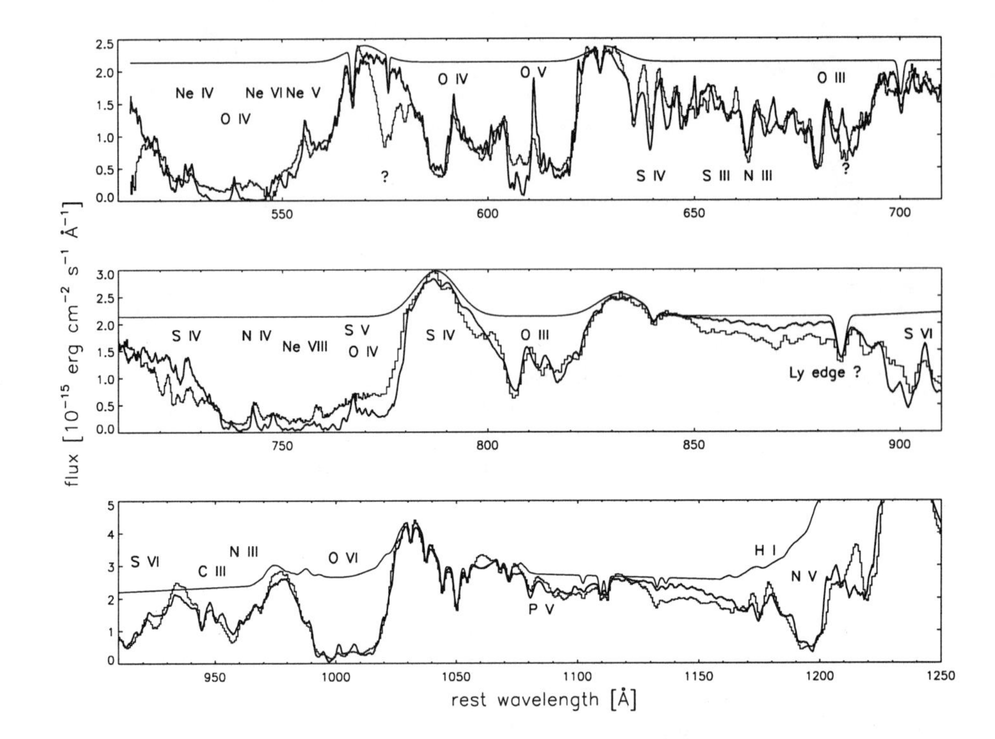

Figure 3. A representative fit to the spectrum, using the fixed effective continuum. The most important BALs are labeled. Intervening and Galactic absorption systems appear in the effective continuum spectrum for identification purposes.

Arav, N., et al. 2002, in Mass Outflow in Active Galactic Nuclei: New Perspectives, eds. D.M. Crenshaw, S.B. Kraemer, & I.M. George (San Francisco: ASP), p. 179

Arav, N., et al. 2001, ApJ, in press

Arav, N., Korista, T. K., de Kool, M., Junkkarinen, V. T. & Begelman, M. C. 1999, ApJ, 516, 27 (FOS99)

Kaiser, M. E., et al. 1998, PASP, 110, 978

Korista, K. T., Weymann, R. J., Morris, S. L., Kopko, M., Turnshek, D. A., Hartig, G. F., Foltz, C. B., Burbidge, E. M., & Junkkarinen, V. T. 1992, ApJ, 401, 529

Lindler, D. 1998, CALSTIS Reference Guide (CALSTIS Version 5.1)

Mathur, S., et al. 2000, ApJ, 533, L79

Weymann, R. J., Morris, S. L., Foltz, C. B., & Hewett, P. C. 1991, ApJ, 373, 23

Zheng, W., Kriss, G. A., Telfer, R. C., Grimes, J. P., Davidsen, A. F. 1997, ApJ, 475, 469

Mass Outflow in Active Galactic Nuclei: New Perspectives
ASP Conference Series, Vol. 255, 2002
D.M. Crenshaw, S.B. Kraemer, and I.M. George

The Polarization Properties of Broad Absorption Line QSOs: Observational Results

D. Hutsemékers[1]

European Southern Observatory, Chile

H. Lamy

University of Liège, Belgium

Abstract. Correlations between BAL QSO intrinsic properties and polarization have been searched for. Some results are summarized here, providing possible constraints on BAL outflow models.

1. Introduction

From 1994 to 1999 we have obtained broad-band linear polarization measurements for a sample of approximately 50 Broad Absorption Line (BAL) QSOs using the ESO 3.6m telescope at La Silla (Chile).
On the basis of this sample plus additional data compiled from the literature, possible correlations between BAL QSO intrinsic properties and polarization have been searched for. Here we present some of our most interesting results, updated with recent data.

2. Analysis and results

A careful distinction between BAL QSO subtypes has been made. In addition to the BAL QSOs with high-ionization (HI) absorption features only, we have distinguished BAL QSOs with strong (S), weak (W), and marginal (M) low-ionization (LI) absorption troughs (Hutsemékers et al. 1998, 2000 for details).
Several indices are used to quantify the spectral characteristics: the balnicity index (BI) which is a modified velocity equivalent width of the C IV BAL, the detachment index (DI) which measures the degree of detachment of the absorption trough relative to the emission line, the maximum velocity $v_{\rm max}$ in the C IV BAL, and the power-law index α of the continuum.
Although most BAL QSOs are radio-quiet, some of them appear radio-moderate, and radio-to-optical flux ratios $R^\star$ were also collected.
Correlations and sample differences were searched for by means of the usual statistical tests. Survival analysis was used for censored data (mainly $R^\star$). While the study of polarization was our main goal, correlations between different in-

[1]Also Research Associate FNRS, University of Liège, Belgium

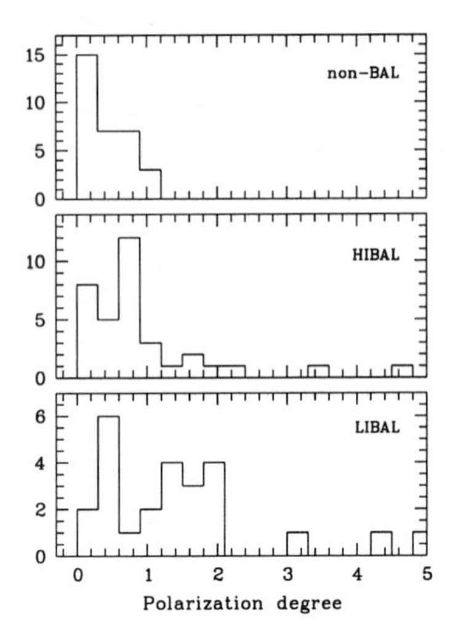

Figure 1. The distribution of the polarization degree p_0 (in %) for the three main classes of QSOs. LIBAL QSOs contain the three sub-categories, i.e. strong, weak and marginal LIBAL QSOs. Data are from H1998, S1999, L2000, O2000 (a LIBAL QSO with p_0=7.5 and the unclassified BAL QSOs are not represented here)

dices have also been considered.
Results presented by Hutsemékers et al. (1998, 2000; H1998, H2000) are updated with polarimetric data from Schmidt & Hines (1999; S1999), Lamy & Hutsemékers (2000; L2000), and Ogle et al. (2000; O2000). Only polarimetric measurements with $\sigma_p \leq 0.4\%$ are taken into account, such that the debiased polarization degree p_0 has a typical uncertainty $\sigma_p = 0.2$-0.3%. The radio-loud BAL QSOs recently discovered in the FIRST survey (Becker et al. 2000; B2000) are included in the present study.

• Evidence for polarization differences between low- and high-ionization BAL QSOs

The distribution of the polarization degree p_0 for the three main classes of QSOs is illustrated in Fig. 1. We can see that the bulk of QSOs with $p_0 > 1.2\%$ belong to the sub-class of LIBAL QSOs. Note that not all LIBAL QSOs are highly polarized. As a class, HIBAL QSOs appear less polarized than LIBAL QSOs and more polarized that non-BAL QSOs. They seem to have intermediate properties. All these differences are statistically significant ($P_{\rm K-S} \geq 99\%$).

• The correlation between the balnicity and the slope of the continuum

In addition to their higher polarization, it is seen from Fig. 2 that most LIBAL QSOs have also larger balnicities and more reddened continua than HIBAL QSOs. Considering the whole BAL QSO sample (i.e. HI+LI BALs), a significant ($P_\tau \geq 99\%$) correlation is found between the balnicity index BI and the slope of the continuum. Since LIBAL QSOs as a class are more reddened and

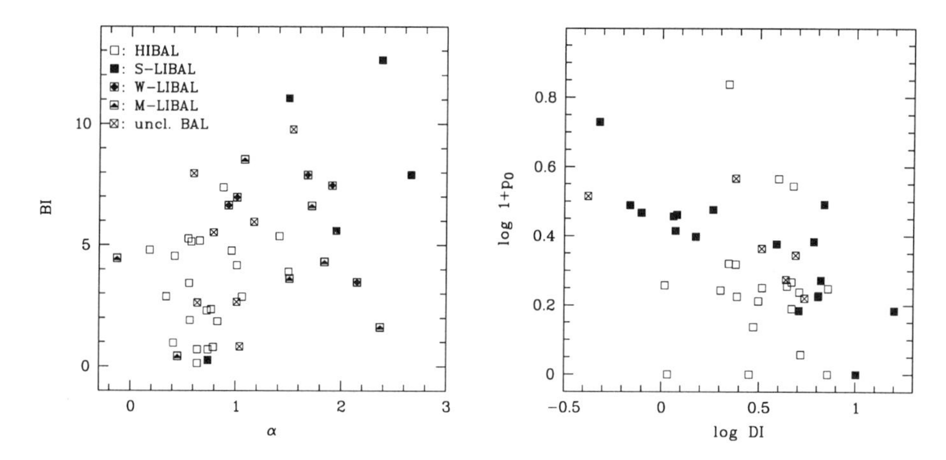

Figure 2. **Left**: the correlation between the balnicity index BI (in 10^3 km s^{-1}) and the power-law index α ($F_\nu \propto \nu^{-\alpha}$). The 3 sub-categories of LIBAL QSOs are distinguished here. Data and objects are from H1998, L2000, H2000. **Right**: the correlation between the polarization degree p_0 (in %) and the line profile detachment index DI. The correlation is especially apparent for the LIBAL QSOs (filled squares). Data from H1998, S1999, L2000

more polarized than HIBAL QSOs, it also results a correlation between the power-law index and the polarization, although less convincing.

• The correlation between the polarization of the continuum and the line profile detachment index

Among several possible correlations of the polarization with spectral indices like the balnicity index, the equivalent width and the velocity width of C IV and C III], the only significant ($P_\tau \geq 99\%$) correlation we found is a correlation with the line profile detachment index, quite unexpectedly. Fig. 2 illustrates the correlation between the polarization degree p_0 and the line profile detachment index DI for all BAL QSOs of our sample. The correlation is especially apparent and significant for the LIBAL QSOs. It indicates that the BAL QSOs with P Cygni-type line profiles (DI$\ll$) are the most polarized.

• The absence of correlation between the polarization and $R^\star$

If the higher polarization of BAL QSOs as a class is due to an attenuation of the direct continuum with respect to the scattered one –at least in some objects– (Goodrich 1997), then we expect the polarization to be correlated with the radio-to-optical flux ratio. In Fig. 3, the BAL QSO polarization p_0 is plotted against the radio-to-optical flux ratio $R^\star$. No correlation is seen, as confirmed by the statistical tests. Note that the distribution of $R^\star$ is not found to differ between the HIBAL and LIBAL subsamples

• The absence of correlation between the terminal velocity and $R^\star$

In order to investigate the claimed anticorrelation between the terminal velocity of the flow and the radio-to-optical flux ratio (Weymann 1997), we have plotted

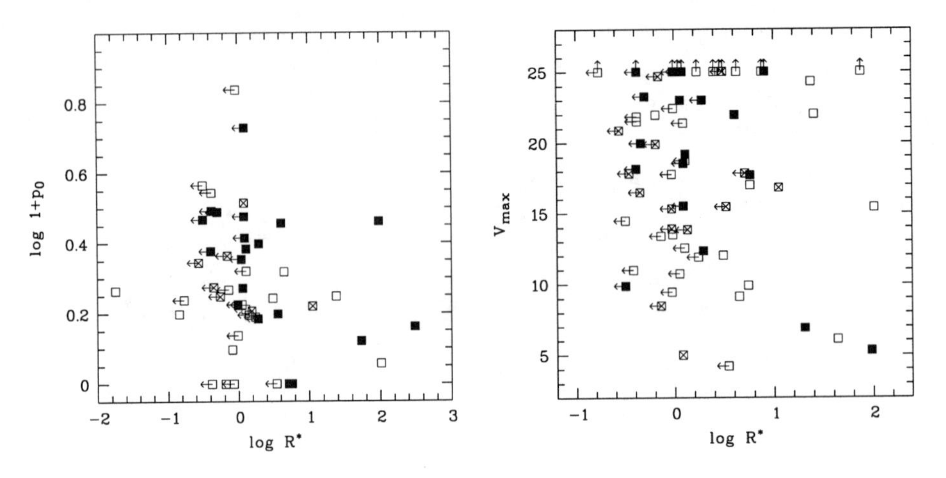

Figure 3. **Left**: The polarization degree p_0 plotted against the radio-to-optical flux ratio $R^\star$. Data are from H2000. **Right**: the maximum velocity $v_{\max}$ in the C IV BAL (in 10^3 km s^{-1}) is plotted against the radio-to-optical flux ratio $R^\star$. Data from H2000, B2000. In both figures, open squares represent HIBAL QSOs, filled squares LIBAL QSOs, and squares with a cross unclassified BAL QSOs, while arrows indicate censored data points

in Fig. 3 the maximum velocity $v_{\max}$ in the C IV BAL against the radio-to-optical flux ratio $R^\star$. No correlation is found, as confirmed by the statistical tests.

3. Conclusions

Our results show that polarization is correlated with BAL QSO line profiles and types, emphasizing the extreme behavior of LIBAL QSOs already reported by several studies. These results could provide constraints on the BAL outflow models and geometry. As discussed by Hutsemékers et al. (1998), they are consistent with the Murray et al. (1995) disk-wind model.

References

Becker, R. et al. 2000, ApJ 538, 72 [B2000]

Goodrich, R. 1997, ApJ 474, 606

Hutsemékers, D., Lamy, H., Remy, M. 1998, A&A 340, 371 [H1998]

Hutsemékers, D., Lamy, H. 2000, A&A 358, 835 [H2000]

Lamy, H., Hutsemékers, D. 2000, A&AS 142, 451 [L2000]

Murray, N. et al. 1995, ApJ 451, 498

Ogle, P. et al. 2000, ApJS 125, 1 [O2000]

Schmidt, G., Hines, D. 1999, ApJ 512, 125 [S1999]

Weymann, R. 1997, ASP Conf 128, 3

Mass Outflow in Active Galactic Nuclei: New Perspectives
ASP Conference Series, Vol. 255, 2002
D.M. Crenshaw, S.B. Kraemer, and I.M. George

Q0059-2735: A Hybrid Starburst / Broad Absorption Line QSO? Clues from Spectropolarimetry

D. Hutsemékers[1]

European Southern Observatory, Chile

H. Lamy

University of Liège, Belgium

Abstract. Spectropolarimetric data support the hybrid BALQSO / starburst model proposed for the iron Lo-BAL QSO Q0059-2735.

1. Q0059-2735 and the Hybrid BALQSO / Starburst Model

Q0059-2735 (z =1.59) is one of the extreme and very rare low-ionization broad absorption line (Lo-BAL) QSOs which exhibit narrow absorption lines (NALs) from metastable levels of Fe II. Q0059-2735 is often considered as the prototype of the "iron Lo-BAL" QSO class.
A hybrid BAL QSO / Starburst model has been proposed for Q0059-2735 by Cowie et al. (1994) and Egami et al. (1996). These authors suggest a model in which the central (BAL) QSO is surrounded by a dusty shell of young stars contributing to the excess of UV continuum, and at the origin of the Fe II NALs. The mixture of starburst and reddened quasar properties appears in varying degree as a function of wavelength, and could be the signature of a younger stage in the quasar evolutionary sequence. A couple of BAL QSOs with characteristics supporting this interpretation were recently found by Becker et al. (1997).
Since Q0059-2735 is polarized like many other low-ionization BAL QSOs, we have investigated its UV rest-frame polarized spectrum.

2. Clues from Spectropolarimetry

Spectropolarimetric data have been obtained for Q0059-2735 with the ESO 3.6m telescope at La Silla, and are presented in Lamy & Hutsemékers (2000).
In agreement with the standard interpretation of BAL QSO spectropolarimetry (Cohen et al. 1995, Schmidt & Hines 1999), the rise of polarization in the BAL troughs (cf. Fig. 1 and Table 2 in Lamy & Hutsemékers 2000) suggests that the scattered continuum is less absorbed in the BAL region (BALR) than the direct unpolarized continuum. The fact that some BALs (e.g. C IV) are also seen in the polarized flux indicates that the scattered flux crosses the BALR in regions of

[1] Also Research Associate FNRS, University of Liège, Belgium

lower opacity. However, the absence of Al III in the polarized flux of Q0059-2735 suggests that the scattered flux misses the region of the BALR where Al III is formed. This behavior is not unique among BAL QSOs: in the polarized flux of Q1246-0542 there is apparently no trace of any BAL, including C IV. In the case of Q0059-2735, this suggests that the low-ionization Al III BALR is less extended than the high-ionization one, and does not cover the scattering region. On the other hand, the Fe II absorption blends are detected in the polarized flux, while they are not significantly more polarized than the continuum. Thus, the iron absorbing gas must intercept both the polarized and unpolarized continua with roughly the same opacity, suggesting a different location and/or geometry of the Fe II absorbing region. Compared to the other low-ionization BALs, the behavior of the Mg II BAL is quite striking since it appears more polarized than the continuum (although not as much as Al III), while it is seen in the polarized flux, like Fe II. These intermediate properties may indicate a hybrid origin of the Mg II BAL.

These results are consistent with the interpretation that the spectrum of Q0059-2735 is a superposition of a BAL QSO spectrum and a starburst one, the starburst being at the origin of the Fe II NALs (Cowie et al. 1994; Egami et al. 1996). In this model, Q0059-2735 is seen along a line of sight close to the dusty equatorial plane. In the framework of the disk-wind model of the BALR, such an orientation could explain the presence of low ionization troughs, the very deep and steep C IV absorption trough, and the high degree of polarization in the continuum (cf. Murray et al. 1995, Hutsemékers et al. 1998). Free electrons and/or dust scatter the continuum photons along lines of sight that cross the BALR where the opacity is still large for C IV, and much smaller for Al III and Mg II, the latter ions being located much closer to the disk as suggested by Murray et al. (1995). Within this model, the Fe II NALs are produced beyond the BALR, in material swept up by the strong winds of supernovae in the starburst (Hazard et al. 1987, Norman et al. 1994), such that the scattered and direct continua are similarly absorbed. The Mg II BAL could be hybrid, partly formed in the QSO disk-wind and partly in the starburst. The fact that the polarization in the Mg II BAL is not the highest at exactly the wavelength at which the BAL profile is the deepest could support this hypothesis, although data with higher spectral resolution are needed.

References

Becker, R.H. et al. 1997, ApJ479, L93

Cohen, M.H. et al. 1995, ApJ448, L77

Cowie, L.L. et al. 1994, ApJ432, L83

Egami, E. et al. 1996, AJ112, 73

Hazard, C. et al. 1987, ApJ323, 263

Hutsemékers, D. et al. 1998, A&A340, 371

Lamy, H., Hutsemékers, D. 2000, A&A356, L9

Murray, N. et al. 1995, ApJ451, 498

Norman, C.A. et al. 1996, ApJ472, 73

Schmidt, G.D., Hines, D.C., 1999, ApJ512, 125

Mass Outflow in Active Galactic Nuclei: New Perspectives
ASP Conference Series, Vol. 255, 2002
D.M. Crenshaw, S.B. Kraemer, and I.M. George

The Importance of Shocks in the Ionization of the Narrow Line Region of Seyferts

H. R. Schmitt

National Radio Astronomy Observatory, P.O. Box 0, Socorro, NM87801

A. L. Kinney

NASA Headquarters, 300 E St., Washington, DC20546

J. B. Hutchings

Dominion Astrophysical Observatory, Victoria, BC V8X 4M6, Canada

J. S. Ulvestad

National Radio Astronomy Observatory, P.O. Box 0, Socorro, NM87801

R. R. J. Antonucci

University of California, Santa Barbara, Santa Barbara, CA93106

Abstract. We discuss the viability of shocks as the principal source of ionization for the Narrow Line Region of Seyfert galaxies. We present the preliminary results of [OIII] λ5007Å and radio 3.6 cm imaging surveys of Seyferts, discuss the effects of shocks in the ionization of two galaxies, and also calculate an upper limit to the Hβ luminosity that can be due to shocks in 36 galaxies with unresolved radio emission. We show that, for favored values of the shock parameters, that shocks cannot contribute more than ~15% of the ionizing photons in most galaxies.

1. Introduction

The close connection between the narrow line and the radio emission in Seyfert galaxies is a well known fact. The first papers in this subject found a correlation between the [OIII] and radio luminosities, and between the radio luminosity and the Full Width Half Maximum (FWHM) of the [OIII] lines (DeBruyn & Wilson 1978, Heckman et al. 1981, Whittle 1985). Emission line and radio imaging of Seyfert galaxies shows that the radio and line emission of these galaxies are aligned (Haniff et al. 1988). Higher resolution imaging shows, for some of these galaxies, that the radio emission is surrounded by line emission (Pogge & De Robertis 1995, Capetti et al. 1996, Falcke et al. 1998), which suggests that the Narrow Line Region gas (NLR) is ionized by shocks due to the interaction of the radio plasma with the gas (see Figure 1 for an example). Another evidence

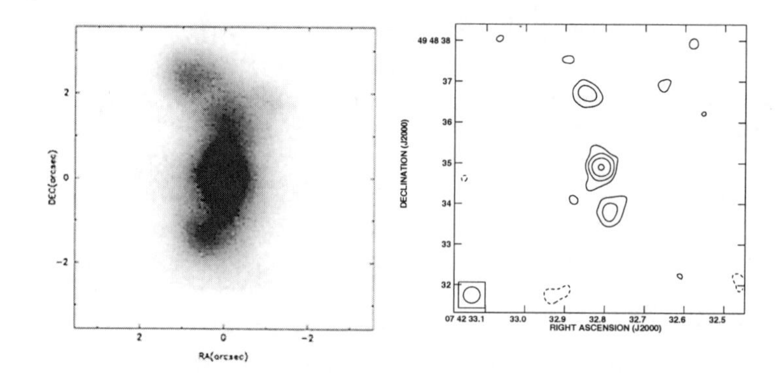

Figure 1. Comparison of the CFHT [OIII] λ5007Å image of Mrk 79 (left) with the VLA A-conf. 3.6 cm image (right), on the same scale.

of shocks in the NLR of Seyferts is the detection of anomalous velocity fields, like double components separated by hundreds of km/s (Whittle et al. 1988, Winge et al. 1997, Hutchings et al. 1998), and "extra" line widths associated with radio jets (Whittle 1992).

These results lead Dopita & Sutherland (1995, 1996) to propose a series of models where the NLR is ionized by fast shocks, with velocities in the range $150 \leq V_{shock} \leq 500$ km/s. According to their models, high energy photons are created behind the shock and ionize the preshock gas upstream (precursor). This model has several predictions, which were analyzed by Morse et al. (1996) and Wilson (1997), and are summarized below. The models predict we should find line emission surrounding radio lobes, and a photon deficit in the NLR. However, this can also be explained by the Unified Model (see Young et al. 2001 for a discussion on X-ray results). Shocks can explain the [OIII] temperature problem in these galaxies, where the measured gas temperature is much higher than the one predicted by simple photoionization models, but this can also be explained combining matter bounded and ionization bounded clouds (Binette et al. 1996).

Here we present the preliminary results of a radio and [OIII] λ5007Å imaging survey of a sample of Seyfert galaxies. We discuss the case of two individual galaxies and calculate a limit to the Hβ luminosity that can be due to shocks in those galaxies without extended radio emission.

2. Radio and [OIII] survey of Seyfert galaxies

Our radio continuum survey was done using the VLA in A-configuration at 3.6 cm, which gives a resolution of 0.25″. The results of this survey were published by Kinney et al. (2000) and Schmitt et al. (2001). The [OIII] survey is underway. It started with the CFHT and is now being done with *Hubble Space Telescope (HST)*.

One of the results from the radio continuum survey is that only 50% of the Seyferts show extended emission. This percentage decreases to 25% if we count only those galaxies where the radio emission is composed of multiple components. Comparing the radio and [OIII] images we find some interesting cases, like NGC 5347 (Figure 2). This Figure shows that the [OIII] emission is resolved, but the radio emission is unresolved, contrary to what would be expected if shocks were important in the ionization of the NLR of this galaxy. On the other hand we also have cases like Mrk 79 (Figure 1), where both the [OIII] and radio emission are extended and cospatial, suggesting that shocks may be an important source of ionizing photon. We discuss these galaxies below.

We will use in the following subsections the equations given by Dopita & Sutherland (1996) to predict the Hβ luminosity produced by a shock and the precursor, which are, respectively,

$$L(H\beta) = 7.44 \times 10^{-6} A (V_s/100 \text{ km s}^{-1})^{2.41} (n/\text{ cm}^{-3}) \text{ erg s}^{-1}$$

$$L(H\beta) = 9.85 \times 10^{-6} A (V_s/100 \text{ km s}^{-1})^{2.28} (n/\text{ cm}^{-3}) \text{ erg s}^{-1}$$

where A is the shock area, V_s is the shock velocity and n is the preshock density.

2.1. NGC 5347

This is a Seyfert 2 galaxy with a radial velocity of 2335 km s^{-1}, so 1″ corresponds to 150 parsecs at the galaxy (we assume H_0 = 75 km s^{-1} Mpc^{-1}). As we can see in Figure 2, the [OIII] image is extended, with a cone shaped structure in the circumnuclear region. At ≈ 3″ NE from the nucleus we can see [OIII] emission detached from the nuclear emission (NE structure), with a total linear extent of 2″. The borders of this structure are well aligned with the borders of the cone structure at the nucleus, suggesting that the NLR is ionized by the nuclear source, and the collimation of the radiation is due to a circumnuclear torus. Figure 2 also shows the archival VLA 6 cm image of this galaxy, which is unresolved. The radio emission is unresolved at 3.6 cm, 6 cm and 20 cm, indicating that we did not resolve out diffuse emission around the nucleus. In fact, the comparison between the 20 cm flux from the nuclear source (3.4 mJy) with that from the entire galaxy (5.6 mJy), obtained from the NVSS, shows that most of the radio emission in this galaxy originates in the nuclear region.

Here we calculate how much of the Hβ emission from the NE structure can be due to shocks. If we assume that the Hβ flux is produced by a shock seen edge-on, with a circular cross section of radius 1″, we get that the shock area is 0.071 kpc^2 (6.75×10^{41} cm^2). González Delgado & Pérez (1996) obtained spectra of this galaxy and showed that the NE structure has L(Hβ) = 2×10^{39} erg s^{-1}, [OIII]/Hβ = 7.6, n_e = 350 cm^{-3}, and the FWHM([OIII])= 80 km s^{-1}. According to Ferruit et al. (1999), the pre-shock density for a shock with this velocity would be of the order of n = 15 cm^{-3}. Using the equations given above, for a shock of velocity V_s = 100 km s^{-1} and preshock density n = 15 cm^{-3}, we get L(Hβ) = 7.5×10^{37} erg s^{-1} and L(Hβ) = 1.0×10^{38} erg s^{-1} for the shock and precursor, respectively. This corresponds to ≈10% of the observed Hβ luminosity. Another problem with the shock ionization of the NE structure is that the [OIII]/Hβ ratio can only be reproduced with a shock of 400 km s^{-1}, which is inconsistent with the observed FWHM of the [OIII] line.

We now use this same approach for the nuclear region of this galaxy. According to González Delgado & Pérez (1996), the nuclear spectrum has L(Hβ) = 1.0×10^{40} erg s^{-1} and n_e = 400 cm^{-3}. They also found that the [OIII] line can

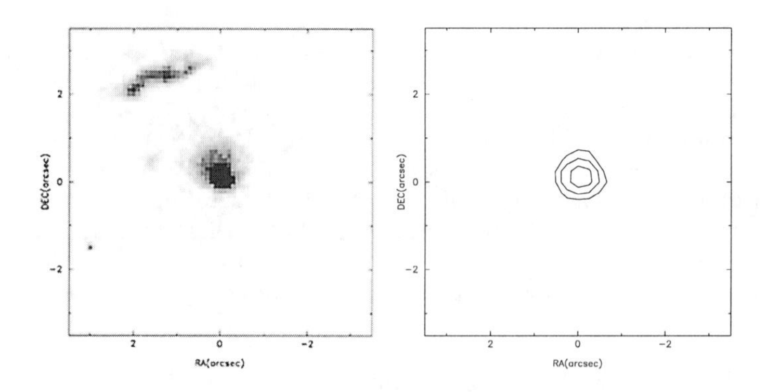

Figure 2. Comparison between the *HST* [OIII]λ5007Å images of NGC 5347 (left) with the archival VLA A-config. 6 cm image (right).

be decomposed into a narrow and a broad component, with the broad component having FWHM([OIII])=500 km s^{-1}. Again, if we assume that we are seeing the shock edge-on, with a circular cross section of radius 0.5″, we get that the shock area is 0.018 kpc^2 (1.7×10^{41} cm^2). We get that for a shock velocity V_s = 500 km s^{-1} and a preshock density n = 10 cm^{-3}, L(Hβ) = 6.3 × 10^{38} erg s^{-1} for the shock and 6.6 × 10^{38} erg s^{-1} for the precursor, which corresponds to ≈ 10% of the observed Hβ luminosity. These results indicate that shocks are not extremely important for the ionization of the NLR of this galaxy.

2.2. Mrk 79

The [OIII] and radio images of Mrk 79 are shown in Figure 1. This Figure shows that the [OIII] emission of this Seyfert 1 galaxy is extended by ≈ 5″ along the N-S direction. The radio emission lies along the same direction, but is extended by only ≈ 3″. The radio structures to the N and S of the nucleus are surrounded by the [OIII] emission, suggesting that shocks may play an important role in the ionization of the gas. Using the same technique used for NGC 5347, we calculate what would be the contribution of shocks to the Hβ luminosity of the N and S regions of the NLR. We assume the shocks have a circular cross section of radius 0.5″, which corresponds to an area of 0.145 kpc^2 (1.4 × 10^{42} cm^2), considering the galaxy has a radial velocity of 6643 km s^{-1}. From Whittle et al. (1988) we have that the N and S lobes have FWHM([OIII])=400 km s^{-1} and 350 km s^{-1}, respectively, which we assume to be the shock velocities. Assuming a preshock density n =10 cm^{-3}, we can calculate that the sum of the shock and precursor Hβ luminosities are 6.1×10^{39} erg s^{-1} for the N component, and 4.5×10^{39} erg s^{-1} for the S component. Using our CFHT Hα image and assuming Hα/Hβ = 3.1, we get the observed L(Hβ)= 2.8 × 10^{40} erg s^{-1} and 1.9 × 10^{40} erg s^{-1} for the N and S components, respectively. This means that ≈25% of the photons ionizing the N and S lobes of this galaxy can be due to shocks.

2.3. Seyferts with unresolved radio emission

Another test of the importance of shocks to the ionization of the NLR of Seyferts was done using those galaxies for which we do not observe extended radio emission. We use the same method used for NGC5347 and Mrk 79. Since these galaxies are unresolved in radio, we make the conservative assumption that the shock has a circular cross section with radius 0.25″, two times larger than the resolution of the observations. We do not have information about the FWHM([OIII]) of these galaxies, so we assume that all the galaxies have shocks of 500 km s^{-1}, at the high end of the parameters modeled by Dopita & Sutherland (1995, 1996), and use a preshock density of $n = 10$ cm^{-3}. Given these assumptions, we should consider these results as upper limits to the contribution of shocks to the ionization of these NLR's.

Table 1. Observed and predicted Hβ luminosities[1]

Name	L(Hβ) Obs	L(Hβ) Calc	R.	Name	L(Hβ) Obs	L(Hβ) Calc	R.
Mrk 1	16.55	1.34	1	I 01475-0740	5.99	1.65	1
Mrk 1040	7.24	1.42	2	UGC 2024	6.72	2.64	1
MCG -2-8-39	20.13	4.61	1	I 03125-0254	10.93	3.03	1
Mrk 607	2.20	0.43	1	I 04385-0828	1.27	1.20	1
MCG -5-13-17	3.37	0.84	3	UGC 3478	1.87	0.86	1
UGC 4155	26.14	3.42	1	Mrk 1239	39.53	2.09	3
NGC 3783	4.99	0.38	4	NGC 4593	10.23	0.43	5
NGC 4704	8.46	3.87	1	MCG -2-33-34	19.67	1.13	3
I 13059-2407	1.49	1.02	1	MCG -6-30-15	1.10	0.32	6
NGC 5347	12.00	1.47	11	I 14434+2714	31.44	4.54	1
UGC 9826	8.49	4.48	1	I 15480-0344	10.55	4.83	1
I 16288+3929	26.41	4.83	1	I 16382-0613	11.64	4.05	1
UGC 10889	7.56	4.15	1	MCG +3-45-3	30.58	3.11	1
UGC 11630	1.22	0.78	1	NGC 7213	3.69	0.19	7
Mrk 915	35.63	3.06	2	UGC 12138	16.22	3.18	1
UGC 12348	26.96	3.37	1	Mrk 590	22.66	3.66	8
Mrk 1058	1.74	1.54	2	Mrk 705	19.16	4.39	9
UGC 6100	39.39	4.51	9	UGC 10683 B	6.54	4.99	10

[1]Luminosities are in units of 10^{39} erg s^{-1}. The Hβ luminosities calculated based on shock models include both the shock and the precursor. References: (1) de Grijp et al. (1992); (2) Dahari & De Robertis (1988); (3) Rodríguez-Ardila et al. (2000); (4) Winge et al. (1992); (5) Clavel et al. (1983); (6) Reynolds et al. (1997); (7) Filippenko & Sargent (1988); (8) Stirpe (1990); (9) Cruz-Gonzalez et al. (1994); (10) Wilson et al. (1981); (11) González Delgado & Pérez (1996).

We show in Table 1 the observed and calculated (shock+precusor) Hβ luminosities of the galaxies with unresolved radio emission, as well as the references from which we obtained the observed values. On average only 15% of the observed Hβ luminosity can be due to shocks, and this number is likely to be smaller, given the favorable shock parameters we assumed.

3. Summary

We presented the preliminary results of a survey of radio and [OIII] images of Seyfert galaxies. About ~50% of our galaxies have unresolved radio emission (smaller than 0.25″). Based on conservative assumptions about the area and velocity of a shock in these galaxies, and on the preshock gas density, we calculated the shock and precursor Hβ luminosities for each one of them and compared these values with the observed ones. This showed that, on average, only 15% of the observed Hβ emission can be due to shocks, which is confirmed by the detailed study of two individual galaxies with resolved [OIII] and radio emission (Mrk 79 and NGC 5347). In summary, in most of the cases shocks are not a viable source of ionizing photons for the NLR of Seyfert galaxies.

Acknowledgments. Support for this work was provided by NASA grants AR-8383.01-97A and GO-08598.07-A. The National Radio Astronomy Observatory is a facility of the National Science Foundation operated under cooperative agreement by Associated Universities, Inc.

References

Binette, L., Wilson, A. S., & Storchi-Bergmann, T. 1996, A&A, 312, 357
Capetti, A. et al. 1996, ApJ, 469, 554
Clavel, J. et al. 1983, MNRAS, 202, 85
Cruz-Gonzalez, I. et al. 1994, ApJS, 94, 47
Dahari, O., & De Robertis, M. M. 1988, ApJS, 67, 249
de Bruyn, A. G., & Wilson, A. S. 1978, A&A, 64, 433
De Grijp, M. H. K., et al. 1992, A&AS, 96, 389
Dopita, M. A., & Sutherland, R. S. 1996, ApJS, 102, 161
Dopita, M. A., & Sutherland, R. S. 1995, ApJ, 455, 468
Falcke, H., Wilson, A. S., & Simpson, C. 1998, ApJ, 502, 199
Ferruit, P., et al. 1999, MNRAS, 309, 1
Filippenko, A. V., & Sargent, W. L. W. 1988, ApJ, 324, 134
González-Delgado, R. M., & Pérez, E. 1996, MNRAS, 280, 53
Haniff, C. A., Wilson, A. S., & Ward, M. J. 1988, ApJ, 334, 104
Heckman, T. M., et al. 1981, ApJ, 247, 403
Hutchings, J. B., et al. 1998, ApJ, 492, L115
Kinney, A. L., et al. 2000, ApJ, 537, 152
Morse, J. A., Raymond, J. C., & Wilson, A. S. 1996, PASP, 108, 426
Pogge, R. W., & De Robertis, M. M. 1995, ApJ, 451, 585
Reynolds, C. S., et al. 1997, MNRAS, 291, 403
Rodríguez-Ardila, A., Pastoriza, M. G., & Donzelli, C. J. 2000, ApJS, 126, 63
Schmitt, H. R., et al. 2001, ApJS, 132, 199
Stirpe, G. M. 1990, A&AS, 85, 1049
Whittle, M. 1985, MNRAS, 213, 33
Whittle, M. et al., 1988, ApJ, 326, 125
Whittle, M. 1992, ApJ, 387, 121
Wilson, A. S. 1997, in Emission Lines in Active Galaxies: New Methods and Techniques, ed. B.M. Peterson, F.Z. Cheng, & A.S. Wilson (San Francisco: ASP), 264
Wilson, A. S. et al., 1981, AJ, 86, 1289
Winge, C., Axon, D. J., Macchetto, F. D., & Capetti, A. 1997, ApJ, 487, L121
Winge, C., et al. 1992, ApJ, 393, 98
Young, A. J., Wilson, A. S., & Shopbell, P. L. 2001, ApJ, in press

Mass Outflow in Active Galactic Nuclei: New Perspectives
ASP Conference Series, Vol. 255, 2002
D.M. Crenshaw, S.B. Kraemer, and I.M. George

Kinematics of the NLR in Mrk 3

Jose R. Ruiz
Catholic University of America, 620 Michigan Ave, NE, Washington, DC 20064

D. Michael Crenshaw
Department of Physics and Astronomy, Georgia State University, Atlanta, GA 30303

Steven B. Kraemer
Laboratory for Astronomy and Solar Physics, NASA's Goddard Space Flight Center, Code 681, Greenbelt, MD 20771

Abstract. We present two sets of observations of the narrow-line region (NLR) gas of the Seyfert 2 galaxy Mrk 3, both using the Space Telescope Imaging Spectrograph (STIS) onboard *HST*. Our analysis of these two datasets produce radial velocity maps of the emission-line gas, which indicate general trends in the gas motion. We fit the radial velocity maps to kinematic models of the NLR gas. The data clearly favor a model in which the gas exists in a partially filled bicone, is accelerated radially away from the nucleus, and is followed by a constant deceleration (possibly due to collision with an ambient medium).

1. Introduction

In many Seyfert 2 galaxies, the NLR clouds appear to lie in a biconical or roughly linear configuration surrounding the nucleus (e.g., Schmitt & Kinney 1996). Various kinematic models have been proposed to explain NLR cloud motion. Capetti et al. (1995) have compared optical and radio measurements of the NLR of Mrk 3 and concluded that the NLR clouds are the result of radio jet plasma expanding away from the bicone axis. Winge et al. (1997, 1999) postulate gravitational motions for the NLR in NGC 4151. Recently, Crenshaw & Kraemer (2000) and Kaiser et al. (2000) have determined radial velocities as a function of position in the NLRs of NGC 1068 and NGC 4151 (the brightest Seyfert 2 and Seyfert 1, respectively) with the STIS on *HST*. Crenshaw et al. (2000) have proposed a model where clouds on the surface of a bicone are radially accelerated from the nucleus by wind pressure or radiation pressure, encounter and collide with an ambient medium, then decelerate to near-systemic values. It explains the general trends seen in the radial velocity as a function of position in the inner kiloparsec around the nuclei of these galaxies. In this paper, we will show that this same model fits the observed kinematics of the NLR clouds in Mrk 3.

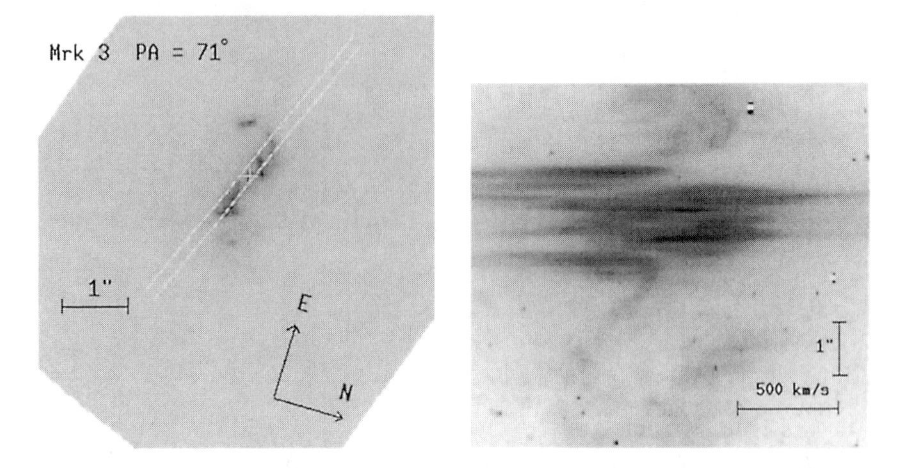

Figure 1. Left-hand panel-The FOC image of the bright NLR clouds of Mrk 3, with the position of the long slit overlaid. Right-hand panel-The STIS slitless spectrum showing the region around [O III] λ5007. The horizontal axis is along the dispersion, while the vertical scale is the spatial axis.

Mrk 3 is a well-studied Seyfert 2 galaxy, whose host is classified as an elliptical or S0 galaxy type. It lies 53 Mpc away (H_0=75km s^{-1} Mpc^{-1}, z=0.0135, 3.82″ kpc^{-1}). Mrk 3 has bright [O III] emission-line clouds that lie in a bent biconical configuration (apex of the two cones coincident with the nucleus) along PA=80°, with a half-opening angle of 22.5° (Schmitt & Kinney 1996; also see the left-hand panel of Figure 1). At the end of the western cone, a large, diffuse knot appears, while on the end of the eastern cone, a bright knot appears out of the bicone, giving the entire structure an 'S' shape (Kukula et al. 1993). Schmitt & Kinney measure the clouds as extending 280 pcs on either side of the nucleus. Radio jets have also been observed (Axon et al. 1998; Kukula et al. 1999) along the same PA; these appear to follow the biconical structure, although the half-opening angle is far less ($\sim$ 8-10°). Though they lie close to the emission-line clouds, they are not exactly coincident; Axon et al. (1998) suggest that they lie along the convex edge of the S-shaped curvature.

2. Observations and Data Analysis

In order to find the NLR knots in the slitless observations, and hence, determine their velocities, a companion [O III] image, as well as a continuum image are required. To that end, the archival images of Mrk 3 were obtained. The first of these is a Wide Field Planetary Camera (WFPC) [O III] observation. This image served to match the bright NLR clouds. A Wide Field Planetary Camera 2 (WFPC 2) continuum image was also retrieved, as well as a Faint Object Camera (FOC) [O III] image. The FOC image is shown in left-hand panel of Figure 1. It served to match the faint extended narrow-line region (ENLR) clouds.

The new slitless observations take advantage of STIS's spatial resolution (0″.1) and the G430M spectral resolution ($\lambda/\Delta\lambda \approx 10{,}000$). The observations were centered at 5093 Å, with a bandwidth of 286 Å. In addition to [O III] λ5007, [O III] λ4959 and Hβ were also observed. The spectral region around [O III] λ5007 is shown in the right-hand panel of Figure 1. The horizontal axis is the dispersion axis, while the vertical axis is the spatial axis, as indicated by the scale. The bright NLR clouds are smeared out along the dispersion axis, indicating large velocity dispersions. They are also shifted along this axis, from which their radial velocities can be calculated. We also obtained long-slit observations of Mrk 3 using the G140L, G230L, G430L and G750L gratings and the 52″ x 0″.1 aperture. The complete observations covered the wavelength range from 1150-10,000 Å. A full analysis of these data will be presented separately (Collins et al. 2001, in preparation). For this work, the single emission line of [O III] λ5007 from these observations was used. The slit had a PA of 71°, and was chosen to pass through the nucleus. The position of the long-slit is shown in the left-hand panel of Figure 1, overlaid on the FOC image for comparison with the [O III] clouds.

3. Results

The NLR dataset is shown in Figure 2, which shows the NLR radial velocities (relative to systemic) plotted against distance from the optical continuum center. The two sets of data points are seen to be compatible, both in terms of location and velocity. The clouds have a wide range of radial velocities, from -1000 km s^{-1} to $+600$ km s^{-1}. There are roughly equal numbers of points that show redshifts and blueshifts. They also show general properties that must be duplicated by any model fit. Now we briefly discuss each of these properties.

Both sets of data points show fast rises in velocities (from systemic values at the nucleus to $\sim\pm600$ km s^{-1}) out to $\sim$0″.25 of the nucleus on each side. The climbs in velocities are seen in both redshifts and blueshifts. All the rises are then followed by shallower velocity downturns, so that at $\sim$0″.7-1″.0, the velocities have returned to near systemic values. The amplitudes of the maximum velocities are not equal. The amplitudes range from $\sim$300 km s^{-1} on the blueshifted west side to $\sim$800 km s^{-1} on the blueshifted east side (ignoring a few high-velocity points). This variation in amplitude implies that any fitted cone is tilted, and in fact, the angle of inclination can be calculated by the amplitude difference. Finally, the range of velocities is fairly narrow. For example, $\sim$0″.30 west of the nucleus, blueshifted velocities are seen exclusively from -200 to -300 km s^{-1}, while redshifted velocities are seen spanning a narrow range from 300 to 500 km s^{-1}. At this distance, there are no velocities seen from -200 to 300 km s^{-1}. Any model fit must be able to match these narrow velocity ranges.

4. Model Fitting

Once the spatial orientations and radial velocities of the clouds were obtained, we attempted to fit these observations using kinematic models. These models calculate the radial velocities and spectral lines of material on the surface of a thin disk, or a bicone, either filled or hollow. Each geometry can assume

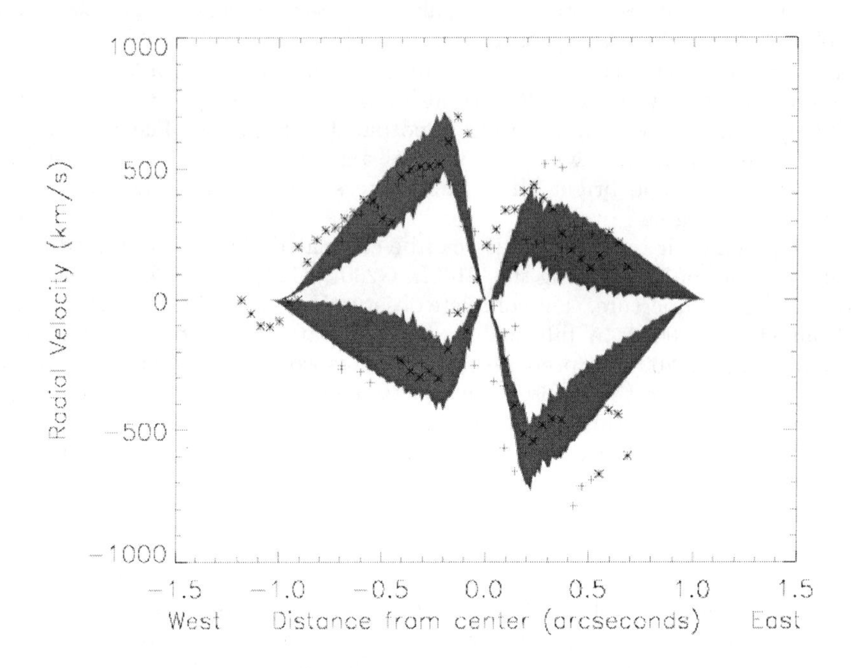

Figure 2. Radial velocity against the distance from the center (slitless: plus signs, long-slit: asterisks). The best fit model of the radial acceleration + constant deceleration model is overlaid.

one of various velocity laws that control the material's movement. The bicone program has been used previously to model the NLR emission-line clouds of NGC 1068 and NGC 4151 (Crenshaw & Kraemer 2000, Crenshaw et al. 2000). The two cones (one on either side of the nucleus) are assumed to possess identical properties, including geometry, size and velocity law. In addition, the cones are assumed to have a filling factor of 1 within the minimum and maximum half opening angle, and not to absorb [O III] photons. We adjust certain parameters, to obtain the best fit. We concentrate here on the NLR out to $\sim 1''.0$ on either side of the nucleus.

We can immediately rule out gravitational rotation models by calculating the mass required to impart radial velocities on the order of 500-1000 km s^{-1} at a distance of 100-200 pcs away from the nucleus. This mass is of the order of 10^{9-10} $M_{\odot}$. Typical masses for black holes in Seyfert nuclei are 10^{6-8} $M_{\odot}$ (Peterson & Wandel 2000). Observationally, the observed morphology of the NLR does not suggest a disk geometry, while the redshifts and blueshifts on either side of the nucleus cannot be the result of simple Keplerian rotation. Thus we are left with outflow models, or models where material flows tangentially outward from the radio axis.

Several model input variables can be constrained from the observations. The first of these was a minimum and maximum half-opening angle. The maximum angle was measured from the images, giving a value of 25°. This agrees with Schmitt & Kinney (1996), who measure a maximum half-opening angle of 22.5°. The minimum half-opening angle is not visible in the [O III] images, so it was varied to match the data. The optimum value for the models was 15°. This value places the emission-line material outside the observed radio jet cone (half-opening angle ~7-8°; Capetti et al. [1995]).

The inclination angle was calculated based on the differences between the radial velocity maxima on the W and E sides of the cones. The maximum blueshifts are higher by ~300 km s^{-1} than the maximum redshifts on the east side. The NLR inclination angle was then calculated as ~5-10°, using simple trigonometry. Finally, the value for the maximum deprojected velocity of the NLR gas was chosen so that it would match the observed NLR radial velocity peak (~-800 km s^{-1}). The model that fit the most data points is one which involves radial acceleration + constant deceleration. The model can be visualized as material first accelerated by wind or radiation pressure from the nucleus, which then impacts an ambient medium and then decelerates at a constant rate. This model implies that the emission-line clouds originate from a region closer to the nucleus and move outward from there. Figure 2 shows the long-slit and slitless data points overlaid with the shading from this model. Obviously, this model does not perfectly fit every point, but it fits the gross features of the observations well.

5. Discussion

The orientation of the host galaxy has been previously reported as 27° out of the plane of the sky (Schmitt & Kinney 1996). If this orientation extends down to kiloparsec scales, then the plane of the galactic disk would lie within the angular range of one side of each cone (15° to 25°, tilted out 5° of the plane of the sky). The situation then resembles NGC 1068 and NGC 4151 (Crenshaw et al. 2000, Crenshaw & Kraemer 2000), which also seem to have the galactic disk and one side of the bicone in the same plane. We propose the same geometry in this galaxy.

The radio jet and the NLR emission share a similar axis, and are nearly coincident. However, other than their spatial coincidence, there do not appear to be any other correlations, as would be expected if the radio plasma's expansion were the source of the NLR velocities. Firstly, in the data itself, there are no bright NLR clouds that correspond to jet flux maxima. This lack of correspondence has been noted in other objects (NGC 4151; Kaiser et al. [2000]). In addition, there is no evidence for peculiar velocities at the positions of the radio lobes. In terms of the modeling, transverse velocity models predict equal blueshift/redshift amplitudes no matter what inclination angle the bicone is tilted. However, the data show a definite difference (200-400 km s^{-1}) in velocity maxima between redshifts and blueshifts.

6. Conclusions

Two STIS spectra were obtained of the NLR of the Seyfert 2 galaxy Mrk 3. Radial velocities were determined of the emission-line gas as a function of position (out to ~1 kpc from the nucleus). The velocity maps indicate general trends in the gas motion. These include: blueshifts and redshifts on either side of the nucleus, steep velocity rises from systemic up to $\sim\pm700$ km s^{-1} taking place in the inner $0''.3$ (0.8 kpc) both east and west of the nucleus, and gradual velocity descents back to near-systemic values from $0''.3$-$1''.0$. In addition, the spectra gave extremely compatible results for individual NLR clouds, though one was a long-slit spectrum and the other was a slitless spectrum of the same region.

The data were then fitted to kinematic models for the NLR gas on the surface of the bicone. The data sets were fit best with a radial acceleration + constant deceleration model. In the model, the cones extend out to a radius of $0''.75$ from the nucleus, with a half-opening angle between 15° and 25°. The modeled material reaches a maximum deprojected velocity of 1750 km s^{-1}, reaching this velocity at a distance of $0''.3$-$0''.43$ from the nucleus, close to the observed distance of $0''.2$-$0''.3$ from the nucleus.

This work was supported by NASA Guaranteed Time Observer funding to the STIS Science Team under NASA grant NAG 5-4103 and by *HST*. Additional support for this work was provided by NASA through grant number HST-GO-08340.01-A from the Space Telescope Science Institute, which is operated by AURA, Inc., under NASA contract NAS5-26555.

References

Axon, D.J., Marconi, A., Capetti, A., Macchetto, F.D., Schreier, E., & Robinson, A. 1998, ApJ, 496, L75

Capetti, A., Macchetto, F., Axon, D.J., Sparks, W.B., & Boksenberg, A. 1995, ApJ, 448, 600

Crenshaw, D.M., & Kraemer, S.B. 2000, ApJ, 532, L101

Crenshaw, D.M., Kraemer, S.B., Hutchings, J.B., Bradley, L.D., II, Gull, T.R., Kaiser, M.E., Nelson, C.H., Ruiz, J.R., & Weistrop, D. 2000, AJ, 120, 1731

Kaiser, M.E., Bradley, L.D. II, Hutchings, J.B., Crenshaw, D.M., Gull, T.R., Kraemer, S.B., Nelson, C.H., Ruiz, J., & Weistrop, D. 2000, ApJ, 528, 260

Kukula, M.J., Ghosh, T., Pedlar, A., Schilizzi, R.T., Miley, G.K., de Bruyn, A.G., & Saikia, D.J. 1993, MNRAS, 264, 893

Kukula, M.J., Ghosh, T., Pedlar, A., & Schilizzi, R.T. 1999, ApJ, 518, 117

Peterson, B.M., & Wandel, A. 2000, ApJ, 540, L13

Schmitt, H.R., & Kinney, A.L. 1996, ApJ, 463, 498

Winge, C., Axon, D.J., Macchetto, F.D., & Capetti, A. 1997, ApJ, 487, L121

Winge, C., Axon, D.J., Macchetto, F.D., Capetti, A., & Marconi, A. 1999, ApJ, 519, 134

Mass Outflow in Active Galactic Nuclei: New Perspectives
ASP Conference Series, Vol. 255, 2002
D.M. Crenshaw, S.B. Kraemer, and I.M. George

Resolving the NLR of NGC 1068 with STIS: Associated Absorbers Seen in Emission?

G. Cecil

University of North Carolina, Dept. of Physics & Astronomy, Chapel Hill, NC 27599-3255

Abstract. Our R=5,000 Space Telescope Imaging Spectrograph (STIS) spectra map and resolve [O III] and Hβ emission-line profiles across most of the narrow-line region (NLR) of NGC 1068, and are matched to line emitting structures in *Hubble Space Telescope (HST)* images. We find emission knots with radial velocities blueshifted by up to 3200 km s^{-1} relative to galaxy systemic, that project $70 - 140$ pc NE of the nucleus up to 40 pc from the radio jet, emit several % of the NLR line flux, coincide with a region of collisionally ionized X-ray and IR coronal-line emitting gas, show velocity gradients of up to 2000 km s^{-1} in 7 pc and total extent averaged over $0''.2 \times 0''.05$ areas of up to 1250 km s^{-1}, ionized masses ~ 250 M$_\odot/n_{e,4}$ (density $n_{e,4} = 10^4$ cm^{-3}), and follow the overall trend of radial acceleration from near the nucleus, yet are kinematically contiguous with brighter clouds moving 200-800 km s^{-1} relative to systemic that are themselves adjacent to the radio jet. In their spatio-kinematic properties the knots are candidates for the associated absorbers projected on the UV continua of other active galactic nuclei (AGN).

1. Introduction

The NLR of NGC 1068 spans a several arcsecond ($1'' = 70$ pc) region that coincides with a twisting radio jet, and has been studied extensively. Its many kinematical components are blended at ground spatial resolution (e.g. Cecil et al. 1990; Arribas et al. 1996) and in Faint Object Spectrograph spectra (Caganoff et al. 1991). During their GTO program with the STIS, Kraemer & Crenshaw obtained a long-slit spectrum along one cut through the NLR, using R=1,000 gratings that span $\lambda\lambda 120 - 1027$ nm. In analyzing continuum and line fluxes (Kraemer & Crenshaw 2000, KC hereafter) and gas kinematics (Crenshaw & Kraemer 2000, CK hereafter) along the slit, they found that flux ratios are reproduced by AGN photoionization, with considerable absorption along those sight-lines from the nucleus that intersect the galaxy gas disk. They confirmed a correlation between ionization potential and line blueshift from galaxy systemic velocity that Marconi et al. (1998) obtained from ground-based IR spectra. Fluxes of these IR collisional lines peak $1 - 2''$ ($70 - 140$ pc) NE of the nucleus, the same region where Ogle et al. (2002) found in the *Chandra X-ray Observatory* High Energy Transmission Grating Spectrometer zero-order image what they argued was evidence for collisional ionization that augments photoionization. (However, Behar et al. (2002) have a photoexcitation interpretation.)

Nearby, the GTO STIS spectra show (CK) that the outflowing clouds must decelerate rapidly and (KC) that additional ionization from ~1000 km s^{-1} shocks is required to fit the observed line flux ratios.

2. Emission-Line Profiles Are Resolved Across the NLR

The GTO R=1,000 spectrum had insufficient velocity resolution to disentangle NLR kinematics. Therefore, we spent 7 of our 14 *HST* orbits building a datacube of [O III] and Hβ profiles across most of the NLR. We used STIS grating G430M, stepped the 0″.2-wide slit, and binned spectra in dispersion before readout to maximize sensitivity in our single-orbit exposures. (We also used L-gratings to map UV line fluxes across the NLR, but do not discuss those data here.)

At ~ 650 points in the NLR we scaled and subtracted the red wing of [O III] λ5007 to that of λ4959, then fit and subtracted the continuum to form composite [O III] and Hβ profiles (Figure 1). We synthesized an [O III] image from the resulting flux map, and registered it spatially to the Faint Object Camera (FOC) *HST* image (Capetti et al. 1997). Our datacube yields line profiles of knots 0 – 7 and clouds *A* – *K* in the FOC image, panel 8 of Figure 1. Gas is blueshifted up to 3200 km s^{-1} relative to galaxy systemic, far higher than seen along the GTO slit (that resembles the spectrum in panel 7 of Figure 1). The [O III] and Hβ line profiles show obvious substructure even at *HST* resolution.

Table 1. Properties of NLR Clouds in NGC 1068

Cloud Name	Ionized H Mass [1] (10^2 $M_\odot/n_{e,4}$)	[O III]λ5007/ Hβ [2]	log KE [3] (erg/$n_{e,4}$)	log momentum [3] (dyne s/$n_{e,4}$)
A	13.4±1	15±1	52.74	44.09
B	47.5±4	16±1	53.48	44.80
C	26.3±2	16±1	52.77	43.41
D	27.9±2	18±1	52.60	44.36
E	15.6±1	17±1	52.16	44.00
F	66.9±5	19±1	53.58	44.96
G	21.6±1	23±1	53.11	44.46
H	11.2±1	15±1	52.87	42.26
J	13.6±1	17±1	52.56	44.16
K	10.2±1	14±1	52.64	44.10
HV0	3.1±0.5	12±1	53.46+0.30	43.68+0.30
HV1	9.6±1	19±1	53.10+0.15	44.34+0.15
HV2	1.8±0.4	12±1	52.26+0.34	43.51+0.34
HV3	2.7±0.4	12±1	52.84+0.20	43.95+0.20
HV4	2.2±0.4	12±1	52.57+0.18	43.79+0.18
HV5	4.5±0.7	12±1	52.70+0.15	43.93+0.15
HV6	0.7±0.2	12±1	51.94+0.15	43.15+0.15
HV7	1.3±0.4	12±1	52.01+0.18	43.30+0.18

[1] HVx masses exclude contribution of broad component near systemic velocity.
[2] Velocity averaged; see Figure 2 for the velocity dependence.
[3] Minimum values, assume that all velocity is along our sight-line; + on HVx denotes contribution of broad component. Density $n_{e,4}$ is in units of 10^4 cm^{-3}.

We dereddened fluxes assuming E_{B-V} ~ 0.35, the mean blue-wing value found by KC from their large wavelength-baseline spectra >0″.2 NE of the nu-

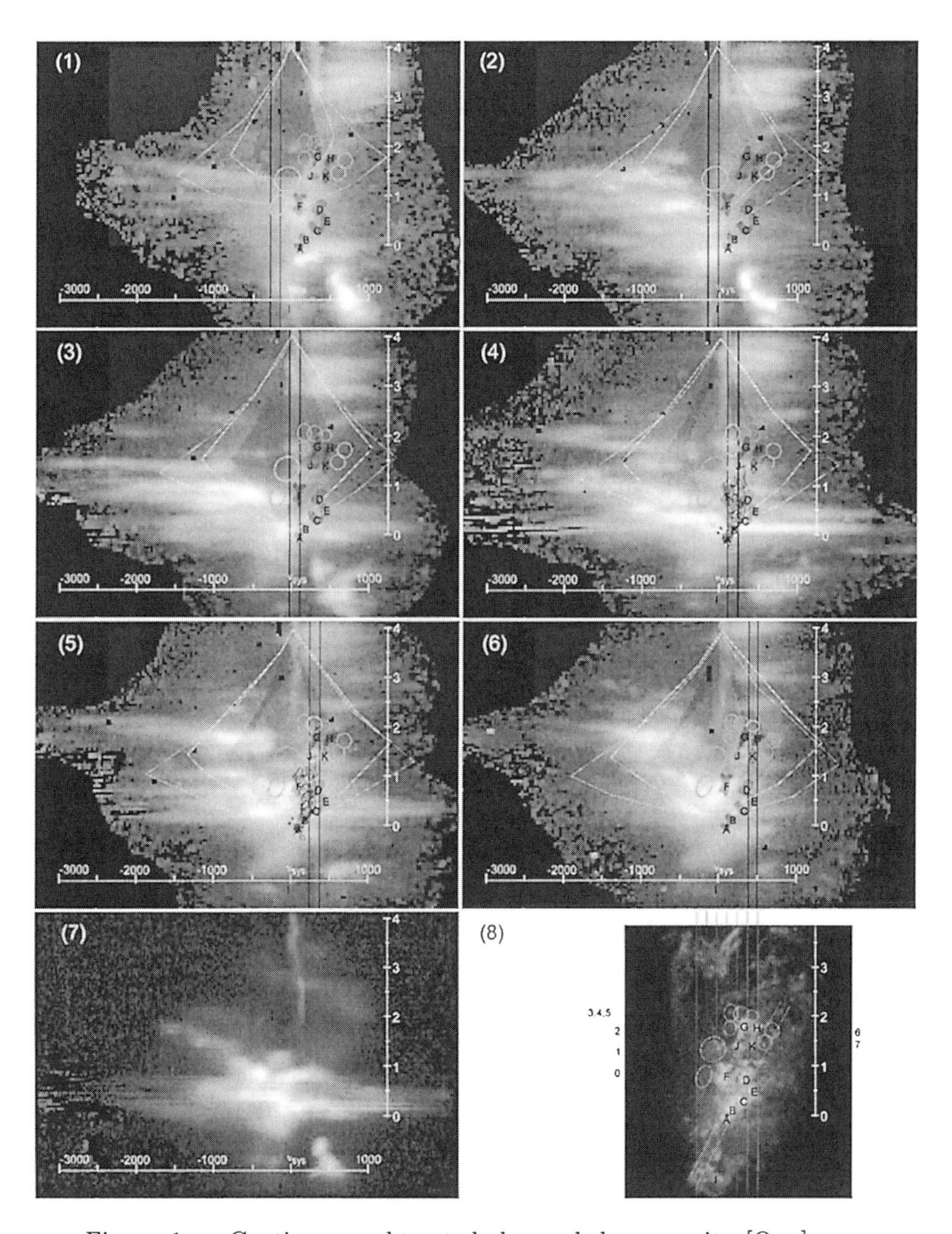

Figure 1. Continuum-subtracted, log-scaled composite [O III] profiles on progressive vertical slices through the NLR. 0″.19-wide slits are spaced 0″.2 apart along P.A. 38° (not panel 7), and are registered to the brightest radio contours and to the FOC [O III] image in panel 8. The base of the NE radio lobe is at top, the SW radio lobe is at bottom, the vertical scale is in arc-seconds from the nucleus, and the horizontal scale is in km s^{-1} relative to galaxy systemic. The spectrum in panel (7) is extracted along the diagonal in panel (8).

cleus. Table 1 lists lower limits on ionized masses and energetics if filaments are optically thin and recombine under case-B conditions (as KC found for the NE red-wing) at density 10^4 cm^{-3}; lower limits were estimated assuming only line-of-sight motions.

The axis of the ionization bicone is only 5° from the sky plane, so our spectra out to 2″ radius in the NE quadrant can be modeled successfully by accelerating flows that are either perpendicular to this axis or radial from the nucleus. In Figure 1 two kinematical models are projected on our spectra; a higher-fidelity version is at `http://www.thececils.org/science/n1068/n1068stis.jpg`. The brightest flux is usually confined within the contours (outer = radial outflow from the nucleus that peaks at 2500 km s^{-1}, twice what CK inferred from their single slit; inner = lateral expansion from the average axis of the twisting jet).

3. Radiative Acceleration of Dusty Clouds and High-Velocity Knots?

Flip Figure 2 to see that the line profiles (and the properties in Table 1) of knots and the more massive clouds resemble those of associated absorbers (e.g. Hamann et al. 1997). Some recent models of AAL quasar spectra (e.g. de Kool et al. 2001) require that the absorbers lie at comparable distances to those we measure rather than being closer to the AGN. Can our spectra of NGC 1068 shed light on how such filaments are accelerated?

KC's slit samples several clouds NE of the nucleus, so their photoionization models may constrain the features we observe. They found that the blue-wing clouds are optically thin above the Lyman limit, and that between $1''.4-1''.8$ another photon source is required to explain strong NUV lines and a larger He II $\lambda4686/H\beta$ ratio. Because their derived velocity field is discontinuous here, CK posit that clouds are decelerating from an outflow of ~1000 km s^{-1}, producing fast shocks that would generate the observed spectrum (e.g. Dopita & Sutherland 1996). They obtain reasonable fits to blue-wing line fluxes if $85-90\%$ of the flux comes from a tenuous component with $(\log\mathcal{U}, n_H, \log N_H) = (-1.45 \pm 0.15, 3\times10^3\ \mathrm{cm}^{-3}, 21.18\pm0.18\ \mathrm{cm}^{-2})$ and the rest from a dense component with $(\log\mathcal{U}, n_H, \log N_H) = (-2.9, 2\times10^4\ \mathrm{cm}^{-3}, 20.7\pm0.1\ \mathrm{cm}^{-2})$, both with dust fractions of 25%.

Dust absorption dominates opacity in a photoionized plasma like this NLR when $\mathcal{U} > \alpha(T_e)/c\kappa$ (c is lightspeed, $\kappa \sim 20^{-21}$ is the opacity per atom assuming the standard Galactic reddening curve, and $\alpha(T_e) \sim 2\times10^{-13}$ is the recombination coefficient). The dust will compete successfully for ionizing photons in the highly ionized zone, suppressing lines from these species. In addition, a strong pressure gradient will be set up that will much enhance the plasma density by the time lower ionization species emit. In consequence, a lower $\mathcal{U} \sim 0.007$ will provide an apparent model fit to the spectrum, close to the value preferred by KC for their denser component. This ionization parameter is closely related to the $\mathcal{U}$ at which radiation pressure starts to dominate either the pressure gradient, or the dynamical acceleration of the ionized plasma. Thus, the flow of dusty clouds falling toward the nucleus is intrinsically unstable above the dust sublimation point, so clouds are driven into outflow in the direction of slowest accretion (i.e. within the ionization cones).

Dopita (2001) details this scenario, while Cecil et al. (2001, in preparation) apply his formulation to NGC 1068. This NLR is ~ 0.5 kpc across, and Figure 1 shows that massive clouds are accelerated up to 2000 km s^{-1} within $0''.5$ of the nucleus. The flux in the ionizing continuum needed to maintain the observed line luminosity is comparable to the FIR luminosity $1.5 \times 10^{11} L_{\odot}$. This flux radiatively accelerates dusty clouds by $\sim 10^{-5}$ cm s^{-2}, which can therefore reach their observed mean velocities of $\sim$500 km s^{-1} over a distance of 35 pc. In Figure 2, the knots are composed of narrower sub-components than are the clouds. In hydrodynamical simulations of cloud acceleration, tenuous gas streams ablate from massive clouds (e.g. Schiano et al. 1995), providing a reservoir for further acceleration of knots from already moving clouds.

This scenario ignores the role of the radio jet in agitating this NLR (Bicknell et al. 1998), but its influence may be localized at specific jet/cloud interactions (e.g. B & C, Gallimore et al. 1996; G & H, Capetti et al. 1995). While the region of collisional line emission coincides with the base of the NE radio lobe, numerical simulations of a breakout lobe do not show (D. Balsara, private comm.) a strong backflow vortex at its base that might form high-velocity shocks. In any event, many of the high-velocity knots have similar line profiles in Figure 2, and are found $\gtrsim 40$ pc from the brighter parts of the jet.

In summary, we have resolved the NLR of NGC 1068, demonstrating that integral-field spectrometers on 8-10m telescopes with adaptive optics *will* be able to constrain dynamics using nebular diagnostics in the optical and near-IR.

Acknowledgments. NASA grants GO-7353 and GO-01153A to UNC-CH

References

Arribas, S., Mediavilla, E. & Garcia-Lorenzo, B. 1996, ApJ, 463, 509

Behar, E. et al. 2002, in Mass Outflow in Active Galactic Nuclei: New Perspectives, ed. D.M. Crenshaw, S.B. Kraemer, & I.M. George (San Francisco: ASP), p. 43

Bicknell, G., et al. 1996, ApJ, 495, 680

Caganoff, S. et al. 1991, ApJ, 377, L9

Capetti, A., Macchetto, F. D., & Lattanzi, M. G. 1997, Ap&SS, 248, 245

Capetti, A., et al. 1995, ApJ, 452, L87

Cecil, G., Bland, J., & Tully, R. B. 1990, ApJ, 355, 70

Crenshaw, D. M. & Kraemer, S. B. 2000, ApJ, 532, L101

de Kool, M. et al. 2001, ApJ, 548, 609

Dopita, M. A. & Sutherland, R. S. 1996, ApJS, 102, 161

Dopita, M. A. 2001, Rev. Mex. AA. (Serie de Conferencias), in press

Gallimore, J. F. et al. 1996, ApJ, 464, 198

Hamann, F., et al. 1997, ApJ, 478, 80

Kraemer, S. B. & Crenshaw, M. 2000, ApJ, 544, 763

Marconi, A. et al. 1998, A&A, 315, 335

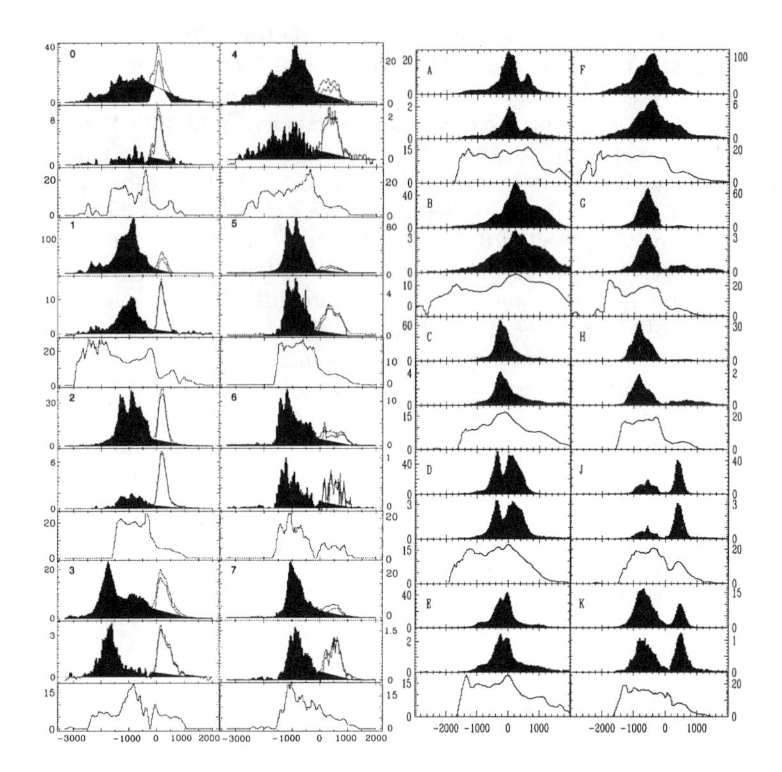

Figure 2. Profiles extracted from the points labeled in panel (8) of Figure 1; top in each trio [O III], middle Hβ, and their ratio below. Velocities are plotted over the range -3200 to 2000 km s^{-1} relative to galaxy systemic, while the vertical scale is line flux in 10^{-13} ergs s^{-1} cm^{-2} or the flux ratio. Left: profiles of high-velocity knots are shaded after subtracting a prominent, broad component seen near systemic velocity. Right: most of the more massive clouds.

Ogle, P. et al. 2002, in Mass Outflow in Active Galactic Nuclei: New Perspectives, ed. D.M. Crenshaw, S.B. Kraemer, & I.M. George (San Francisco: ASP), p. 13

Schiano, A. V. R., et al. 1995, ApJ, 439, 237

Mass Outflow in Active Galactic Nuclei: New Perspectives
ASP Conference Series, Vol. 255, 2002
D.M. Crenshaw, S.B. Kraemer, and I.M. George

Revealing the Energetics and Structure of AGN Jets

Eric S. Perlman[1], Herman L. Marshall[2], and John A. Biretta[3]

Abstract.
Until very recently, few constraints existed on the physics of jets, even though they represent the first known evidence of mass outflow in active galactic nuclei (AGN). This has begun to change with *Hubble Space Telescope (HST)* and *Chandra X-ray Observatory (CXO)* observations, which allow us to observe short-lived, dynamic features, and compare their spectra and morphology to those of longer-lived particles seen in the radio. We examine *HST* and *CXO* observations of M87 and 3C273 which reveal that these two prototype objects seem radically different.

1. Introduction

Collimated jets occur in many settings, including galactic nuclei, binary star systems, and star formation. AGN jets are relativistic flows composed of high-energy particles and magnetic fields, which emit synchrotron radiation in the optical and radio. Jets were the first AGN outflows observed, discovered by Curtis (1918), who noticed a 'curious, straight ray' emanating from M87.

Prior to the launch of *HST*, optical emissions had been observed from two AGN jets: M87 and 3C273 (Schmidt 1963). *ROSAT* and *Einstein* observed X-ray emissions from the jets and/or lobes of these (Schreier et al. 1982, Stern & Harris 1986) plus four others: Cen A (Burns et al. 1983), Cygnus A (Harris et al. 1994), 3C390.3 (Harris et al. 1998) and 3C120 (Harris et al. 1999). *HST* observations, particularly the 3CR Snapshot Survey (Martel et al. 1999), have drastically increased the number of known X-ray or optical jets to about 20, representing a fair cross-section of radio-loud AGN properties.

Because of the featureless nature of synchrotron spectra, the jump from morphology to physics is large for jets. Some progress has been made through numerical modeling and multi-frequency radio mapping, but this elucidates only a small part of the energy spectrum, and details can be obscured by the long particle lifetimes (10^{5-6} yr). Due to their short radiative lifetimes, optical and X-ray synchrotron emitters ($\sim 1 - 100$ yr), represent much more dynamic characteristics. Thus to obtain the tightest constraints, multiband data are required.

[1]Joint Center for Astrophysics, University of Maryland, Baltimore County, 1000 Hilltop Circle, Baltimore, MD 21250 USA

[2]Center for Space Research, Massachusetts Institute of Technology, 77 Massachusetts Ave., Cambridge, MA 02139 USA

[3]Space Telescope Science Institute, 3700 San Martin Drive, Baltimore, MD 21218 USA

2. An issue of stratification

One of the surprises provided by *HST* observations was that the optical and radio morphologies of jets can be quite different. For example, in 3C273 the optical jet appears much narrower than seen in the radio, with a twisted, even 'braided' morphology (Röser et al. 1997). And in M87, there is a large-scale correspondence of features, but detailed comparison reveals a narrower, knottier jet in the optical (Sparks et al. 1996). Therefore it should not have been a surprise when the first *CXO* data revealed yet more differences. For example, in Cen A, the X-ray and radio maxima of several knots appear offset by up to an arcsecond (Kraft et al. 2000). What do these represent: different physical conditions, different emission mechanisms, or both?

Perlman et al. (1999) proposed that the radio-optical differences in the M87 jet could be explained by stratification. What led to this conclusion was *HST* and VLA polarimetric images, which revealed large differences in knots, where the magnetic field vectors seen in the optical (but not radio) become perpendicular to the jet upstream of flux maxima, followed by sharp decreases in optical polarization at flux maxima (their Figures 3-6). Under the Perlman et al. model, high energy electrons are concentrated along the axis and in knots, where the magnetic field is compressed by shocks, while the sheath is dominated by lower-energy particles, with a more static, parallel magnetic field.

Naively, the idea of a stratified jet is not novel: in a fire hose, the fastest part of the flow is in the center. In fact, a decade ago some authors proposed 'two-fluid' jet models (cf. H. Sol, comments at this meeting). Stratification has far-reaching implications. For example, instabilities need not involve all components of the jet flow. Moreover, a compressed, perpendicular magnetic field in knot regions is a recipe for particle acceleration. It instructive to look at the X-ray and optical flux and spectral morphology, where several of these features should show up.

3. Spectral and X-ray Morphology of the M87 and 3C273 Jets

M87 (d=16 Mpc) and 3C273 (z=0.158) are the prototype jets, having both the largest angular extents and high surface brightness. They are very different objects, differing by a factor 100 in luminosity and jet power, and by a factor 30 in physical jet length. One might therefore expect a detailed comparison to show interesting differences.

Looking first at Figure 1, we can see that in the M87 jet, there is an excellent overall correlation between optical flux and optical spectral index α_o ($F_\nu \propto \nu^\alpha$): in high surface brightness regions, one sees flatter spectra. Near knot maxima one sees exactly the kind of hardening, followed by softening, that one would expect if particle acceleration is occurring in the knots. Interestingly, the optical and radio-optical (α_{ro}) spectral indices do not vary together. In each knot region in the inner jet, we observe α_o either leading or lagging α_{ro}. One can understand this in the context of particle acceleration (Kirk et al. 1999): if the acceleration timescale is much less than the cooling timescale for optical synchrotron emitters, one expects α_o to lead α_{ro}; however if the acceleration and cooling timescales are of similar order, one would expect α_{ro} to lead. A different situation is seen

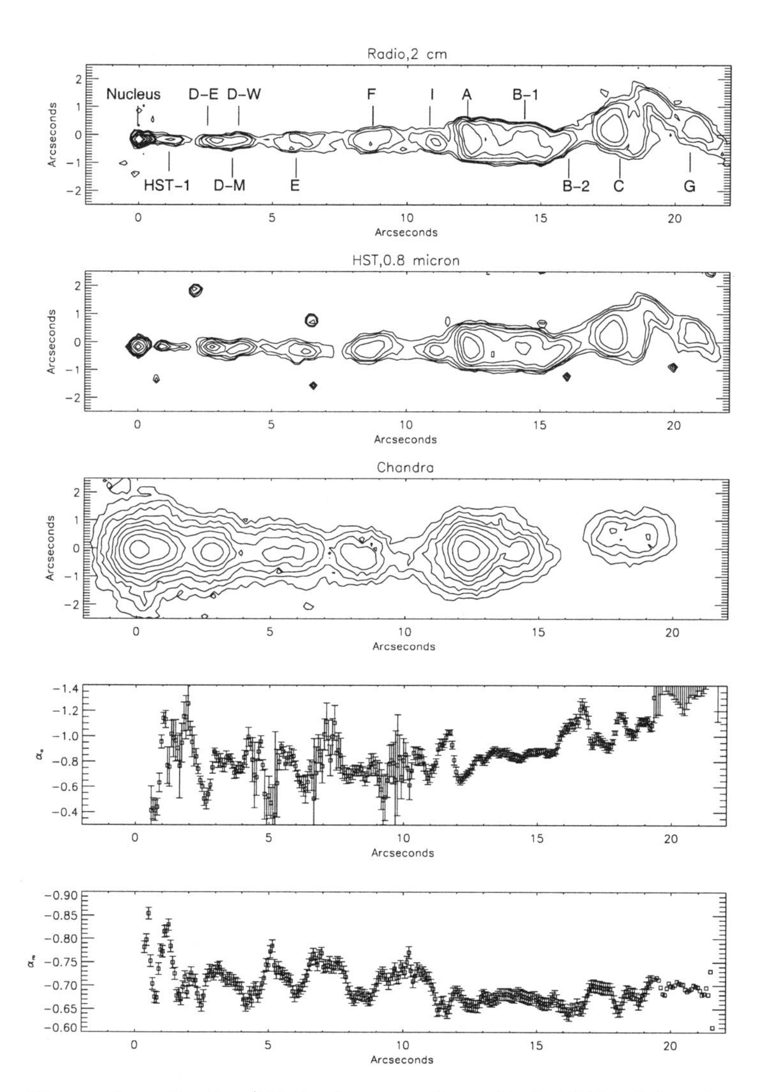

Figure 1. Radio (VLA, 2cm, top), optical (*HST*, I band, second panel), X-ray (*CXO*, third panel), α_o (fourth panel) and α_{ro} (bottom) images of the M87 jet. The *CXO* Advanced CCD Imaging Spectrometer+High Energy Transmission Grating Spectrometer image was adaptively smoothed. Contours represent (3, 5, 8, 16, 32, 64, 128, 256...) $\times$ rms noise. Note the strong correlation between optical flux and spectral index. This, along with polarimetric imaging, strongly argues for energetic stratification. The X-ray and optical peaks of most knots are located in the same places, but there are two optically faint regions where X-ray peaks are seen, in the D-E and E-F transitions.

in the outer jet, where there is an anticorrelation between α_o and α_{ro}. This might indicate that the jet core has a steeper injection index than the sheath.

X-ray emission is seen from all regions of the M87 jet, and several optical and X-ray knot maxima are located at the same θ_{nuc}, e.g., HST-1 (nuclear distance $\theta_{nuc} = 1''$), A ($\theta_{nuc} = 12''$) and C ($\theta_{nuc} = 18''$). Knot A's X-ray and optical maxima are at the same θ_{nuc}, not displaced by $0.5''$, as had been claimed by Neumann et al. (1997) and Böhringer et al. (2000) The most likely explanation is the unexpected brightness of HST-1, which *CXO* only partially resolves from the core, and is hopelessly blended with it in *ROSAT* and *XMM-Newton* images. There are, however, significant morphological differences. In knot E ($\theta_{nuc} = 5-6''$), the X-ray bright region begins $1''$ upstream of the optical peak, in an optically faint region. Also, in knot F ($\theta_{nuc} = 8''$), the X-ray bright region begins $0.6''$ upstream of the optically bright region, and the maxima are displaced. The difference in knot D ($\theta_{nuc} = 2.5'' - 4''$) is more subtle: the decline in X-ray flux following maximum is much steeper than in the optical.

The X-ray spectral indices and broadband SEDs indicate that the X-ray emissions of the M87 jet are due to synchrotron radiation (Böhringer et al. 2001, Marshall et al. 2001a). Since the lifetimes of X-ray synchrotron emitting electrons are only a few years, particle acceleration is required. Spectral fits and variability timescales constrain these regions to be a small fraction of the volume of each knot (Harris et al. 1997, Perlman et al. 2001, Marshall et al. 2001a).

Turning to Figure 2, we see that in the 3C273 jet there is no correlation between the optical flux and spectral index. Instead, there is a gradual steepening in α_o with θ_{nuc}, overlaid with small variations, e.g., flattenings in the A-B1, B1-B2 and C1-C2 transitions. Thus physical conditions in the 3C273 jet change remarkably smoothly over scales of many kiloparsecs. A detailed spectral index map also reveals evidence of superposed emission regions in some knots (Jester et al. 2001a). This is very different from the M87 jet, illustrated above.

Jester et al. (2001a) note that the consistency of the ground and *HST*-observed runs of α_o, the gradualness of the spectral changes and the lack of correlation between optical flux and α_o, are consistent with no radiative cooling over scales of many kiloparsecs. But optical synchrotron emitting particles have lifetimes $\sim$ hundreds of years. Jester et al. (2001a) conclude that the only way to escape this paradox is to have continuous reacceleration, throughout the jet's length. This is not inconsistent with stratification, but it is very different from what we see in M87. Optical polarimetry would be invaluable to examine these issues further. Unfortunately the current *HST* polarimetry (Thomson et al. 1993) is too low signal to noise; reobservation is required.

Comparing optical and X-ray images of 3C273 (Marshall et al. 2001b, Sambruna et al. 2001), one sees further differences. Knot A is by far the most powerful in the X-rays, whereas the optical fluxes of all knots are within a factor of two, and the radio maximum occurs in knot H. In addition, the X-ray and optical peaks in knot B appear offset, and no X-ray emission is seen in knot H2. There is debate over the X-ray emission mechanism for knot A: Marshall et al. (2001b) and Röser et al. (2000) favor synchrotron radiation, while Sambruna et al. (2001) claim that Comptonization is required. But these authors agree that the X-ray emissions of the other knots is due to Comptonization of Cosmic

Microwave Background photons. Deeper observations are needed to distinguish between mechanisms.

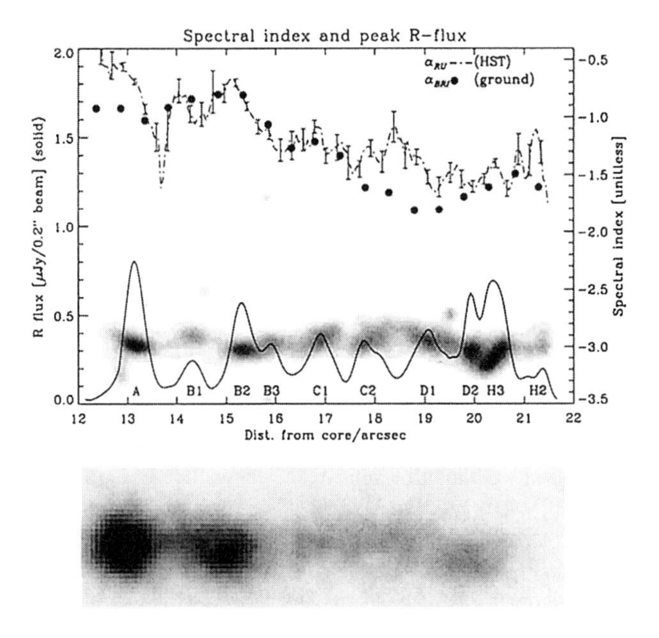

Figure 2. Optical (Jester et al. 2001b, top) and X-ray emissions (bottom) of the 3C273 jet. The top panel shows runs of α_o (points and dashed curve at top) and optical flux (solid curve). The *CXO* image is adaptively smoothed. By contrast to M87, in the 3C273 jet optical flux and α_o are uncorrelated. Also, the X-ray and optical maxima of knot B are offset by $\sim 0.5''$ and X-ray emission is not seen at $\theta_{nuc} \gtrsim 21''$. This implies different physics than seen in the M87 jet.

4. Discussion

There are significant differences between the observed morphologies of the M87 and 3C273 jet, which probably translate to differences in structure and physics. Unfortunately, not all of the observations are in place even for these objects. For 3C273, better *HST* polarimetry is required, and several more bands of deep optical imaging would be very helpful to pin down the α_o map better (compare the error bars in the runs of α_o in Figures 1 and 2). For M87, the magnetic and energetic structures are better constrained, but there are other issues, for example the observed superluminal motion in several knots (Biretta et al. 1999a,b) and the evolution of physical conditions in those components.

We are only scratching the surface of the range of properties in jets. Many new observations are needed. Through cycle 9, four of the twelve known jets brighter than 20 mag/arcsec^2 have only snapshot *HST* observations, and four more have only shallow exposures and poor spectral coverage. Moreover, through Cycle 9, polarimetry had been done only for M87 and 3C273 (3C264 and 3C78 are scheduled for Cycle 10). *CXO* is doing somewhat better; deep observations have been obtained or scheduled for 7 of the 12 brightest optical jets.

References

Biretta, J., Perlman, E., Sparks, W., & Macchetto, F. 1999a, in The Radio Galaxy M87, ed. H.-J. Röser & K. Meisenheimer (Springer), p. 210

Biretta, J. A., Sparks, W. B., & Macchetto, F. 1999b, ApJ, 520, 621

Böhringer, H., et al. 2001, A & A, 365, L181

Burns, J. O., Feigelson, E. D., & Schreier, E. J. 1983, ApJ, 273, 128

Curtis, H. D. 1918, Pub. Lick Observatories, 13, 55

Harris, D. E., Carilli, C. L., & Perley, R. A. 1994, Nature, 367, 713

Harris, D. E., Hjorth, J., Sadun, A. C., Silverman, J. D., & Vestergaard, M. 1999, ApJ, 518, 213

Harris, D. E., Leighly, K. M., & Leahy, J. P. 1998a, ApJL, 499, L149

Harris, D. E., Biretta, J. A., & Junor, W. 1997, MNRAS, 284, L21

Jester, S., Röser, H.-J., Meisenheimer, K., Perley, R., & Conway, R. 2001a, A & A, in press, astro-ph/0104393

Jester, S., Röser, H.-J., Meisenheimer, K., Perley, R., & Garrington, S. 2001b, in Particles and Fields in Radio Galaxies, ed. R. Laing & K. Blundell, (sn Francisco: ASP), in press, astro-ph/0011413

Kirk, J. G., Rieger, F. M., & Mastichiadis, A. 1998, A & A 333, 452.

Kraft, R. P., et al. 2000, ApJL, 531, L9

Marshall, H. L., et al. 2001a, ApJL, submitted

Marshall, H. L., et al. 2001b, ApJ 549, L167

Martel, A. R., et al. 1999, ApJS, 122, 81

Neumann, M., Meisenheimer, K., Röser, H.-J., & Fink, H. H. 1997, A & A, 318, 383

Perlman, E. S., Biretta, J. A., Sparks, W. B., Macchetto, F. D., & Leahy, J. P. 2001, ApJ, 551, 206

Perlman, E. S., Biretta, J. A., Zhou, F., Sparks, W. B., and Macchetto, F. D. 1999, AJ, 117, 2185

Röser, H.-J., et al. 1997, Rev. Mod. Ast., 10, 253

Röser, H.-J., et al. 2000, A & A 360, 99

Sambruna, R. M., et al. 2001, ApJL, 549, 161

Schmidt, M. 1963, Nature, 197, 1040

Schreier, E. J., Gorenstein, P., & Feigelson, E. D. 1982, ApJ, 261, 42

Sparks, W. B., Biretta, J. A., & Macchetto, F. 1996, ApJ 473, 254

Stern, C. P., & Harris, D. E. 1985, BAAS, 17, 831

Thomson, R. C., Mackay, C. D., & Wright, A. E. 1993, Nature, 365, 133

Mass Outflow in Active Galactic Nuclei: New Perspectives
ASP Conference Series, Vol. 255, 2002
D.M. Crenshaw, S.B. Kraemer, and I.M. George

Cygnus A Revisited

Ilse M. van Bemmel

Kapteyn Astronomical Institute, P.O.Box 800, NL–9700 AV Groningen

Abstract. Deep spectro-polarimetry observations of Cygnus A show a redshift of the emission lines in polarized flux on the far side of the galaxy. This shift is interpreted as an outflow of the scattering medium with respect to the nucleus of the object. On the other hand, the coronal lines indicate the matter to flow towards the nucleus, i.e. in the opposite direction. They are only detected on the far side and clearly blueshifted. This apparent contradiction can be explained by assuming that the coronal lines are created on the surface of dusty narrow-line clouds that faces the active nucleus. However, some unsolved problems still remain.

1. The background

Cygnus A has over the past been extensively studied, as it is the closest double-lobed radio galaxy to Earth. So far this has revealed important insights in the detailed physics of active galaxies. Deep *Hubble Space Telescope* images of Cygnus A have shown its optical structure to be quite complex. There are two ionization cones aligned with the radio axis, in which the interstellar medium (ISM) is illuminated and ionized directely by the central active nucleus. In addition a ring of young blue stars is observed, oriented perpendicular to the radio axis, and there is a diffuse older stellar population. Finally, many more small scale structures can be recognized in the form of dust clouds or highly ionized regions (Fosbury 1999).

In this contribution I will present the results obtained with a deep spectropolarimetry observation on the Keck telescope with the Low Resolution Imaging Spectrograph (Oke et al. 1995). Data were obtained in two sets, resulting in a total of over 2 hours observing. More details of the observations are described by Ogle et al. (1997). The spectro-polarimetry is obtained using a slit aligned on the radio axis, which will provide spatial information by subtracting the spectra of the nucleus and two equidistant apertures on the east and the west of it. All apertures are of the same size in order to compare the fluxes and polarization data. From the orientation of the radio jet, it is known that the eastern aperture corresponds to the far side of the galaxy, where the ionization cone is tilted away from the observer. The western aperture is then the near side of the galaxy, where the ionization cone is tilted towards the observer. Datareduction was done using standard IRAF routines and a Monte-Carlo code to obtain unbiased values for polarization, as described by Vernet et al. (2001).

In this paper I will focus on two main issues. First I will discuss the polarization profile of the forbidden emission lines. Second I will present the detection

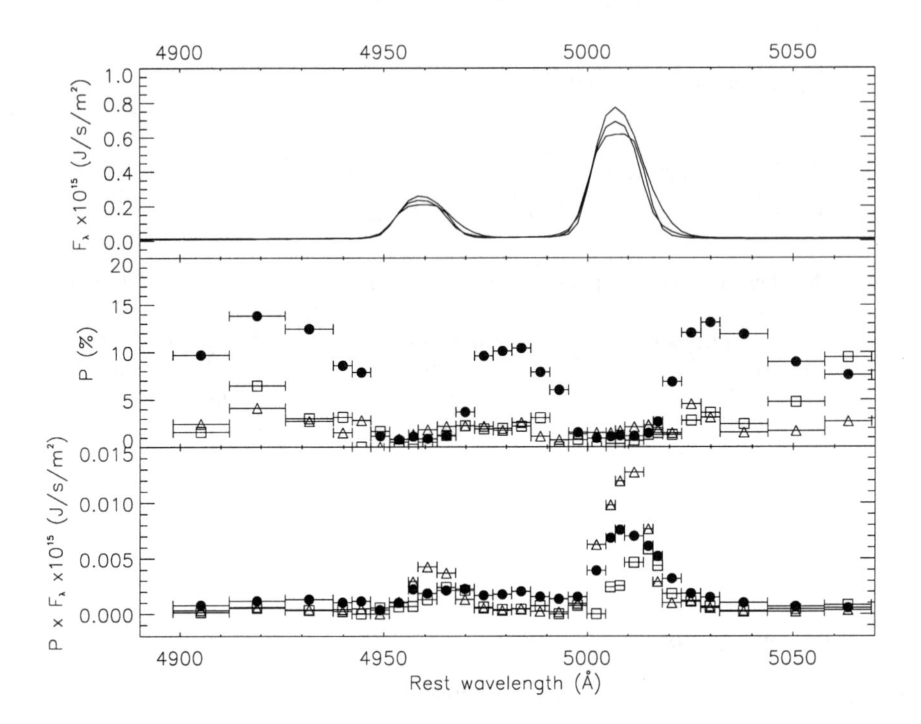

Figure 1. Polarization structure in the [OIII] doublet. From top to bottom is shown the total flux, the fractional polarization and the polarized flux. The apertures are coded as follows: triangles - east, squares - nucleus, circles - west. The total intensity spectrum shows nucleus (top line), east (middle line) and west (lower line) at the emission peak.

of coronal lines in the total flux spectrum. For details on the continuum polarization, see Ogle et al. (1997).

2. Emission line polarization

In order to study the polarization of the emission lines, I will focus on the [OIII] $\lambda\lambda$4959, 5007Å doublet, which are the strongest lines in the spectrum in all three apertures. Thus it provides the best signal to noise to study the polarization over the line.

As shown in Figure 1, three main effects are found:

1. A strong polarized red wing in the nucleus. In the total intensity spectrum it is evident that the nucleus shows a strong red wing. In the polarized flux spectrum this wing appears to be highly polarized.

2. In the fractional polarization a peak is observed on the red side of the lines in all three apertures. The signature is most evident in the western aperture, due to the higher polarization of the continuum. As expected, the polarization is low in the center of the lines.

3. Comparing east and west in the polarized flux, it is clear that an offset to the red is present in the eastern polarized line, while the western polarized line is unshifted.

The nature of the polarized emission in the nucleus is unclear. Because it is so highly polarized and only observed in the nuclear aperture, the emission must come from a region very close to the active nucleus. In that case there are two explanations for the redshift, either it is a broader component of which the blue part is absorbed, or the emission is gravitationally redshifted. So far the real nature of this emission remains unclear.

The peak in fractional polarization has been observed before in planetary nebula NGC 7027 (Walsh & Clegg 1994), and there it is interpreted as an outward movement of the scattering medium. Assuming Cygnus A has an expanding medium in the ionization cones, it is relatively straight-forward to explain the observed behaviour of the polarized flux in the eastern and western apertures.

Because the dust is moving away from the central source, the scatterers always see the photons redshifted. Scattering is equivalent to a quick absorption and re-emission of the photon. On the east the observer sees the dust moving away from him, thus the observed photon obtains a double shift with respect to the observer: first from the movement of the scatterer with respect to the source of emission and second from the movement of the scatterer with respect to the observer. This causes a redshift of the line that is related to twice the velocity of the dust. On the west the observer sees the dust moving towards him, and then the two effects cancel. Now the redshift in the absorption of the photon adds up to a blueshift in the re-emission and no shift in wavelength of the line should be observed. Modelling this process also reproduces the observed red peak in the fractional polarization.

From the observed shift in wavelenght of the polarized line on the east, the velocity of the scattering medium can be obtained. It is measured to be around $100\,\mathrm{km\,s^{-1}}$. This is much smaller than the velocities measured from absorption line complexes in the ultra-violet and X-ray spectra. The outflowing matter should thus be unrelated to these absorbing clouds.

3. Coronal emission lines

Coronal lines are generally defined as emission lines having an ionization potential higher than 100 eV. In most active galaxies they are observed close (~10–100 pc) to the nucleus, with a clear blueshift and a somewhat broader line profile than the narrow lines. The coronal lines in active galaxies are generally assigned to an outflow on the border of the narrow- and broad-line region (Oliva 1996).

Cygnus A shows a different behaviour. Coronal lines are detected up to [ArXI] and [SXII], which have ionization potentials of more than 500 eV. So far [SXII] has only been detected in the Sun and NGC 1068 (Kraemer & Crenshaw 2000). In Figure 2, the relations between the velocity of the line, the width

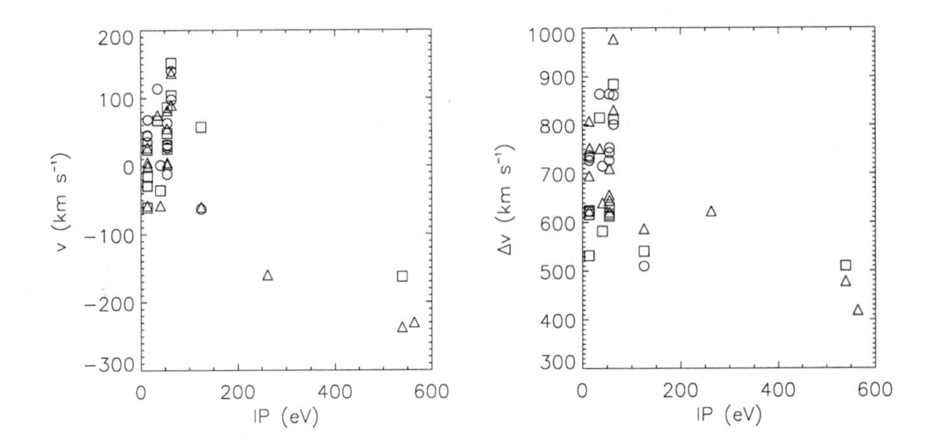

Figure 2. Left panel: relation between observed line velocity with respect to the systemic velocity as measured from [OIII]5007Å versus ionization potential. Right panel: idem, but now for the full width at 70% of the maximum intensity of the line. The apertures are coded as follows: triangles = east, squares = nucleus, circles = west.

of the line and ionization potential are shown. Although the coronal lines are blueshifted, they are not broadened. The latter might be related to the fact that in Cygnus A the coronal lines are observed at $\sim 1\,\mathrm{kpc}$ from the nucleus. This lies well within the narrow-line region and thus it is expected that all lines emerging from this region have comparable widths.

However, the highest ionization lines are observed only in the eastern aperture, and some are detected in the nucleus. No coronal line emission is observed in the western aperture. This contradicts both the explanation for coronal line creation in an outflow, and the detection of an outflowing medium in the ionization cones. Since the east is the far side, the blueshift of the coronal lines indicates the hot gas is moving towards us, while the general motion of the scattering medium is away from us. The general symmetry of Cygnus A requires all processes observed on one side to be present on the other. Thus the absence of coronal line emission on the near side of the galaxy also needs to be explained.

Assuming that the coronal lines are photo-ionized, as are the other narrow emission lines in Cygnus A, there is an explanation that circumvents the inconsistency. If the coronal lines are created on the back side of dusty narrow-line clouds, the hot gas that emits them is evaporating from the surface of the cloud. This motion will go in the opposite direction than the general flow of the ISM. In addition this model will not break the symmetry of Cygnus A, since the illuminated side of the narrow-line clouds is not visible in the western aperture. If the clouds are slightly dusty, the coronal line emission can only escape on the hot side of the clouds and will be absorbed when travelling in the opposite direction. Thus on the near side of the galaxy, where we observe the cool side of the narrow-line clouds, no coronal line emission is expected in this model.

Using the observed flux and assuming the medium producing the coronal lines is photo-ionized, one can calculate the filling factor of the coronal-line clouds. Using literature values for some of the parameters in this calculation, it is found that this model requires the whole space to be filled with clouds in order to produce the observed flux. From detailed observations of narrow-line regions in other active galaxies, it is clear that no more than 10% of the space can be filled with clouds.

There are some assumptions within this model that can cause problems. The main problem lies in the determination of the ionizing spectrum, which relies on an interpolation over the extreme ultra-violet, where no observations can be made. Generally, this interpolation is done with a relatively flat spectrum. However, this method might not reproduce the total of the observed ionizing emission, and thus underestimate the real ionizing flux. As a consequence, the abundance of the atoms is underestimated and thus the filling factor will be overestimated.

Another solution to the apparent anistropy can be that the coronal lines are generated by shocks due to jet-cloud interactions on the eastern side. In case of pure collisional ionization of the coronal lines, their spectrum is clearly different from a photo-ionized spectrum. In addition most jet-cloud interactions result in a bending of the jet, which is not observed in Cygnus A. At this point no further study of this possibility has been made.

4. Conclusions

I have shown with the use of optical spectropolarimetry that an outflow of matter is present in the ionization cones of Cygnus A. This method can be applied to any other astronomical object to determine the motion of matter, as long as the object is polarized. This no longer requires the use of absorption lines to study outflows, although this flow is clearly unrelated to the high velocity clouds in broad-absorption line quasars.

In addition I have discussed the presence of coronal line emission solely on the far side of the host galaxy and at $\sim$1 kpc from the nucleus. The blueshift of the detected coronal lines is inconsistent with models for coronal line creation in active galaxies, but it can possibly be explained by heating of dusty narrow-line clouds.

Acknowledgments. I would like to thank Bob Fosbury, Joël Vernet, Jeremy Walsh, Marshall Cohen & Montse Villar-Martin for their motivation, support and comments. Also I would like to thank Pat Ogle, Hien Tran, Bob Goodrich & Joe Miller for their initial involvement.

References

Fosbury, R.A.E. 1999, in KNAW Coll. Procs., The Most Distant Radio Galaxies, ed. H.J.A. Röttgering, P.N. Best & M.D. Lehnert

Kraemer, S.B., & Crenshaw, D.M. 2000, ApJ, 532, 256

Ogle, P.M., Cohen, M.H., Miller, J.S, et al. 1997, AJ, 452, 371

Oke, J.B., Cohen, J.G., Carr, M., Cromer, J., Dingizian, A., Harris, F.H., Labrecque, S., Lucinio, R., Schall, W., Epps H., & Miller, J. 1995, PASP, 107, 375

Oliva, E, 1996. in Emission Lines in Active Galaxies: New Methods and Techniques, ed. B.M. Peterson, F.Z. Cheng & A.S. Wilson (San Francisco: ASP), 288

Vernet, J., Fosbury, R.A.E., Villar-Martìn, M., Cohen, M.H., Cimatti, A. & di Serego Alighieri, S. 2001, A&A, 366, 7

Walsh, J.R, & Clegg, R.E.S. 1994, MNRAS, 268, L41

Mass Outflow in Active Galactic Nuclei: New Perspectives
ASP Conference Series, Vol. 255, 2002
D.M. Crenshaw, S.B. Kraemer, and I.M. George

Black-Hole Mass and the Formation of Radio Jets in Quasars

Marek J. Kukula

Institute for Astronomy, University of Edinburgh, Royal Observatory, Edinburgh EH9 3HJ, U.K.

James S. Dunlop

Institute for Astronomy, University of Edinburgh, Royal Observatory, Edinburgh EH9 3HJ, U.K.

Ross J. McLure

Department of Astrophysics, Nuclear & Astrophysics Laboratory, University of Oxford, Keble Road, Oxford OX1 3RH, U.K.

Abstract. We summarise the results of a major *Hubble Space Telescope* imaging study of the host galaxies of radio-loud and radio-quiet quasars. We find that for absolute magnitudes $M_R < -24$ the hosts of both types of quasar are massive elliptical galaxies. By applying the relation between galaxy spheroid mass and black-hole mass derived for nearby inactive galaxies we estimate the masses of the central black holes in the quasars and find the first signs of a difference between radio-loud and radio-quiet objects. The black-hole masses of the radio-loud quasars (RLQs) are systematically larger than those of radio-quiet quasars (RQQs) of similar optical luminosity. The implications of this result for the origins of radio loudness are discussed.

1. Introduction

Less than 10% of quasars are strong sources of radio emission and it is now generally accepted that there are two distinct populations of RLQs and RQQs. In radio-loud objects such as radio galaxies and RLQs the radio emission is produced by relativistic jets of plasma, which are ejected from the vicinity of the central engine (e.g., Junor & Biretta 1995) and extend to scales much larger than the host galaxy itself. Recent studies have shown that the central engines of RQQs are also capable of producing collimated outflows of radio-emitting plasma (e.g., Blundell et al. 1996; Kukula et al. 1998), but these structures are typically confined to the nuclear regions of the host and, by definition, are orders of magnitude less luminous than the jets in RLQs. At other wavelengths the spectra of RLQs and RQQs are very similar, and all the evidence suggests that their central engines - believed to be an accretion disk around a supermassive black hole - are essentially the same. The favoured energy source for the radio jets in RLQs and radio galaxies is the extraction of mechanical energy from a spinning black hole (e.g., Blandford 2000, and references therein) and an appealing aspect of this theory is that the radio power is not directly related to the accretion of material (and hence the optical luminosity of the quasar), but direct observational support has been hard to come by.

2. Radio loudness and host galaxy morphology

Nearby, low-luminosity active galactic nuclei (AGN) display marked preferences in terms of host galaxy type, with (radio-quiet) Seyferts favouring spiral hosts whilst (radio-loud) radio galaxies are exclusively associated with massive ellipticals. This suggested that the radio luminosity of an AGN was largely determined by the morphology of its host galaxy, and lead to the expectation that radio-quiet quasars would be found to lie in disk galaxies and radio-loud quasars in elliptical galaxies.

Detailed observations of the galaxies which host quasar activity finally became possible in the last decade, with the launch of the *Hubble Space Telescope (HST)* and improvements in ground-based observing facilities, and a great many studies of quasar hosts have now been carried out (eg Véron-Cetty & Woltjer 1990; Hutchings & Morris 1995; Disney et al. 1995; Bahcall et al. 1997; McLeod, Rieke & Storrie-Lombardi 1999; Hamilton, Casertano & Turnshek 2001). The picture which has emerged demolishes the simple 'radio quiet AGN = disk host' paradigm: whilst the hosts of RLQs are indeed always found to be elliptical galaxies, RQQs occur in both disk and elliptical hosts.

In order to investigate further, we carried out our own *HST* study of low redshift ($0.1 \leq z \leq 0.25$) quasar hosts using carefully matched samples of 11 RLQs and 12 RQQs, all with luminosities in the range $-23 \leq M_V \leq -26$ (McLure et al. 1999; Dunlop et al. 2001). We also included a sample of 10 radio galaxies, matched to the RLQ sample in terms of redshift and steep-spectrum 5GHz luminosity. Observations were made using the $F675W$ filter (roughly equivalent to R-band) on the Wide Field & Planetary Camera; for objects at these redshifts the filter neatly excludes strong emission lines such as [OIII] λ5007 and Hα which might otherwise mask or confuse the underlying distribution of starlight in the host galaxies. Two-dimensional modelling of the images allowed us to separate the contribution of the quasar point source from the extended galaxy light, and to determine the size, luminosity and morphology of the host.

In common with other host galaxy studies we find that the hosts of RLQs are all massive elliptical galaxies, indistinguishable from the radio galaxies in our sample, whilst the RQQs are found in both elliptical and disk-dominated hosts. However, perhaps the most important result to emerge from this work is that for quasars brighter than $M_R \sim -24$ the hosts are invariably massive elliptical galaxies, regardless of radio power (Figure 1). In fact, only the two least luminous RQQs are found to lie in disk-dominated hosts, but the nuclear luminosity of these two objects means that technically they should be classified as Seyfert galaxies rather than quasars. This is an exceptionally clean result when compared to other studies of quasar hosts, some of which find evidence for very luminous RQQs with disk-dominated hosts. (At higher redshifts the situation is still uncertain, though see Kukula et al. 2001). However, the weight

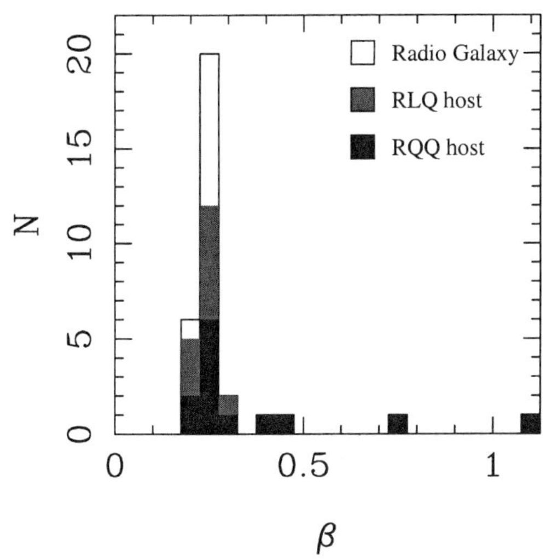

Figure 1. Histogram showing best-fit galaxy morphologies from our modelling of the hosts of RQQs, RLQs and radio galaxies. The model fits the radial surface-brightness profile of the galaxy with a function of the form $I(r) \propto exp(-r^{\beta})$, where $\beta = 1$ corresponds to an exponential disk profile and $\beta = 0.25$ is a de Vaucouleurs spheroidal-galaxy profile. The radio-loud objects and all but two of the RQQ hosts are dominated by a spheroidal component. The two disk-dominated RQQs are the least luminous objects in the sample.

of evidence at low redshifts is that elliptical hosts are increasingly prevalent for RQQs of higher nuclear luminosity.

In our sample the elliptical galaxies containing RQQs are very similar to the hosts of RLQs and radio galaxies; all are luminous ($L \geq 2L^{\star}$), large (typical scalelengths $\simeq$ 10 kpc), and display a Kormendy relation identical to that of normal massive elliptical galaxies in the local universe. A slight caveat to this is that, despite the fact that the RLQ and RQQ samples were designed so that their total optical luminosity distributions were statistically indistinguishable, there is a tendency for the elliptical hosts of RQQs to be slightly less luminous than the RLQ hosts. However, in the current (rather small) samples the trend is not statistically significant, and does not alter the fact that, in this study, the most powerful nuclear activity is overwhelmingly associated with massive elliptical galaxies, regardless of radio luminosity.

In retrospect this result can be seen as an inevitable consequence of the correlations between black-hole mass & galaxy-spheroid luminosity (e.g., Magorrian et al. 1998) and black-hole mass & host velocity dispersion (Ferrarese & Merritt 2000; Gebhardt et al. 2000) derived for local, inactive galaxies. According to these relationships, black holes of the mass necessary to power luminous quasars would only be expected in galaxies with bulge components so massive that they must in turn be classed as giant ellipticals.

3. Black-hole mass and radio loudness

Our 2-D modelling of the images allows us to reliably separate the contribution of the quasar nucleus from that of the host, and to measure the luminosity of the host galaxy's spheroidal component. This information, combined with the relation $M_{bh} = 0.0025 M_{sph}$ (from Gebhardt et al. 2000) and assuming a suitable mass-to-light ratio for the host, enables us to estimate the black hole masses for the RLQs and RQQs in our sample. Due to the large scatter involved in these relationships it is clearly desirable to obtain an independent estimate of the black-hole masses. This has already been done by McLure & Dunlop (2001), using Hβ line widths and assuming simple Keplerian orbits for the clouds in the broad-line region.

We plot these two estimates of black-hole mass against one another in Figure 2. This diagram demonstrates two points. Firstly, despite the large uncertainties inherent in both the host-luminosity/black-hole mass and the Hβ-line width/black-hole mass calculations, the masses derived for the black holes by the two (entirely independent) methods are remarkably consistent. In only three objects (two visible on the diagram) is the difference in the two estimates greater than a factor of ~ 3, and this can easily be understood as an underestimate on the part of the Hβ-M_{bh} relation in objects in which the broad-line clouds orbit close to the plane of the sky. Secondly, the non-linear form of the mass-to-light ratio relation has amplified the slight difference in host galaxy luminosities, with the result that the black-hole masses of the RLQs are systematically larger than those of the RQQs.

This is the first strong evidence for a difference between the RLQs and RQQs in our host-galaxy study. Though the samples are small the difference is significant at the 3-sigma level, and a similar result was obtained by Laor (2000) from a study of the virial masses of a much larger sample of Palomar Green quasars. From our data it seems that a massive black hole ($M_{bh} > 5 \times 10^8 M_\odot$), is necessary to power any quasar of absolute magnitude $M_R < -24$. But black holes capable of producing powerful radio jets are apparently confined to masses greater than $\sim 10^9 M_\odot$. However, the fact that there is a degree of overlap between the mass distributions of RLQ and RQQ black holes shows that black-hole mass cannot be the sole cause of radio loudness. Some other physical parameter must also be involved. In the following section, we consider the possibility that the relevant parameter is black-hole angular momentum.

4. Black-hole angular momentum and radio loudness

The spin of the central black hole has long been proposed as the most plausible energy source for relativistic radio jets. The fact that the angular momentum of the black hole is independent of the current accretion rate would naturally explain why the RQQs and RLQs in our sample can have similar optical luminosities but different radio properties. But why should a high rate of spin be preferentially associated with the most massive black holes, as suggested by our study? If the angular momentum of the black hole was simply a result of the material accreted during periods of quasar activity then in order to maintain a

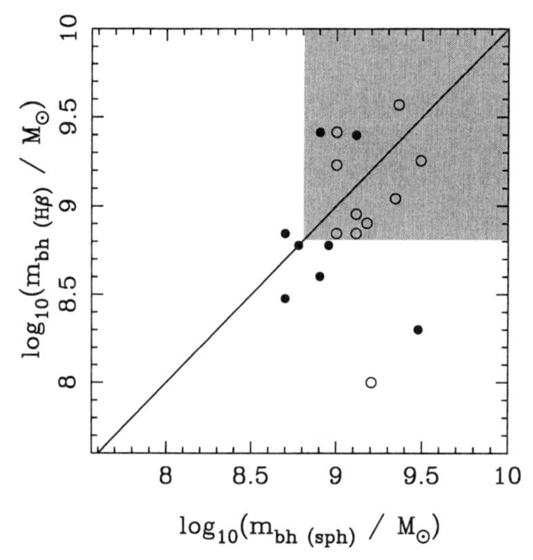

Figure 2. A comparison between of the black-hole masses predicted from host galaxy spheroidal luminosity and those determined from $H\beta$ line-width by McLure & Dunlop (2001). Open circles are RLQs and filled circles RQQs. The shaded area demonstrates that there is a region in which both approaches agree that $M_{bh} > 10^9 M_{\odot}$, and that this region contains all except one of the RLQs, but excludes all except two of the RQQs

high spin rate the angular momentum vector of the accretion disk would have to remain stable over the entire lifetime of the black hole - this seems unlikely.

However, a recent model proposed by Wilson & Colbert (1995) evokes galaxy mergers to forge a link between the mass of the central black hole and its spin. The model assumes that in a merging event involving two galaxies the merger product will retain the morphology of the most massive galaxy, but the spin of the coalesced central black hole will depend critically on the ratio of the masses of the two progenitors. Only a merger between black holes of roughly equal mass will produce a rapidly spinning black hole, capable of launching powerful relativistic jets.

Thus, low mass-high mass galaxy mergers should result in a high mass elliptical galaxy with a high-mass, low-spin black hole. These objects will be capable of forming an optically-luminous quasar, but not of producing powerful radio jets. In contrast, high mass-high mass mergers produce a high mass elliptical galaxy with a high-mass, rapidly spinning black hole. These are the rarest merger events, but they account for the majority of the most massive black holes, and the galaxies they give rise to would form the parent population for RLQs.

5. Conclusions

Clearly our finding that the most massive black holes are more likely to be found in RLQs is consistent with the Wilson & Colbert scenario. Interestingly the quasar hosts in our study occupy the same region of the surface brightness - half-light radius projection of the fundamental plane as massive inactive ellipticals with 'boxy' central isophotes (Genzel et al. 2001), which may indeed suggest that they are the product of massive-galaxy mergers (Naab & Burkert 2001). The scatter in the mass distribution of merging galaxies also provides a natural explanation for the observed overlap in RQQ and RLQ black-hole masses. It is too soon to say whether this picture is correct; the current sample is relatively small, and there are many other uncertainties which need to be resolved, including the exact role of mergers in galaxy formation, timescales for the coalescence of the central black holes, and the relationship between the quasar phenomenon and the evolution of the parent galaxy population. However, black-hole mass distributions are providing the first signs of a fundamental difference between RLQs and RQQs which might begin to explain the radio-loud/radio-quiet dichotomy. A more detailed discussion of these results and their implications can be found in Dunlop et al. (2001).

References

Bahcall, J.N., et al. 1997, ApJ 479, 642
Blandford, R. 2000, astro-ph/0001499
Blundell, K.M., et al. 1996, ApJ, 468, L91
Disney, M.J., et al. 1995, Nature, 376, 150
Dunlop, J.S., et al. 2001, MNRAS submitted
Ferrarese, L., & Merritt, D. 2000, ApJ, 539, L9
Gebhardt, K., et al. 2000, ApJ, 539, L13
Genzel, R., et al. 2001, astro-ph/0106032
Hamilton, T.S., Casertano, S., & Turnshek, D.A. 2001, astro-ph/0011255
Hutchings, J.B., & Morris, S.C. 1995, AJ, 109, 1541
Junor, W., & Biretta, J.A. 1995, AJ, 109, 500
Kukula, M.J., et al. 1998, MNRAS, 297, 366
Kukula, M.J., et al. 2001, MNRAS in press (astro-ph/0010007)
Laor, A. 2000, ApJ, 543, L111
Magorrian, J., et al. 1998, AJ, 115, 2285
McLeod, K.K., Rieke, G.H., & Storrie-Lombardi, L.J. 1999, ApJ, 511, L67
McLure, R.J., et al. 1999, MNRAS, 308, 377
McLure, R.J., & Dunlop, J.S. 2001, MNRAS, in press (astro-ph/0009406)
Naab, T., & Burkert, A. 2001, astro-ph/0103476
Véron-Cetty, M.-P., & Woltjer, L. 1990, A&A, 236, 69
Wilson, A.S., & Colbert, E.J.M. 1995, ApJ, 438, 62

Mass Outflow in Active Galactic Nuclei: New Perspectives
ASP Conference Series, Vol. 255, 2002
D.M. Crenshaw, S.B. Kraemer, and I.M. George

General Relativistic Simulations of Jet Formation by a Rapidly Rotating Black Hole

Shinji Koide

Toyama University, 3190 Gofuku, Toyama 930-8555, Japan

David L. Meier

Jet Propulsion Laboratory, 4800 Oak Grove Dr. Pasadena, CA 91109, USA

Kazunari Shibata

Kwasan and Hida Observatories, Kyoto University, Yamashina, Kyoto, 607-8471, Japan

Takahiro Kudoh

National Astronomical Observatory, Mitaka, Tokyo 181-8588, Japan

Abstract. We have performed general relativistic magnetohydrodynamic simulations of jet formation in accretion disks around rapidly rotating black holes. Here we compare the case where the disk rotates in the same sense as the black hole with the opposite, counter-rotating, case.

Radio observations of active galactic nuclei and "microquasars" in our Galaxy (Mirabel & Rodriguez 1994) have revealed compelling evidence of existence of relativistic jets. It is believed that the cores of these objects contain a rapidly rotating black hole and that magnetic phenomena occurring near the hole produce the relativistic jets (Blandford & Znajek 1977; Blandford & Payne 1982; Shibata & Uchida 1986; Kudoh & Shibata 1997). To simulate jet formation in the magnetosphere of a black hole, we have newly developed a Kerr general relativistic magnetohydrodynamic (KGRMHD) code (Koide et. al. 1998; Koide et. al. 1999; Koide et. al. 2000). We report here on the numerical results of our jet formation simulations.

Our study is based on the general relativistic conservation laws of mass, momentum, and energy in conducting fluids and on Maxwell equations. We use the Kerr metric with spin parameter $a = 0.95$, which describes the space-time around the nearly maximally rotating black hole. We employ the *simplified total variation diminishing* (TVD) method to integrate the partial differential equations. For the initial conditions, we assume a transonic free fall corona and relativistic Keplerian disk with a uniform magnetic field.

Figure 1 shows the simulation results when the accretion disk is rotating *counter* to the black hole rotation at a time $t = 47\tau_S$. Here τ_S is the unit of time, $\tau_S = r_S/c$, where r_S is Schwarzschild radius and c is speed of light. The disk falls toward the black hole rapidly because no stable orbit exists for $R < 4.5r_S$, allowing part of the magnetized disk to enter the *ergosphere* ($R < r_S$) quickly. The jet has two layers: the plasma beta of the head and skin of the jet is high and the magnetic field azimuthal component B_ϕ is negative, while that of the root and the center of the jet is low and B_ϕ is positive. The high beta jet is a gas pressure driven jet produced by a shock that develops in the disk, while the low beta jet is a magnetically-driven jet caused by the strong magnetic field and rapid field rotation due to frame dragging near and inside the ergosphere. This acceleration mechanism is thought to be identical to the Blandford-Znajek mechanism (Blandford & Znajek 1977). At $t = 47\tau_S$, the jet of the counter-rotating disk case is accelerated mainly by the electromagnetic force; in contrast, in the case of a co-rotating disk, the jet is due to gas pressure alone (in other words, no low beta jet is formed at this time).

We thank M. Koide for her important comments on this study and also thank NIFS for the use of their super-computers.

References

Mirabel, I.F., & Rodriguez, L.F. 1994, Nature, 371, 46
Blandford, R.D., & Znajek, R. 1977, MNRAS, 179, 433
Blandford, R.D., & Payne, D.G. 1982, MNRAS, 199, 883
Shibata, K., & Uchida, Y. 1986, PASJ, 38, 631
Kudoh, T., & Shibata, K., 1997 ApJ, 476, 632
Koide, S., Shibata, K., & Kudoh, T. 1998, ApJ, 495, L63
Koide, S., Shibata, K., & Kudoh, T. 1999, ApJ, 522, 727
Koide, S., Meier, D. L., Shibata, K., & Kudoh, T. 2000, ApJ, 536, 668

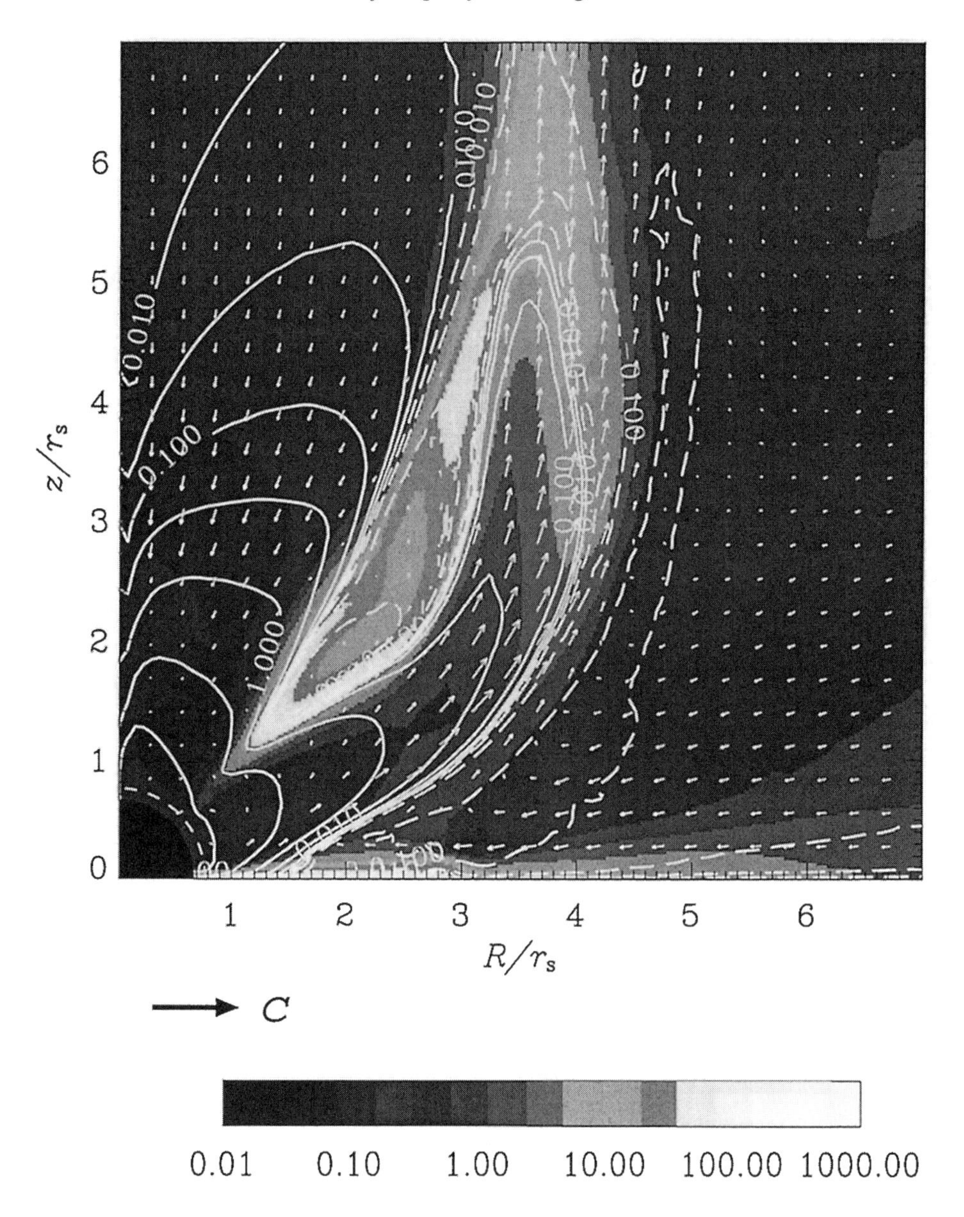

Figure 1. The plasma beta (gray-scale) and the azimuthal component of the magnetic field B_ϕ (contour) of the counter-rotating disk case at $t = 47\tau_S$. The solid lines show contours of positive B_ϕ and the dashed lines negative B_ϕ. The arrows indicate velocity vectors, and the black fan-shaped region at the origin shows the horizon of the Kerr black hole. The dashed line near the horizon is the inner boundary of the calculation region.

Mass Outflow in Active Galactic Nuclei: New Perspectives
ASP Conference Series, Vol. 255, 2002
D.M. Crenshaw, S.B. Kraemer, and I.M. George

Spectropolarimetry of Broad Hα Lines and Geometry of the BLR

Marshall H. Cohen

California Institute of Technology, Pasadena, CA 91125

André R. Martel

Johns Hopkins University, 3400 N. Charles Street, Baltimore, MD 21218

Abstract. In a small fraction of Broad Line Radio Galaxies (BLRG) and Seyfert 1 galaxies, the polarization position angle rotates across the broad emission lines, especially at Hα. An understanding of this behavior can potentially yield important information on the scattering geometry in the nucleus. We show two examples of this phenomenon, 3C 445, a BLRG, and Mrk 231, a Seyfert 1, and present an equatorial scattering model that explains some of its features in a straightforward way.

1. Introduction

The literature contains a number of examples of rotation of the polarization position angle (*PA*) in Hα in Seyfert 1 galaxies and in broad-line radio galaxies (BLRG). This has usually been explained as radiation from the nuclear sources (continuum and broad line region [BLR]) being scattered from separated clouds, so that different Doppler-shifted wavelengths are seen at different *PA*. The most elaborate discussion of this is by Martel (1998) who decomposed Hα in NGC 4151 into components each with its own *PA*. Several of the components were able to be associated with various features in the galaxy. Such an interpretation, dominated by radial motions, suffers from a lack of knowledge of the relative location and velocities of the emitters and scatterers.

In this paper we investigate a different possibility, that the rotations are caused by scattering of Hα radiation on nearby clouds which see both red- and blue- shifted radiation but from different directions. The scattering planes are different for the red- and blue- shifted rays and an integration gives a *PA* that can rotate through Hα. A discussion of this idea is in Cohen et al. (1999; hereafter C99) and it is mentioned by Goodrich & Miller (1994).

2. Examples

2.1. Mrk 231

Figure 1 shows (a) total flux, (b) polarization *P*, and (c) position angle *PA* for the Hα region of Mrk 231 (Martel 1996; hereafter M96) and 3C 445 (C99). Note that *P* is S-shaped in Mrk 231, decreasing on the blue side and increasing on

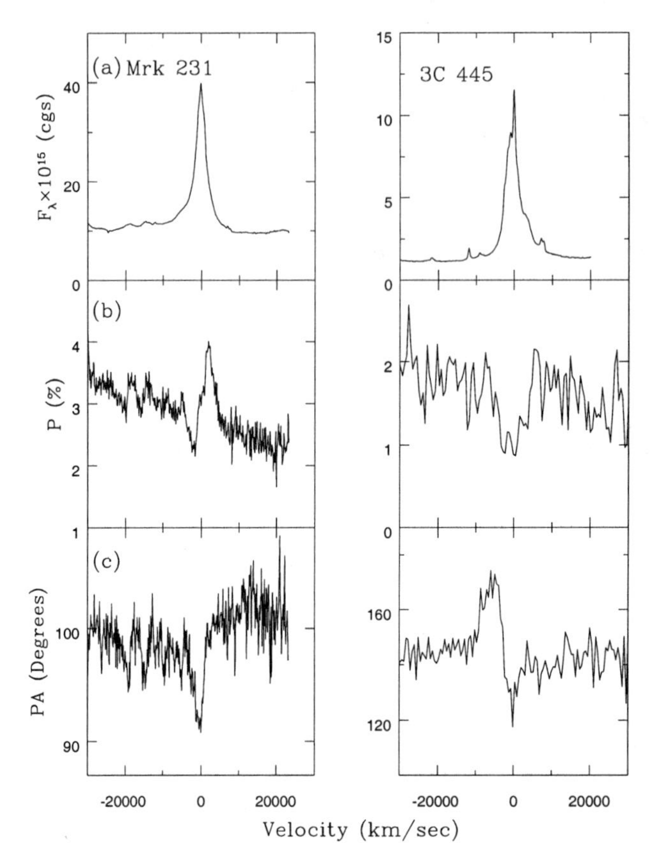

Figure 1. Hα spectral region of Mrk 231 and 3C 445 : (a) total flux (b) percentage polarization (c) *PA*. Note the S-shaped changes in *P* in Mrk 231 and in *PA* in 3C 445.

the red side of Hα. This shows that the *PA* of the Hα flux is different in the two wings; and that either *P* is higher in Hα than in the continuum, or (more likely) that the dilution of *P* by unpolarized components is lower in (continuum + Hα) than in the continuum alone. In (c), it can be seen that the *PA* of Hα is similar to that of the continuum in most of the red wing, but *PA*(blue) is smaller. In fact the intrinsic (continuum-corrected) shift in *PA*(blue) must be rather substantial, because in (b) the dip in the blue is nearly as big as the rise in the red. M96 modelled this line with 5 Gaussian components, each with its own wavelength and *PA*. His result is that there is a trend in *PA* corresponding to velocity, and the *PA* difference between the red and blue wings is about 30°.

2.2. 3C 445

Spectra for the BLRG 3C 445 are shown in the right-hand panels of Figure 1 (C99). The polarization fraction *P* has a weak S pattern, but *PA* has striking changes across Hα. A decomposition of the flux into line and continuum components (C99) shows that $PA(\mathrm{H}\alpha)$ has about a 45° change from the red to the blue wing, and much of the change occurs about 2500 km s^{-1} to the blue of the line center.

2.3. Others

The Hα polarization in seven Seyfert 1 galaxies is discussed in M96. For most, the continuum and line polarizations are low, $\sim$ 1%, and they all show structure across their Hα profiles in both P and PA. Mrk 304 and NGC 3516 are similar to Mrk 231 in having a strong S shape in P, and the S pattern is offset to the blue by perhaps 2500 km s^{-1} in Mrk 304. As for Mrk 231, the PA also changes where P dips. In Mrk 704, the PA rotates about 8° across the central 5500 km s^{-1} of Hα. A few BLRG, in addition to 3C 445, show polarization changes in Hα. 3C 227 has both Hα and Hβ offset by $\sim$ 20° from the continuum; in this case the PA of the line radiation itself is approximately constant (C99).

Many broad-absorption-line quasars show PA rotations (Ogle 1998, Ogle et al. 1999), especially in the deep absorption troughs. It is not clear if these rotations are generically similar to those seen in the emission lines of BLRG and Seyfert 1's.

3. An Equatorial Model

Figure 2 shows the equatorial plane of an idealized active galactic nucleus (AGN). Hα-emitting clouds orbit with velocity v on the inner circle of radius r, and scattering clouds lie on the outer circle of radius R ($\rho = r/R$). Both the Hα and scattering clouds are uniformly distributed around the circles, and the former radiate isotropically. The observer is in the $y - z$ plane at angle θ from the z-axis. AGN unification scenarios generally assume that BLRG are seen near the boundary of an obscuring dusty torus, which shields the central quasar from direct view. In Figure 2, the projection of the torus' edge on the equatorial plane is represented by arc AC; the dashed region is hidden, and the observer has an asymmetric view of the equatorial plane. The asymmetry is necessary to obtain rotation of the PA, and also for the model to mimic the observed spread between the optical continuum PA and the radio axis, which we assume to be projected onto the y axis. Point B receives both red- and blue-shifted radiation but the scattering planes are different so the red and blue wings of the scattered line will be at different position angles. When this is integrated over the visible arc ABC, the PA can rotate as a function of velocity.

Dust is much more efficient than electrons at scattering and so we take R to be at the inner wall of the dusty torus. If this is near 1 pc and the BLR is at a few light months, then $\rho \sim 0.1$. This picture is very simplified since in reality the torus will be irregular (think of the Galactic Center). It would be more realistic to put clouds randomly near the outer circle and to vary R; and also of course to add structure in the z direction for both the emission and scattering clouds. These elaborations are not warranted at present.

In the examples shown below, we have taken $\rho = 0.1$ and the only free parameters are θ and the points A and C which define the visible arc. In fact, a wide range of results can be obtained by varying A and C, essentially because the integration adds together many vectors with a wide range of angles, and the

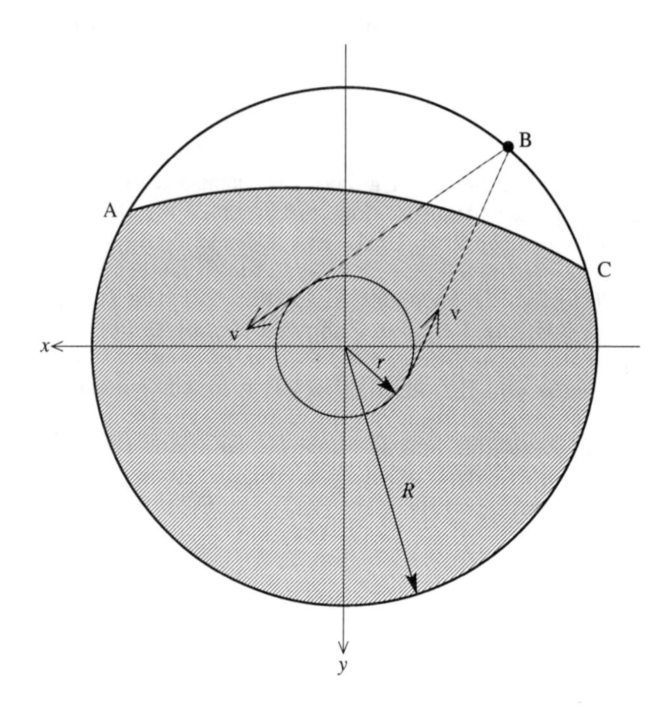

Figure 2. Equatorial plane of an AGN viewed along the z-axis. Hα clouds orbit with velocity v on the inner circle. Emission from the clouds is scattered by material on the outer circle and is seen by an observer at angle θ from the z-axis. A dusty torus hides the hatched region from direct view. At point B, the scattering planes are different for the blue and red rays, and the observed blue and red polarizations are at different *PA*'s.

(often small) residual vector can depend critically on the details of cancellation. For the results shown here we weigh radiation in the equatorial plane by the inverse square of the distance and with a cosine that makes the opposite point invisible.

Light from the central continuum source, which has no motions in our model, will also be scattered from the arc ABC, and its *PA* will be the same as that of the zero-velocity point of Hα. However, in some AGN these *PA*'s differ. This can be accommodated in our model by adding another velocity, an outflow of Hα, or circular or radial motions of the scatterers. This opens up a wide range of parameter space which we have explored only in a limited fashion. Here, we only discuss models where the Hα clouds have circular motions and the scatterers are stationary.

If the arc ABC is strictly symmetric around the x or y axis then the continuum polarization will be parallel or perpendicular to the radio axis, respectively.

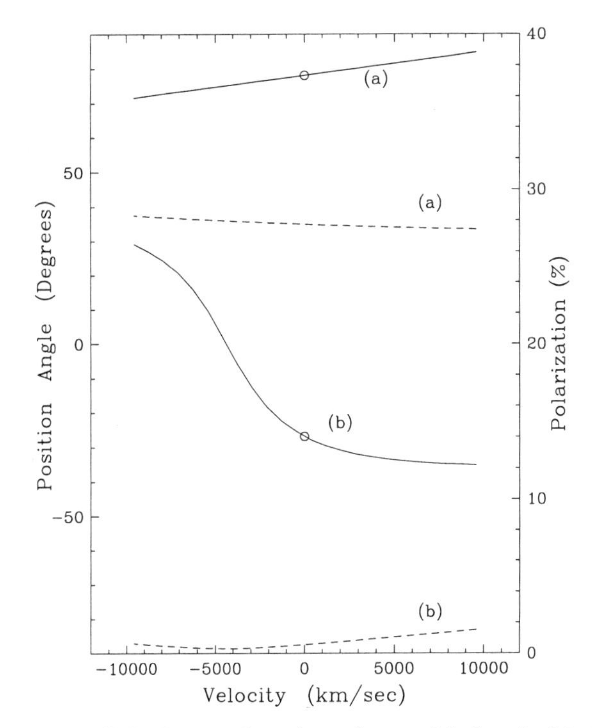

Figure 3. Calculations based on the model sketched in Figure 2 with $\rho = 0.1$, $\theta = 20^\circ$, and $v = 10,000$ km s^{-1}. Solid curves are the position angle, and dotted curves are the percentage polarization, of the scattered Hα radiation. (a) arc ABC is from 1.1π to 1.8π; (b) arc ABC is from 1.05π to 2.0π. The open circles show the *PA* of the continuum.

But we expect that in general the x axis will be obscured and the arc will be more or less as shown. This means that radio-optical *PA* differences will avoid 0°, but values near 90° will not be common either. This is in accord with observations of BLRG; for five BLRG in C99 the median difference is 31°. If we take the radio axis to be given by the smallest-scale observations with the VLBA, and the optical *PA* to be that of the continuum near Hα, then the radio-optical differences for Mrk 231 and 3C 445 are 33° and 31°, respectively (M96; Ulvestad et al. 1999; C99)

4. Results

Figure 3 shows two examples of the many possibilities that can be obtained with the model. To compare them with observations (Figure 1) it is necessary to combine Hα and continuum components. In case (a), the *PA* rotates uniformly and the continuum *PA* is in the middle of the Hα range. This results in an

S-shaped *PA* curve as in 3C 445. However, two features of the observation are inconsistent with the model: the measured curve is symmetric around -2500 km s^{-1} instead of 0 km s^{-1}, and the radio-optical *PA* difference is 31°, rather far from the model 75°. These discrepancies, especially the first, appear to be rather difficult to resolve; but adding a velocity (e.g., circular motion of the scatterers) introduces many possibilities including offsetting the *PA* of the zero-velocity component from the continuum. The polarization in the scattered light is high, nearly 30%, which is good because there also could be non-scattered components of both Hα and continuum, which would dilute the polarization.

In case (b) the continuum *PA* is close to that of the red wing of Hα, and this gives *P* and *PA* curves like those in Mrk 231. The model radio-optical *PA* difference is 27°, close to the observed value of 33°. However, the cancellation of the vectors is nearly complete, and the net polarization is only about 1%. The observed polarization is 3%, and even without dilution the model produces insufficient polarization. This is a serious consideration, since we do expect that there will be other components to the continuum and line radiation. There might be attenuated and reddened light seen directly fron the BLR and the central continuum source; C99 estimate that for 3C 445 the central source is seen through about 2 magnitudes of extinction.

We conclude that the equatorial scattering model in Figure 2 can explain some of the features seen in Type 1 AGN polarization data. Discrepancies remain but these may be resolved in the future by exploring the parameter space of radial and circular motions of the emitters and scatterers.

References

Cohen, M.H., Ogle, P.M., Tran, H.D., Goodrich, R.W. & Miller, J.S. 1999, ApJ, 118, 1963 (C99)

Goodrich, R.W. & Miller, J.S. 1994, ApJ, 434, 82

Martel, A.R. 1996, Ph.D. thesis, University of California, Santa Cruz (M96)

Martel, A.R. 1998, ApJ, 508, 657

Ogle, P.M. 1998, Ph.D. thesis, California Institute of Technology

Ogle, P.M., Cohen, M.H., Miller, J.S., Tran, H.D., Goodrich, R.W & Martel, A.R. 1999, ApJS, 125, 1

Ulvestad, J.S., Wrobel, J.M. & Carilli, C.L. 1999, ApJ, 516, 127

Mass Outflow in Active Galactic Nuclei: New Perspectives
ASP Conference Series, Vol. 255, 2002
D.M. Crenshaw, S.B. Kraemer, and I.M. George

How Much of the Broad-Line Region is Outflowing?

C. Martin Gaskell and Victoria Y. Mariupolskaya

Dept. Physics & Astronomy, Univ. Nebraska, Lincoln, NE 68588-0111

Abstract. The blueshifting of the high-ionization broad emission lines in AGN is probably caused by outflow of high-ionization gas coupled with obscuration by a disk but there has been no consensus on the overall kinematics of the BLR gas and the relationship of the low-ionization BLR gas to the high-ionization BLR gas. We have studied the widths of high- and low-ionization emission lines in *Hubble Space Telescope*, *IUE*, and ground-based spectra. We confirm that there is considerable scatter between the C IV and Mg II FWHMs, but nonetheless there is a highly-significant correlation and the profiles can be identical. The widths of these lines and Hβ are correlated over the full range of widths from Narrow-Line Seyfert 1s (NLS1s) to Broad-Line Radio Galaxies (BLRGs). The correlations imply that the kinematics of the high and low-ionization gas are related, and support an underlying unity among the different classes of AGN. A significant part of the low-ionization BLR gas is thus probably outflowing with the high-ionization BLR. One problem any theory of BLR structure and kinematics must explain, however, is the absence of lags between the blue and red wings of lines as they vary.

1. Introduction

Intrinsic absorption lines provide incontrovertible evidence for outflow in active galactic nuclei (AGN). The kinematics and dynamics of broad *emission*-line region (the BLR) remain uncertain but acceleration of BLR clouds by radiation pressure has long been given serious consideration (e.g., Mathews & Blumenthal 1975). Since the broad absorption line (BAL) gas is clearly being accelerated by radiation pressure, radiation pressure must be an important factor for BLR gas too. The similarity of absorption-line and BLR velocities (hundreds to thousands of km s^{-1}) provides some circumstantial evidence in favor or outflow of the BLR. The high-ionization lines in quasars are blueshifted relative to the lines arising from the low-ionization gas (Gaskell 1982). For convenience we will call the gas most of the high-ionization BLR lines come from "BLR I" and the gas most of the low-ionization lines come from, "BLR II". The blueshifting requires radial motion of the BLR I gas and some opacity source (absorption or scattering). One possibility is that BLR I is outflowing and our view of the gas receding on the far side of the quasar is blocked by the accretion disk (Gaskell 1982). This fits in naturally with the disk-plus-wind outflow models discussed elsewhere in these proceedings.

BLR I and BLR II have a number of differences (see Gaskell 2000). The blueshifting of the BLR I lines suggests that they arise from a different kinematic component from BLR II. The main questions we want to address here are, what is the relationship between the kinematics of BLR I and BLR II? In particular, if BLR I is outflowing, is BLR II outflowing also? (or, how much of BLR II is outflowing?) If BLR I arises predominantly from the same wind or bi-conical outflow as the absorbers while BLR II arises predominantly in a disk, their line widths could be quite unrelated and would show different behaviors with orientation: the BLR I lines would be broadest in a face-on orientation, while the BLR II lines would be narrowest face-on.

There is no ideal way to compare line profiles. Here we simply investigate the relationship between the full widths at half maxima (FWHMs). Although there are obvious problems with FWHMs (see below) comparing them is nonetheless interesting. We have chosen to study the FWHMs of the high-ionization C IV line and the low-ionization Mg II and Hβ lines. These lines were chosen because of their strength and relative freedom from blending.

Previous studies have produced differing results. Mathews & Wampler (1985) and Brotherton et al. (1994) found only a weak correlation between Mg II and C IV FWHMs. Corbin (1991) found a good correlation between the Hβ and C IV widths, but Ganguly et al. (2001) found almost none. Given these differences we decided to enlarge the samples and investigate the relationships between the FWHMs of all three lines.

2. Full Widths at Half Maximum

We have compared the FWHMs of C IV, Mg II, and Hβ in *Hubble Space Telescope* (*HST*), *International Ultraviolet Explorer* (*IUE*), and ground-based spectra of over 200 quasars. Where possible we favored *HST* spectra over *IUE* spectra, but the lines are almost always broad enough that the lower resolution of the *IUE* has no effect on our conclusions. The objects cover all major classes of AGN and range from Narrow-Line Seyfert 1s (NLS1s) to broad-line radio galaxies (BLRGs).

There are several problems with using FWHMs. We believe that blending of different components with differing line ratios is the most fundamental one. For lines showing displaced-peaks (disk-like profiles) the FWHM is a poor characteristic of the line width. In some AGN intrinsic absorption also has to be allowed for.

3. Results

3.1. Effect of Line Variability

Simultaneous spectra of different wavelength regions are available for very few objects and emission line variability can potentially cause measured FWHMs to vary, either because the intrinsic line widths actually change, or because the ratio of contributions changes. We found no systematic difference between the FWHMs of RMS-difference spectra, and average spectra. For the 35 objects presented in Wandel, Peterson, & Malkan (1999) and Kaspi et al. (2000) we

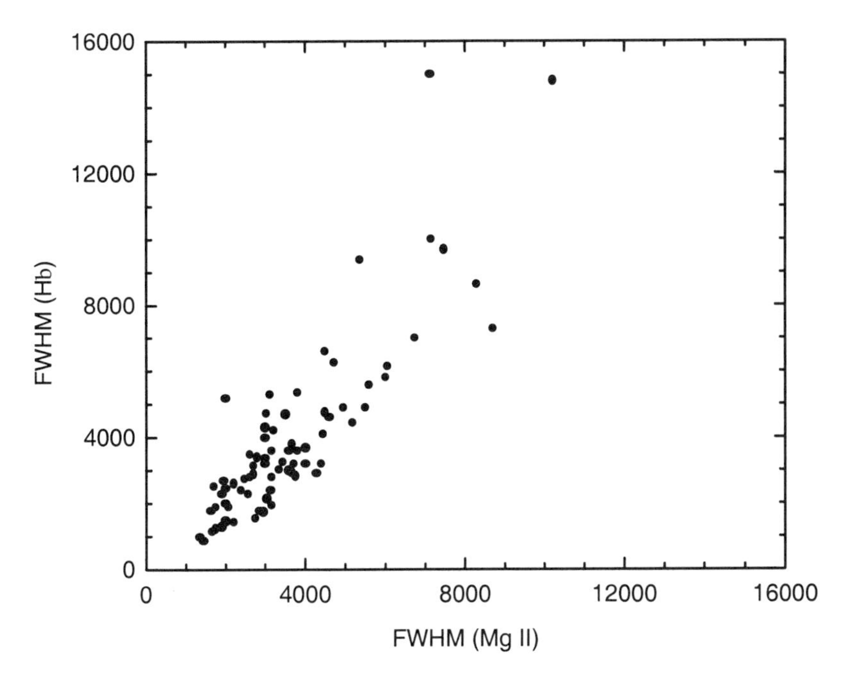

Figure 1. The FWHMs of Mg II and Hβ

found FWHM_{avg} - $\mathrm{FWHM}_{RMS} = 13 \pm 140$ km s^{-1}. For the well-studied AGN NGC 5548 (Wanders & Peterson 1996) we also found no change in the mean FWHM over many observing seasons (again after allowance for the NLR component). Lack of simultaneity is therefore not a major source of scatter when comparing FWHMs of different lines.

3.2. Mg II and Hβ

For all classes of object the Mg II and Hβ FWHMs are well correlated (see Figure 1). For radio-quiet and radio-compact AGN the FWHMs are, on average, identical and the scatter (24%) is consistent with the measuring errors so in these objects the intrinsic profiles of the two lines are probably nearly identical. The similarity of profiles strongly supports the idea that these lines arise from the same clouds. Grandi (1980) and Persson (1988) similarly found O I λ8446 and the Ca II IR triplet to have the same width as Hβ.

For the broader line objects (FWHM > 5000 km s^{-1}) Hβ is $\sim$ 50% broader than Mg II. Inspection of spectra shows that this is due to the disk component in some AGN being stronger in Hβ. These are usually AGN with extended radio structure (e.g., BLRGs). Even for these objects, however, detailed profile comparisons (e.g., Grandi & Phillips 1979) show the similarity of the profiles.

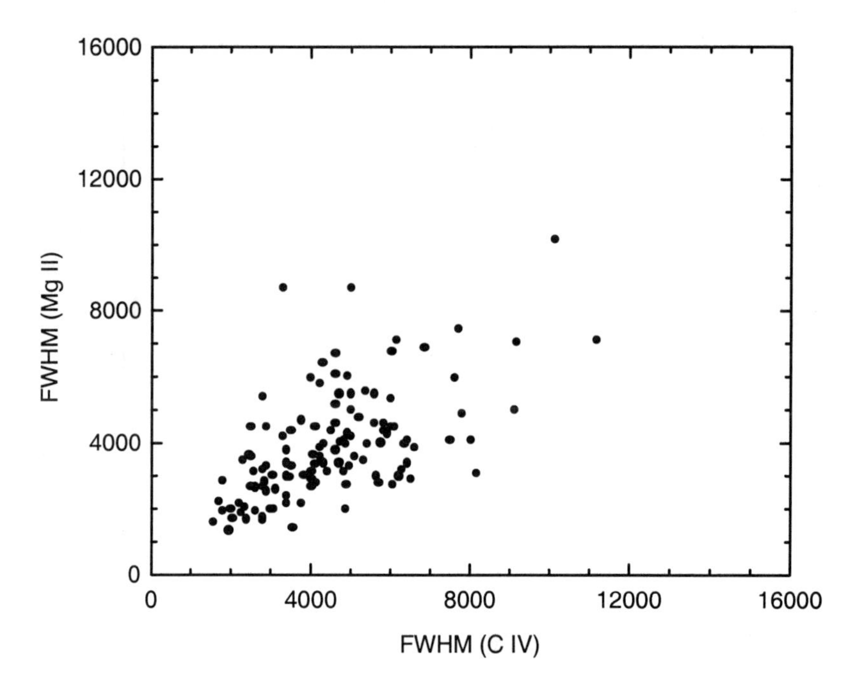

Figure 2. The FWHMs of C IV and Mg II

3.3. High-Ionization Lines

Given the general agreement between the Mg II and Hβ FWHMs, especially for radio-quiet and radio-compact AGN, we believe that either can be regarded as the BLR II FWHM. We confirm that there is considerable scatter between the C IV and Mg II FWHMs (see Figure 2), but nonetheless we find that there is a highly-significant correlation ($>$ 99.99% significance). We believe we find a much more significant correlation than previous studies because of *(i)* our larger sample size, and *(ii)* the presence of NLS1s and BLRGs in our sample.

The C IV and Hβ FWHMs show a similar degree of scatter (Figure 3) as the C IV and Mg II FWHMs, but also a similarly significant correlation. Since we believe that the uncertainties in the FWHM estimates are reflected by the scatter in Figure 1, the greater scatter in Figures 2 and 3 represents intrinsic scatter between the FWHMs of BLR I and BLR II.

Mathews & Wampler (1985) discovered a tendency for C IV to be broader than Mg II. We confirm that C IV is broader for radio-quiet and radio-compact AGN by 35%. We find C IV similarly to be broader than Hβ (by 20%). Brotherton et al. (1994) found the FWHM(C IV)/FWHM(Mg II) ratio to be larger for radio-quiet AGN than for radio-loud ones. We confirm this and find a similar difference for the widths of C IV and Hβ. However, for the radio-extended AGN we find the mean C IV and Mg II FWHMs to be identical. For the extended radio sources Hβ can be much wider than C IV. We believe this is because of the disk component of Hβ.

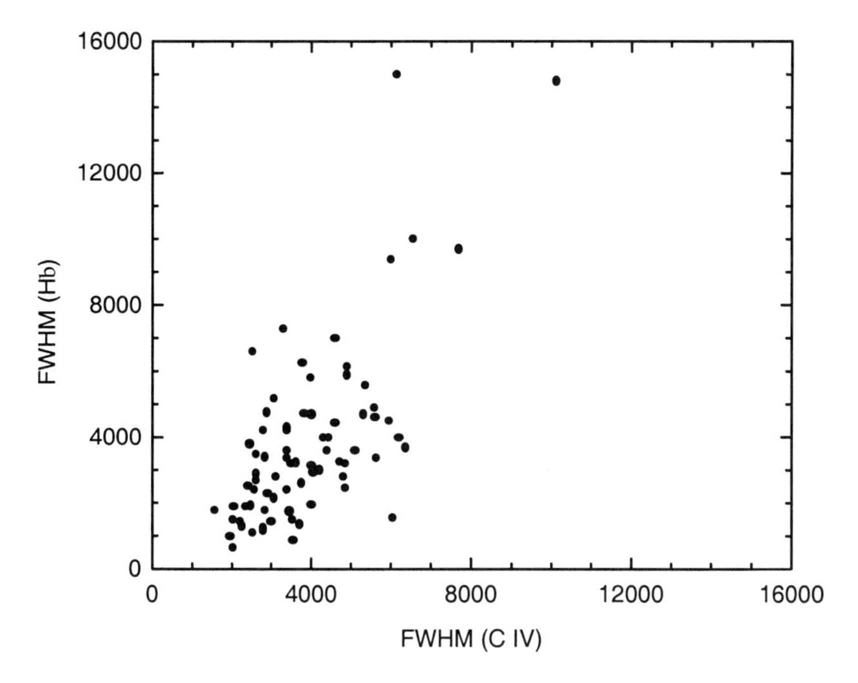

Figure 3. The FWHMs of C IV and Hβ

While detailed profile comparisons often show that BLR I and BLR II lines have different profiles (e.g., Shuder 1982; Crenshaw 1986; Zheng 1992), it is important to note that the C IV and Mg II profiles can sometimes be *identical* even in radio-quiet objects. NGC 3516 is a notable example of this (Goad et al. 1999).

4. Discussion

The highly-significant correlations of the FWHMs of BLR I and BLR II lines for all classes of object, and for a very wide range of luminosity, strongly support the idea that the kinematics of the BLR I and BLR II gas are related. If much (or all) of BLR I is outflowing, as the evidence seems to suggest, then a significant part of the gas producing the BLR II emission is probably *also* outflowing. Some BLR profiles, however, contain the clear signature of emission from a disk, and Gaskell & Snedden (1999) argue that a disk component is present to some degree in the profiles of all broad lines in all AGN. It can be seen in Figure 3 that the AGN with the broadest Hβ (because of a disk component) also have the broadest C IV. This can be explained by the presence of a disk contribution to C IV.

One major observational problem for outflow models must be considered. When broad emission lines vary, reverberation mapping studies of both high- and low-ionization lines do not show the expected leading of the blue-shifted wings relative to the red-shifted wings (Gaskell 1988; Koratkar & Gaskell 1989, 1991; Crenshaw & Blackwell 1990). Given the compelling evidence that much of the BLR *is* outflowing, the solution must be that the red-wing/blue-wing delay does not give the direction of motion in the simple manner one would expect. The models of Chiang & Murray (1996), Bottorff et al. (1997), and Taylor (1999) offer solutions of this sort.

Acknowledgments. We wish to thank all the conference organizers and especially Mike Crenshaw and Steve Kraemer for their efforts in organizing this stimulating and timely workshop. This work has been supported in part by grant AR-05796.01-94A from the Space Telescope Science Institute, which is operated by AURA, Inc., under NASA contract NAS5-26555

References

Bottorff, M., Korista, K. T., Shlosman, I., Blandford, R. D. 1997 ApJ, 479, 200

Brotherton, M. S., Wills, B. J., Steidel, C. C., Sargent, W. L. W. 1994, ApJ, 423, 131

Chiang, J. & Murray, N. 1996, ApJ, 466, 704

Corbin, M. R. 1991, ApJ, 371, L51

Crenshaw, D. M. 1986, ApJS, 62, 821

Crenshaw, D. M. & Blackwell, J. H., Jr. 1990, ApJ, 358, L37

Ganguly, R., Bond, N. A., Charlton, J. C., Eracleous, M., Brandt, W. N., Churchill, C. W. 2001, ApJ, 549, 133

Gaskell, C. M. 1982, ApJ, 263, 79

Gaskell, C. M. 1988, ApJ, 325, 114

Gaskell, C. M. 2000, New Astron. Rev., 44, 563

Gaskell, C. M. & Snedden, S. A. 1999, in Structure and Kinematics of Quasar Broad Line Regions, ed. C. M. Gaskell, W. N. Brandt, M. Dietrich, D. Dultzin-Hacyan, & M. Eracleous (San Francisco:ASP), 157

Goad, M. R., Koratkar, A. P., Kim-Quijano, J., Korista, K. T., O'Brien, P. T., Axon, D. J. 1999, ApJ, 524, 707

Grandi, S. A. 1980, ApJ, 238, 10

Grandi, S. A. & Phillips, M. M. 1979, ApJ, 232, 659

Kaspi, S, Smith, P. S., Netzer, H., Maoz, D., Jannuzi, B. T., & Giveon, U. 2000, ApJ, 533, 631

Koratkar, A. P. & Gaskell, C. M. 1989 ApJ, 345, 637

Koratkar, A. P. & Gaskell, C. M. 1991, ApJS,

Mathews, W. G. & Blumenthal, G. R. 1975, ApJ, 198, 517

Mathews, W. G. & Wampler, E. J. 1985, PASP, 97, 966

Persson, S. E. 1988, ApJ, 330, 751

Shuder, J. M. 1982, ApJ, 259, 48

Taylor, J. A. 1999, Ph.D. thesis, Univ. Maryland

Wandel, A., Peterson, B. M. & Malkan, M. A. 1999, ApJ, 526, 579

Wanders, I. & Peterson, B. M. 1996, ApJ, 466, 174

Zheng, W. 1992, ApJ, 385, 127

Mass Outflow in Active Galactic Nuclei: New Perspectives
ASP Conference Series, Vol. 255, 2002
D.M. Crenshaw, S.B. Kraemer, and I.M. George

Outflow in NGC 3516

J. B. Hutchings

Herzberg Institute of Astrophysics, NRC of Canada, 5071 W.Saanich Rd, Victoria, B.C. V9E 2E7,Canada

Abstract.
Spectra over 900–3000 Å in the 2000 low flux state of the nucleus of the Seyfert 1 galaxy NGC 3516 show P Cygni line profiles, particularly in O VI λ1032, compared with data from an earlier higher state. The profiles are discussed in terms of an accelerating wind driven by a disk continuum source, in which the ionisation radii have changed.

1. Introduction and data

The Seyfert 1 galaxy NGC 3516 has strong and variable absorption lines. It is also one of the rare Seyfert 1s whose narrow-line region shows a bipolar morphology.

We observed NGC 3516 with *Far Ultraviolet Spectroscopic Explorer (FUSE)* on 2000 April 17, and with the *Space Telescope Imaging Spectrograph (STIS)* on April 20. *FUSE* obtains high-resolution (R$\sim$20,000) far-UV spectra covering the 905–1187 Å spectral range. The *STIS* data were obtained in slitless mode: a G140L spectrum (1150–1700 Å R $\sim$ 1000), and a G230L spectrum (1650–3100 Å). As the active galactic nucleus was in an unusually low state, we have compared UV line profiles with the normal high state observed earlier (Kriss et al. 1996; Crenshaw et al. 1998; Goad et al. 1999).

The line profiles in the new low-state data strongly suggest that they are dominated by outflow at velocities not seen in the high state. The absorption and line width in O VI indicate outflow velocities of some 600 km s^{-1} in the hottest (highest ionisation) regions, with the bulk at about 400 km s^{-1}. There is absorption (outflow gas) at all velocities down to zero. The C IV and Lα profiles show the same velocity of maximum absorption and absorption to zero velocity, but the outflow occurs to higher velocity, with a second minimum at 1600 km s^{-1} and a terminal velocity of 2000 km s^{-1}. Much, but not all, of the low-velocity outflow is also present in the high-state profiles, and we know that these can be resolved into several sharp components, from higher resolution data from the *Hubble Space Telescope.*

2. A wind scenario for the BLR

The scenario we suggest is an accelerating wind like a stellar wind. The velocity increases and the ionisation decreases outwards. As the central source

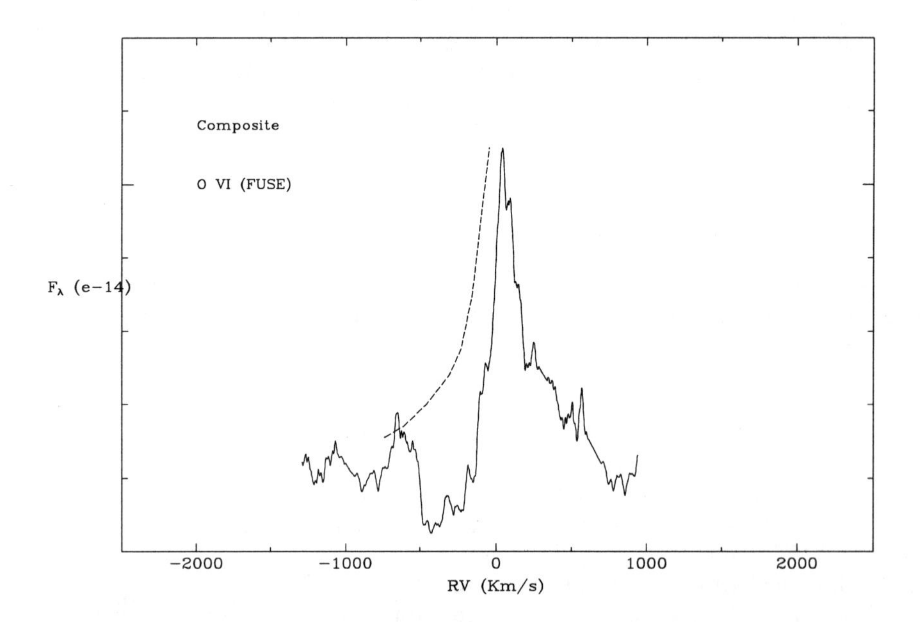

Figure 1. Composite O VI profile from NGC 3516 low state. Dashed line is profile longward wing reflected about systemic velocity. Note narrow emission and P Cygni absorption component.

flux drops, the ionisation boundaries move inwards. Wide profiles with weak or non-detectable P-Cygni absorptions arise if lines are formed in a large radius region at terminal velocity. In the low state, the high ionisation lines are formed at much smaller radii from the nucleus, where P-Cygni absorptions are more prominent, and can be seen through the whole absorbing column to the continuum source. The highest ionisation lines are narrower and have lower velocity absorptions.

This simple model is for spherically symmetric expansion, as expected for a star. If the wind is driven by a disk, geometry is a further parameter. Disk winds are seen in some cataclysmic binaries, and are seen over a broad angle on each side of the disk, but not near the disk plane. Since jets are also driven perpendicular to the disk, there may be a cone of wind avoidance where the jet lies, if there is one. The Seyfert 1 paradigm is that we view the central disk and BLR directly without obscuration by the opaque torus: thus within the expected sightline of a central wind. The wind profiles for such a disk sightline are similar to those for a spherical wind, with slightly less peaky emission profiles. Thus, the simple wind cartoon is generally applicable to the expected geometry, but dependent in detail on the solid angle filled by the wind.

3. Discussion

It appears that the change of nuclear flux allows us to probe the velocity structure of a broad-line wind, by varying the radii of the shells in which different lines are formed. We think that the density structure may not change significantly. The geometry of the wind, what happens to its material as it moves outward and cools, and its relationship with the other mass flows need to be considered and tested with observations in other objects, as well as NGC 3516 itself as it continues to vary.

This work is written up fully by Hutchings et al. (2001).

References

Crenshaw, D. M., Maran, S. P., & Mushotzky, R. F. 1998, ApJ, 496, 797

Goad M.R., Koratkar A.P., Axon D.J., Korista K.T., & O'Brien P.T., 1999, ApJ, 524, 707

Hutchings J.B., Kriss G.A., Koratkar A., Green R.F., Brotherton M., Kaiser M.E., & Zheng W., 2001, ApJ, in press

Kriss G. A., et al. 1996, ApJ, 467, 622

Mass Outflow in Active Galactic Nuclei: New Perspectives
ASP Conference Series, Vol. 255, 2002
D.M. Crenshaw, S.B. Kraemer, and I.M. George

Line Profile Variations in Mrk 110: Probing the Velocity Field in the BLR

Wolfram Kollatschny[1]

Universitäts-Sternwarte, Geismarlandstr. 11, D-37083 Göttingen, Germany

Abstract. We present continuum and emission line light curves of the narrow line Seyfert 1 galaxy Mrk 110. The data set has been obtained with the 9.2m Hobby-Eberly Telescope at McDonald Observatory at 26 epochs from 1999 November through 2000 May. The light curves of the integrated Balmer and helium emission lines are delayed by 3 to 33 light days respectively to the optical continuum variations. The outer line wings respond much faster to continuum variations than the inner regions of the line profiles. The delays of the red wings are slightly shorter than those of the blue wings. This indicates a radial outflow velocity component in the BLR of Mrk 110.

1. Motivation

Mrk 110 is a well known variable Seyfert galaxy (Peterson et al. 1998; Bischoff & Kollatschny 1999). We detected extreme intensity variations in the continuum, and in the Balmer and Helium lines (Bischoff & Kollatschny 1999) in a long-term variability study over 8 years. The variations of the optical emission lines were delayed by 9 to 80 days with respect to continuum variations. Furthermore, we detected strong line profile variations in our first variability campaign. We started a new variability campaign with the 9.2 m Hobby-Eberly Telescope in 1999 to look in detail for variations on time scales of days to weeks only. Besides the study of integrated line variations we are interested in detailed line profile variations to investigate the velocity field in the broad-line region. This requires a homogeneous data set with spectra of very high signal-to-noise ratio (S/N).

2. Observations

All optical spectra of our new variability campaign of Mrk 110 were taken with the 9.2 m Hobby-Eberly Telescope (HET) at McDonald Observatory. The campaign lasted from 1999 November 14 until 2000 May 14 covering 26 epochs. All

[1]Department of Astronomy and McDonald Observatory, University of Texas at Austin, Austin, TX 78712, USA

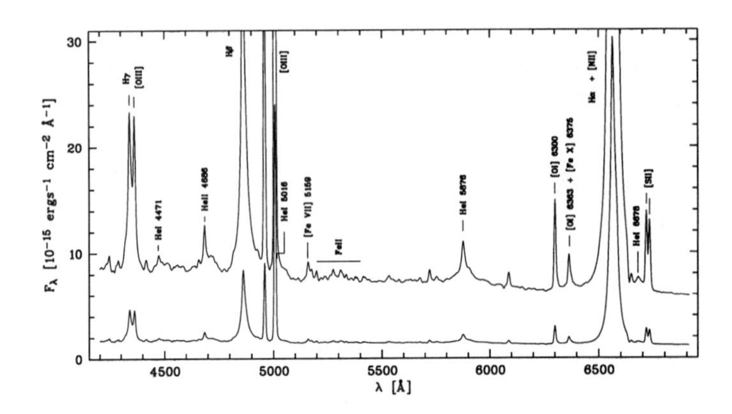

Figure 1. Mean rest frame spectrum of Mrk 110. The upper spectrum is plotted with a vertical magnification of 10 (zero level is shifted by -10) to show both strong and weak lines.

observations were made under identical instrumental conditions with the low-resolution spectrograph (LRS) at prime focus. Exposure times were 10 to 20 minutes yielding a high S/N of > 100 in the continuum. We reduced the spectra in a homogeneous way with IRAF reduction packages. The spectra were calibrated to the same flux of the constant [O III] λ5007 line of 2.26 x 10^{-13} erg s^{-1} cm^{-2}. In this way we got a relative flux accuracy of better than 1 %. Details will be given in a separate paper (Kollatschny et al. 2001).

3. Results

The mean optical spectrum of Mrk 110 of our variability campaign is shown in Figure 1 with two different scalings. This spectrum has been derived from 24 epochs (from November 1999 through May 2000).

Mrk 110 is a narrow-line Seyfert 1 galaxy. Some strong emission lines including the Balmer and He I lines as well as the strongest forbidden lines are labeled. Furthermore, one can clearly identify individual Fe II lines in the Fe II λ5200 line blend.

3.1. Continuum and Hβ Light Curves

Light curves of the continuum flux at λ=5100 Å and of the integrated emission line of Hβ are shown in Figure 2. Our Mrk 110 variability campaign started on 1999 November 14 and ended in 2000 May 14. The continuum light curve reached a maximum stage at the beginning of January 2000 and declined to an extreme low stage in spring 2000.

The size of the broad-line region can be estimated from the cross-correlation function of the continuum light curve with the Hβ light curve. Even without a detailed cross-correlation function analysis one can see a delay of 24 days of the

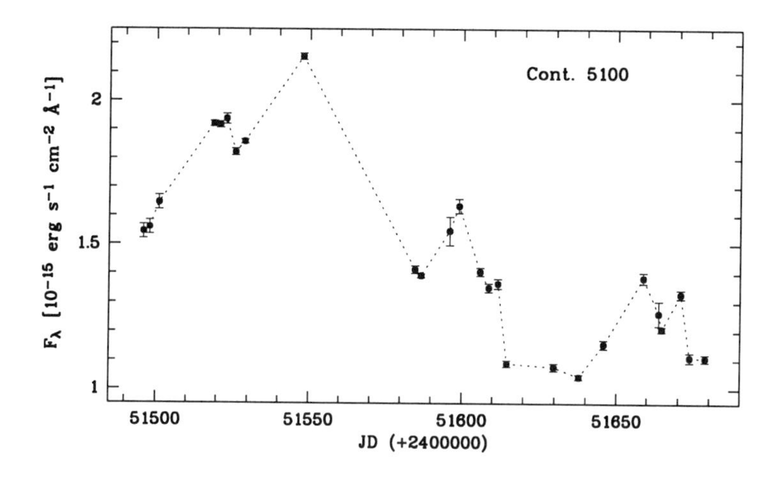

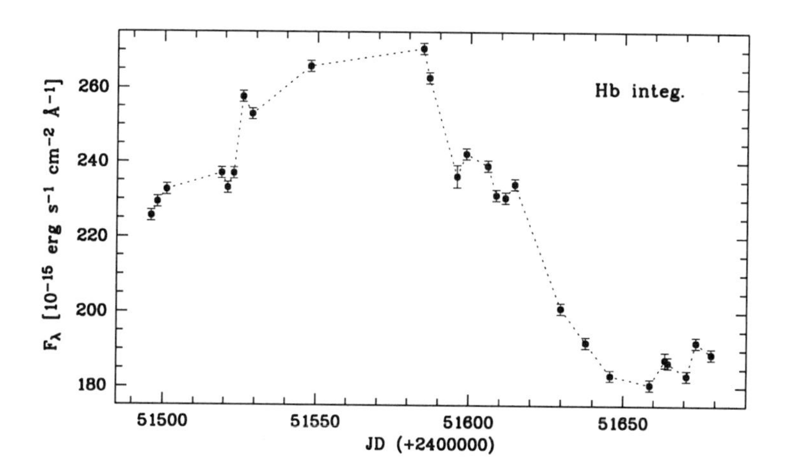

Figure 2. Mrk 110 light curves of the continuum flux at λ_{rest}=5100 Å and of the integrated Hβ emission line flux. The points are connected to aid the eye.

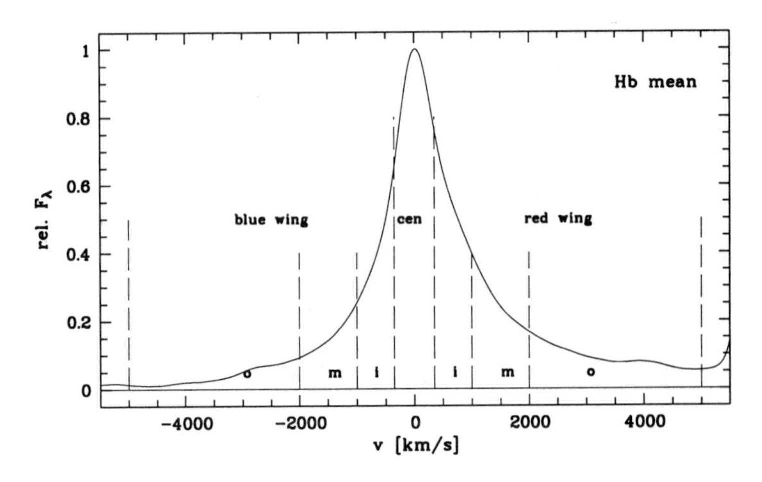

Figure 3. Mean Hβ line profile in velocity space and the subdivision into central and line wing segments.

Hβ light curve with respect to the continuum light curve. In our earlier long-term variability campaign of Mrk 110 (Bischoff & Kollatschny 1999) we have shown that the individual emission lines are delayed by different lags (9 to 80 days) with respect to the continuum as a function of ionization degree.

3.2. Light Curves of Hβ Line Segments

In a next reduction step we divided the strongest emission lines in seven segments (a central component and three components in each wing) for studying the internal line variations in more detail. Figure 3 shows the mean Hβ line profile of our variability campaign and the line segments we used for the segmented light curves. In Figure 4 we plot the light curves of the central Hβ line segment 'cen' and of the two intermediate wing segments 'm'. The light curves of individual line segments are remarkably different. On the other hand the light curves of the red and blue segments do not differ that much. The light curve of the central line region shows the largest delay with respect to the continuum of 30 days. The outer line wings respond much faster to continuum variations than the inner line regions. The shortest delays - of 5 days only - were calculated for the outer line wings.

Figure 5 shows the delays of the individual Hβ line segments of our Mrk 110 variability campaign.

4. Discussion

The integrated emission lines respond with different delays to continuum variations as a function of ionization degree, indicating a stratification in the broad-line region (BLR) of Mrk 110 (Bischoff & Kollatschny 1999). Investigating in more detail the variations of high S/N line profiles, one can notice a strong trend, indicating that the outer emission line wings originate closer to the galac-

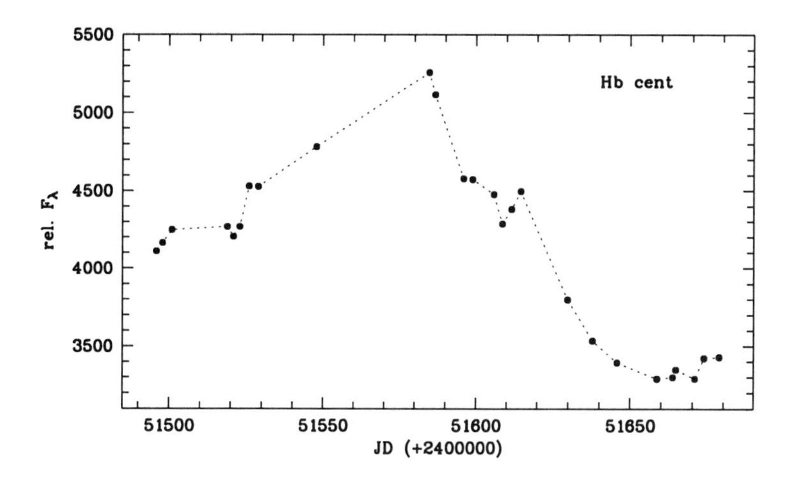

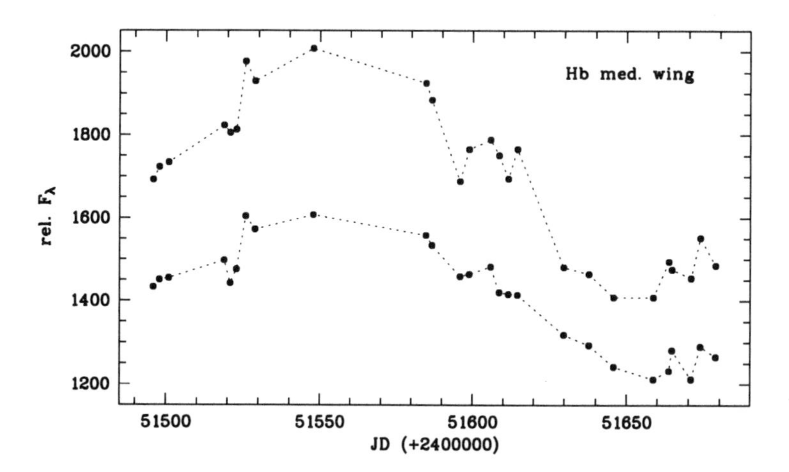

Figure 4. Light curves of the central Hβ line segment and of the intermediate wing segments 'm'. The upper light curve in Figure 4b is that of the red wing. The points are connected to aid the eye.

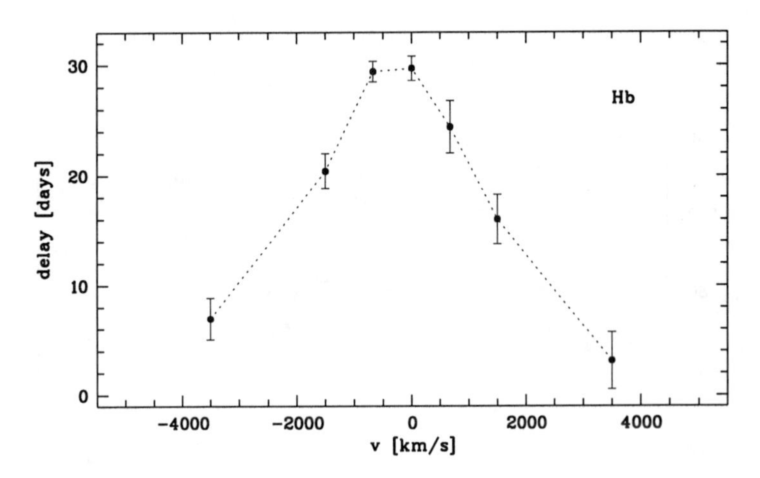

Figure 5. Delays of individual Hβ line segments with respect to continuum variations.

tic center than the inner line cores. This implies that rotational motions and/or turbulent motions are dominant in the BLR of Mrk 110 (e.g., Welsh & Horne 1991; Bottorff et al. 2000).

Additionally, the red emission line wings respond slightly faster than the blue line wings to continuum variations in Mrk 110. This trend is to be seen in all Balmer (Hα, Hβ, Hγ) and Helium (He II λ4686, He I λ4471, He I λ5876) lines. The disk-wind model of Chiang & Murray (1996) and Murray & Chiang (1997) explains an earlier response of the red line wing relative to the blue wing as a result of a radial velocity component (in their model the line emitting gas shows a radial outward velocity component in addition to rotation). But the observed effect, that both outer line wings in Mrk 110 respond simultaneously much faster to continuum variations than the inner line regions, is not yet fully explained in their model.

Acknowledgments. I thank K. Bischoff for valuable comments. Part of this work has been supported by DFG grant Ko 857/24.

References

Bischoff, K., & Kollatschny, W. 1999, A&A 345, 49
Bottorff, M., Ferland, G., Baldwin, J., & Korista, K. 2000, ApJ 542, 644
Chiang, L., & Murray, N. 1996, ApJ 466, 704
Kollatschny, W., et al. 2001, submitted to A&A
Murray, N, & Chiang, L. 1997, ApJ 474, 91
Peterson, B.M. et al. 1998, ApJ 501, 82
Welsh, W.F., & Horne, K. 1991, ApJ 379, 586

Mass Outflow in Active Galactic Nuclei: New Perspectives
ASP Conference Series, Vol. 255, 2002
D.M. Crenshaw, S.B. Kraemer, and I.M. George

Stellar Envelopes Confined by Hot AGN Outflows as Sources of BLR Emission

G. Torricelli-Ciamponi

Osservatorio Astrofisico di Arcetri, Largo E. Fermi 5, 50125-Firenze, Italy

P. Pietrini

Dipartimento di Astronomia, Largo E. Fermi 5, 50125-Firenze, Italy

Abstract. In Active Galactic Nuclei (AGNs) the presence of a star cluster around the central black hole can have several effects on the dynamics and the emission of the global system. In this paper we analyze the interaction of stellar atmospheres with a wind outflowing from the central region of the AGN nucleus. Even a small mass loss from stars, as well as possible star collisions, can give a non-negligible contribution to the feeding of matter into the AGN nuclear wind; moreover, stellar mass loss can produce envelopes surrounding stars that turn out to be suitable for reproducing the observed Broad Line Region (BLR) emission. In this framework, the envelope can be confined by the bow shock arising from the interaction between the expanding stellar atmosphere and the AGN nuclear wind.

1. Introduction

In this paper we re-consider the scenario first suggested by Scoville & Norman (1988) and then re-analyzed by Kazanas (1989) and Alexander & Netzer (1994, 1997), in which stellar extended atmospheres are the source of BLR emission. The main improvement of our treatment lies in the fact that we account for the presence of a background medium, which in turn involves the formation of a shock front limiting the extension of each stellar envelope. The identification of the medium where stellar atmospheres expand can be a fundamental step in understanding the physics of broad line emitting dense gas. Indeed the nature of the medium surrounding BLR envelopes is not evident. Recent observations support the existence of radial outflows even in radio-quiet AGN's, such as Seyfert 1s, (Crenshaw et al. 1999, Weymann et al. 1997, see also this conference). The existence of outflowing gas is a strong constraint for the "intercloud" medium properties, as we have deduced in a previous paper (Pietrini & Torricelli 2000), where we have analyzed the physical structure and characteristics of an AGN nuclear outflow, originating in the very central regions and expanding out to large distances, as a kind of background for various observational components of the AGN structure (BLR, UV absorbers, X-ray "warm absorbers"). The wind integration has proved to be not straightforward in the AGN context and possible only under specific conditions. One of the constraints deriving from such integration is that the outflowing wind may exist only if its density is rather low. The only way to make the wind density more substantial is to feed matter into the wind itself on a certain range of distance from the nucleus (r). A possible source for this can be a stellar cluster around the AGN nucleus. In fact, a cluster of stars orbiting around the central black hole ejects matter into the interstellar medium both because of stellar collisions and of mass loss from individual stars; the global amount of matter fed into the AGN wind can be computed once the star distribution function $\rho_*(r)$ is known.

In conclusion, a consistent solution of the AGN wind, taking into account mass deposition, can determine the environment in which BLR emission occurs and, in particular, the AGN wind density, i.e. the density of the confining gas (see the upper panels of Figures 1 and 2).

2. The model

2.1. BLR confinement

When a mass losing star moves rapidly into the interstellar medium, its wind interacts with the surrounding gas, giving rise to an elongated bow shock. That can be the case for stars orbiting around an AGN nucleus where a nuclear wind is outflowing. In this framework a "high" density region confined by the shock forms around the star; this envelope can be the source of BLR emission.

The order of magnitude of the distance from the star, $R = R_{\rm bow}$, at which the bow shock is formed can be determined by using the standard methods, i.e. from the two equations describing the stellar mass loss rate and the balance between the total pressure (thermal+ram pressure) of the stellar wind and that of the ambient medium:

$$\begin{cases} \dot{M}_* = 4\pi \; m_{\rm H} \; R^2_{\rm bow} n_*(R_{\rm bow}) v_*(R_{\rm bow}) \\ P_* + m_{\rm H} n_*(R_{\rm bow}) v_*^2(R_{\rm bow}) = P_{\rm ext} + P_{\rm ram}. \end{cases}$$

Here $P_* \simeq 2n_*kT_*$ and $P_{\rm ext} \simeq 2n_{\rm ext}kT_{\rm ext}$ are the thermal pressures of the star and of the external medium, respectively; v_* is the expansion velocity of the stellar atmosphere and the ram pressure of the external medium can be identified with the higher between that due to the star Keplerian motion around the central black hole, $V(r) = \sqrt{GM_{\rm BH}/r}$, and that due to the AGN wind:

$$P_{\rm ram} = Max[m_{\rm H} n_{\rm ext} V^2, m_{\rm H} n_{\rm ext} v^2_{\rm ext}] = m_{\rm H} n_{\rm ext} v^2_{\rm Max}.$$

We refer to Torricelli & Pietrini (2001) for details.

The comparison of $R_{\rm bow}$ with the envelope radius derived from other competing mechanisms defining the envelope extension, such as tidal disruption, Comptonization, etc. (see Alexander & Netzer 1994, 1997) shows that in the range of interest the shock front is the most efficient one.

Once derived the thermal properties of the external confining medium, i.e. those of the AGN wind, and prescribed the stellar mass loss, $\dot{M}_*$ and the stellar wind expansion velocity, v_*, the two equations shown above can be solved to derive the envelope extension $R_{\rm bow}$ and its density n_*.

2.2. BLR properties

From the observational and interpretative points of view, stellar atmospheres can contribute to the BLR emission only if $R_{\rm bow}(r)$ and $n_*(r)$ satisfy the following conditions:

a) $R_{\rm bow}(r) \geq \Delta R = 10^{23}U/n_*$ cm, since photoionization equilibrium requires that emission lines are generated in shells of thickness ΔR, which is at least $\Delta R = 10^{23}U/n_*$ cm (see e.g. Peterson 1997);

b) $n_*(r) \geq 10^8$ cm^{-3}, since broad forbidden lines are not observed (Netzer 1990, Peterson 1997);

c) $n_*(r) < 1 \times 10^{12}$ cm^{-3} to avoid line thermalization, *i.e.*, for line emission to be effectively significant (see e.g. Rees, Netzer, & Ferland 1989).

In our model $R_{\rm bow}(r)$ and $n_*(r)$ will attain different values at different distances from the central black hole, hence the above conditions will be satisfied only in a specific range of distances. We call r_1 and r_2 the distances which delimit this interval. Stellar envelopes satisfying points a) to c) contribute to build up the total covering factor $C_{\rm v} = (C_{\rm v})_{\rm SG} + (C_{\rm v})_{\rm RG} + (C_{\rm v})_{\rm MS}$. Here we define

$$(C_{\rm v})_{\rm i}(r) = \int_{(r_1)_{\rm i}}^{r} \pi (R_{\rm bow})_{\rm i}^2 f_{\rm i} \rho_* dr,$$

evaluated at $r = (r_2)_{\rm i}$, as the covering factor due to type star "i" ("i"= super giant (SG), red giant (RG), main sequence (MS)) and we set $f_{\rm SG} = 10^{-4}$, $f_{\rm RG} = 10^{-2}$ and $f_{\rm MS} \simeq 1$.

We present hereafter the results obtained for an AGN ionizing radiation field luminosity $L = 10^{44}$ ergs s^{-1}and with two different sets of stellar mass loss values. The first scenario, with standard mass loss rates, namely $(\dot{M}_*)_{\rm SG} \simeq 10^{-6} M_\odot/yr$, $(\dot{M}_*)_{\rm RG} \simeq 10^{-8} M_\odot/yr$, $(\dot{M}_*)_{\rm MS} \simeq 10^{-14} M_\odot/yr$, turns out to form envelopes suitable for BLR emission mainly around red giant stars. In the second scenario enhanced mass loss rates due to the strong illumination of the stars by the hard central radiation field are allowed (Hameury et al. 1993). Since main sequence stars are especially influenced by this specific environment, we assume in this case $(\dot{M}_*)_{\rm SG} \simeq 10^{-6} M_\odot/{\rm yr}$, $(\dot{M}_*)_{\rm RG} \simeq 10^{-8} M_\odot/{\rm yr}$, $(\dot{M}_*)_{\rm MS} \simeq 10^{-10} M_\odot/{\rm yr}$. With these values, envelopes around MS stars turn out to be the main source of BLR emission.

3. Model results and comparison with observational requirements

The outflow density resulting from the AGN wind integration and the stellar envelope parameters are shown in Figures 1 and 2 for the case of red giant stars (case 1) and for that of main sequence stars (case 2). We can compare them with the constraints coming from observations, getting to the following conclusions (see also Torricelli & Pietrini 2001).

- The typical distance of BLR, $r_{\rm BLR} \simeq 0.1 L_{46}^{1/2}$ pc, lies in the intervals $[r_1, r_2]_{\rm i}$ relative to the stellar types contributing to the build up of $C_{\rm v}$. In fact, for $L = 10^{44}$ ergs s^{-1}, $r_{\rm BLR}$ is $\simeq$ 12 light days, and the intervals in which conditions for BLR emission are fulfilled are $[r_1, r_2]_{RG} = [1.8, 182]$ light days and $[r_1, r_2]_{MS} = [1.4, 82]$ light days, for case 1 and 2, respectively.

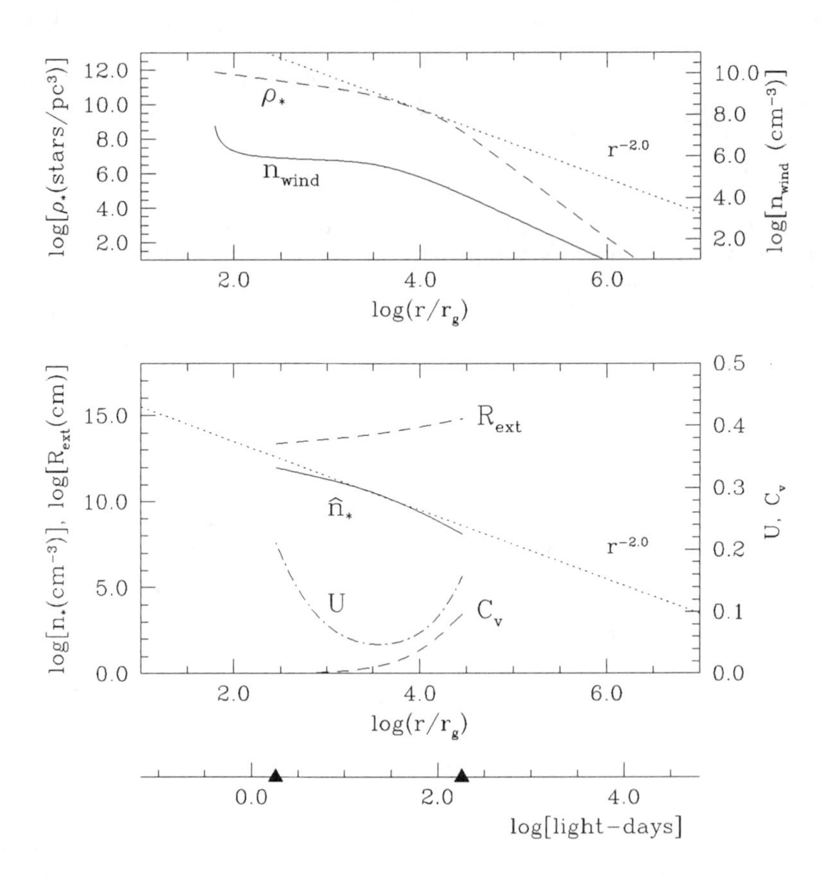

Figure 1. In the upper panel the density in the AGN wind, $n_{\rm wind}(= n_{\rm ext})$, (continuous line) and the star number density, ρ_*, (dashed line) of the central cluster are shown. These quantities have been derived from the AGN wind model by Pietrini & Torricelli (2000). Note the presence of a plateau in the wind density corresponding to a region of high stellar density. The plateau is, in fact, due to a high rate of mass deposition into the wind both from stellar mass loss and star collisions. In the lower panel the mean stellar envelope number density, $\hat{n}_*$, (continuous line), the stellar envelope extension, $R_{\rm ext}$, (dashed line), the resulting covering factor, $C_{\rm v}$, (dashed line) and ionization parameter, U, (dot-dashed line) are shown. The last four quantities are drawn only in the interval $[r_1, r_2]$ (marked by two triangles on the lower axis) where conditions a) to c) are fulfilled (see text). The dotted lines, appearing in both panels, just show reference slopes. This figure refers to the case of red giant stars.

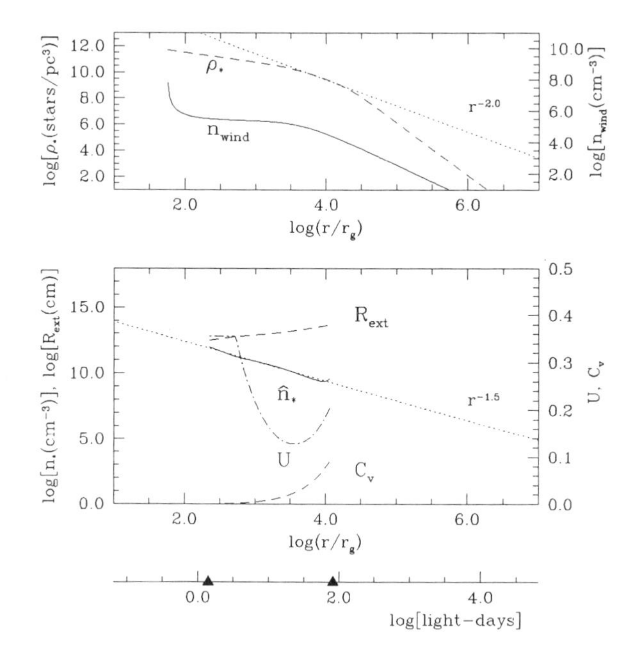

Figure 2. The same as in Figure 1 for the case of main sequence stars.

• U, the ionization parameter $\equiv L/[4\pi c r^2 n_* < h\nu >]$ (where $< h\nu >$ is the average photon energy of the ionizing radiation field) is in the range $U \simeq 0.1 - 1$, as required (Alexander & Netzer 1994, Peterson 1997).

• The covering factor defined above, referring to broad permitted line emission, meets the observed values, i.e. $C_{\rm v} \simeq 0.1$, at r_2, end of the interval.

• The covering factor, $\mathbf{C}_{\rm forb}$,

$$C_{\rm forb} = \sum_{\rm i} (C_{\rm forb})_{\rm i} = \sum_{\rm i} \left[\int_{r_0}^{\infty} \pi (R_{\rm bow})_{\rm i}^2 f_{\rm i} \rho_* dr \right]_{(n_{*\rm i})<10^8 {\rm cm}^{-3}},$$

relative to low density ($n_* < 10^8$ cm^{-3}) envelopes, possibly emitting "broad" forbidden lines, is negligible (see Peterson 1997, Krolik 1999).

• The number of stars contributing to BLR emission is $N_{\rm RG} \simeq 7 \times 10^5$ and $N_{\rm MS} \simeq 2 \times 10^7$.

• ρ_*, stellar density distribution, is $\propto r^{-2}$ in the region where the contribution to the covering factor is substantial.

• $\hat{n}_*$, the mean density in the photoionized shell of the stellar envelope, is $\hat{n}_* \propto r^{-2.0}$ for red giant stars and $\hat{n}_* \propto r^{-1.5}$ for main sequence stars.

These last two points are important to compare our results to Kaspi & Netzer (1999) deductions (for the case of the well studied Seyfert 1 NGC 5548). In fact, translating in our notation Kaspi & Netzer results, their conclusions imply:

i) $\rho_*(r) \propto r^{-2}$, ii) $\hat{n}_*(r) \propto r^{-\rm s}$ with $1 \leq \rm s < 1.5$ and

iii) $\begin{cases} r_1 \approx 3 \text{ light} - \text{days} \\ r_2 \approx (70 \div 100) \text{ light} - \text{days} \end{cases}$

(but models are not sensitive to the exact value of r_1).

It is evident that, while other observational requirements are equally satisfied by red giant stars and main sequence stars, the main sequence star scenario better fits Kaspi & Netzer conditions, since r_2 results in the "right" range and $\hat{n}_*$ has the "expected" slope $\hat{n}_* \propto r^{-1.5}$.

4. Conclusions

The comparison of observational requirements with the results of our analysis is strongly encouraging; in particular, main sequence star scenario better fits Kaspi & Netzer conditions (see for details Torricelli & Pietrini 2001).

In addition, we note that a significant difference between our treatment and those of previous authors is the possibility of a more efficient confinement, due to the presence of an AGN outflow, which allows the formation of smaller envelopes. In this case a larger number of condensed regions is necessary to reproduce the observed BLR emission, but in our scenario this possibility is realistic, since the contribution does not come only from the rare giant stars, but also from the more common main sequence stars (in fact we obtain $N_{\rm MS} >> N_{\rm RG}$). The advantage of such a picture is that the large number of line emitting regions is in accordance with the cross-correlation analysis by, e.g., Arav et al. (1997).

References

Alexander, T., & Netzer, H. 1994, MNRAS, 270, 781

Alexander, T., & Netzer, H. 1997, MNRAS, 284, 967

Arav, N., et al. 1997, MNRAS, 288, 1015

Crenshaw, D.M., et al. 1999, ApJ, 516, 750

Hameury, J.M., et al. 1993, A&A, 277, 81

Kazanas, D. 1989, ApJ 347, 74

Kaspi, S., & Netzer, H. 1999, ApJ, 524, 71

Krolik, J.H., 1999, Active Galactic Nuclei, (Princeton: Princeton University Press)

Netzer, H. 1990, in Active Galactic Nuclei, ed. R.D. Blandford, H. Netzer, & L. Woltjer, (Berlin: Springer Verlag), 57

Peterson, B.M. 1997, in Introduction to Active Galactic Nuclei, (Cambridge: Cambridge University Press)

Pietrini, P., & Torricelli-Ciamponi, G. 2000, A&A, 363, 455

Rees, M.J., Netzer, H., & Ferland, G.J. 1989, ApJ, 347, 640

Scoville, N., & Norman, C. 1988, ApJ, 332, 163

Torricelli-Ciamponi, G., & Pietrini, P. 2001, submitted to A&A

Weymann, R.J., et al. 1997, ApJ, 483, 717

Low-Ionization BAL QSOs in Ultraluminous Infrared Systems

Mass Outflow in Active Galactic Nuclei: New Perspectives
ASP Conference Series, Vol. 255, 2002
D.M. Crenshaw, S.B. Kraemer, and I.M. George

Strong Shocks and Supersonic Winds in Inhomogeneous Media

A.Y. Poludnenko[1], A. Frank[2], and E.G. Blackman [3]

Department of Physics and Astronomy,
University of Rochester, Rochester, NY 14627-0171

Abstract. Many astrophysical flows occur in inhomogeneous media. The broad-line regions (BLR) of active galactic nuclei (AGN) are one of the important examples, where emission-line clouds interact with the outflow.

We present results of a numerical study of the interaction of a steady, planar shock / supersonic postshock flow with a system of embedded cylindrical clouds in a two-dimensional geometry. Detailed analysis shows that the interaction of embedded inhomogeneities with the shock / post-shock wind depends primarily on the thickness of the cloud layer and the arrangement of the clouds in the layer, as opposed to the total cloud mass and the total number of individual clouds. This allows us to define two classes of cloud distributions: thin and thick layers. We define the critical cloud separation along the direction of the flow and perpendicular to it. This definition allows us to distinguish between the interacting and non-interacting regimes of cloud evolution. Finally we discuss mass-loading in such systems.

1. INTRODUCTION

Mass flows are important in many astrophysical systems from stars to the most distant active galaxies. Virtually all mass flow studies focus on homogeneous media. However, the typical astrophysical medium is inhomogeneous. The presence of inhomogeneities can introduce not only quantitative but also qualitative changes to the overall dynamics of the flow.

Practically all current models of AGN agree that the emission-line clouds in BLRs are essential for explaining the observed properties of AGN (Urry & Padovani 1995; Elvis 2000). However, despite the fact that the nature of BLR clouds is an open question, any self-consistent model of the AGN should properly account for properties of BLR cloud interaction with mass outflow. Such self-consistency is important in the context of cloud survival time and cloud displacement prior to destruction. Even when clouds are magnetically confined

[1]wma@pas.rochester.edu

[2]afrank@pas.rochester.edu

[3]blackman@pas.rochester.edu

and stabilized against evaporation (Rees 1987), disruptive action of outflows may prevent cloud survival over the dynamically significant AGN timescales.

Klein, McKee, & Colella (1994; hereafter KMC) addressed the problem of shock-cloud interaction, providing a detailed description of the dynamics of a single, dense, unmagnetized cloud interacting with a strong, steady, planar shock. That work gives an excellent introduction to the subject, in particular the description of the astrophysical significance of the problem of a shock wave interacting with a dense cloud (see also Gregori et al. [2000]). In this work we investigate the general properties of strong shock / supersonic wind interaction with a system of embedded clouds and determine the key quantities governing the evolution of such systems.

2. RESULTS

We have numerically investigated the interaction of a strong, planar shock wave with a system of dense inhomogeneities, embedded in a more tenuous and cold ambient medium. Our code is the AMRCLAW package, which implements an adaptive mesh refinement algorithm for the equations of gas dynamics (Berger & Jameson 1985; Berger & Colella 1989; Berger & LeVeque 1998) in two dimensions. We have assumed constant conditions in the global postshock flow constraining the maximum size of the clouds only by the condition of the shock front planarity. Our results are applicable to strong global shocks with Mach numbers $3 \lesssim M_S \lesssim 1000$. The range of the applicable cloud - unshocked ambient medium density contrast values is 10 – 1000. Figure 1 shows a case of a strong shock and a supersonic post-shock flow interacting with an inhomogeneous system of fourteen identical clouds in regular distribution.

Cloud evolution due to the interaction with the global shock and the post-shock flow has four major phases, namely *the initial compression phase, the re-expansion phase, the destruction phase,* and finally *the mixing phase.* Each image in Figure 1 roughly illustrates each of those phases.

The timescale we use to define time intervals in our numerical experiments is the time it takes for the incident shock wave to sweep across an individual cloud. This is called the *shock-crossing time,* $t_{SC} = (2a_{max}/v_S)$, where a_{max} is the maximum cloud radius in the distribution.

We define the *cloud destruction time* t_{CD} as the time when the largest cloud fragment contains less than 50% of the initial cloud mass. Typically, in our simulations $t_{CD} \approx 24t_{SC}$.

A simple model for the cloud acceleration during the first three phases, i.e. prior to its destruction, can be developed. We find the cloud velocity

$$v_C(t) = \begin{cases} v_{PS}\left(1 - \left(\frac{t}{t_{SC}}a_1 + a_2\right)^{-1}\right), & t \leq 12t_{SC} \\ v_{PS}\left(1 - \left(\left(\frac{t}{t_{SC}} - 12\right)^2 b_1 + \frac{t}{t_{SC}}a_1 + a_2\right)^{-1}\right), & t \leq t_{CD} \end{cases} \quad (1)$$

Here, v_{PS} is the unperturbed postshock flow velocity. For the cases of infinitely strong shocks, i.e. $M_S \to \infty$, the above coefficients have the values

$$a_1 = 1.83 \cdot 10^{-3};\ a_2 = 1.09;\ b_1 = 8.51 \cdot 10^{-5}. \quad (2)$$

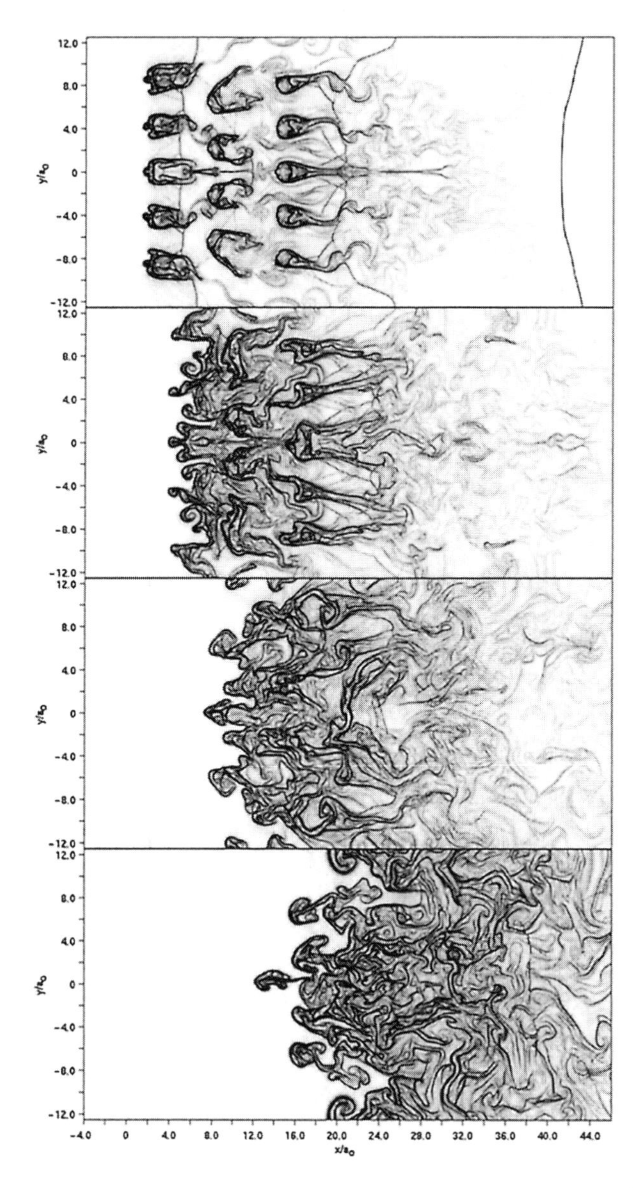

Figure 1. Run $M14$. Time evolution of a system, containing fourteen identical clouds in a regular distribution and interacting with a $M_S = 10$ shock wave. Shown are synthetic Schlieren images of the system at times 22 t_{SC}, 35 t_{SC}, 50 t_{SC}, 69 t_{SC}.

Equation (1) shows that the maximum cloud velocity is not more than 10% of the global shock velocity and not more than 13% of the postshock flow velocity.

Cloud displacement prior to its destruction can be described as

$$L_C(t) = \begin{cases} a_0 c_1 \left(\frac{t}{t_{SC}} - \frac{1}{a_1} \, ln\left(\frac{t}{t_{SC}} \left(\frac{a_1}{a_2} \right) + 1 \right) \right), & 0 \leq t \leq 12 t_{SC} \\ a_0 c_1 \left(\frac{t}{t_{SC}} - c_2 \, tan^{-1} \left(\frac{\frac{t}{t_{SC}} - 12}{\frac{t}{t_{SC}} c_4 + c_5} \right) - c_3 \right), & 12 t_{SC} \leq t \leq t_{CD} \end{cases} \quad (3)$$

Here a_0 is the cloud radius. In the limiting case $M_S \rightarrow \infty$ the values of the coefficients a_1 and a_2 are defined in (2) and $c_1 = 1.5$; $c_2 = 103.22$; $c_3 = 10.9$; $c_4 = 9.43 \cdot 10^{-2}$; $c_5 = 112.56$.

We find the maximum cloud displacement prior to its destruction, L_{CD}, is

$$L_{CD} = L_C(t_{CD}) \leq 3.5 a_{max}. \quad (4)$$

The results of our model are in excellent agreement with the numerical experiments. The difference in the values of cloud velocity and displacement between analytical and numerical results is $\lesssim 10\%$.

The principal conclusion of the present work is that the set Λ of all possible cloud distributions can be subdivided into two large subsets Λ_I, *thin-layer systems*, and Λ_M, *thick-layer systems*, defined as

$$\begin{array}{c} \Lambda_I : (\Delta x_N)_{max} \leq L_{CD}, \\ \Lambda_M : (\Delta x_N)_{max} > L_{CD}, \end{array} \quad (5)$$

where $(\Delta x_N)_{max}$ is the maximum cloud separation in the system along the direction of the flow, or the cloud layer thickness.

Distributions from each subset exhibit striking similarity in behaviour (e.g., Figure 2). The systems containing from one to five clouds, arranged in a single layer, exhibit exactly the same rate of momentum transfer from the global flow to the clouds. The two fourteen cloud runs M_{14} and M_{14r} have different cloud distributions, different total cloud mass, different cloud sizes. Nevertheless, the rate of the kinetic energy fraction increase during compression and re-expansion is different from the single layer cases but is still the same for both fourteen cloud runs. Other global properties exhibit the same behaviour. We conclude that the evolution of a system of shocked clouds depends primarily on the total thickness of the cloud layer and the cloud distribution in it, as opposed to the total number of clouds or the total cloud mass present in the system.

The key parameters determining the type of evolution, are *the cloud destruction length* L_{CD}, defined above in (3), and *the critical cloud separation* transverse to the flow d_{crit}. The latter is defined by the condition that the time for adjacent clouds to expand laterally and merge into a single coherent structure is equal to the cloud destruction time t_{CD}. It can be described by the expression

$$d_{crit} = 2 a_0 \left\{ \frac{t_{CD} - t_{CC}}{t_{SC}} \left(\frac{F_{c1} F_{st}}{\chi} \right)^{\frac{1}{2}} \left(\frac{3\gamma(\gamma - 1)}{\gamma + 1} \right)^{\frac{1}{2}} + 1 \right\}. \quad (6)$$

Here a_0 is the cloud radius; t_{SC} and t_{CD} are the shock-crossing time and cloud destruction times, correspondingly, defined above; t_{CC} is the cloud crushing

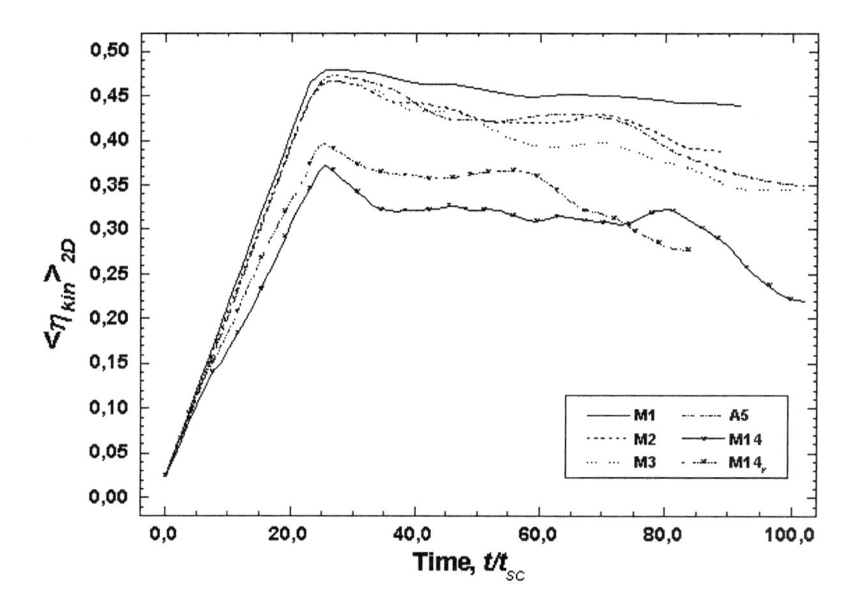

Figure 2. Time evolution of the global average of the kinetic energy fraction $\langle \eta_{kin} \rangle_{2D}$ for the following runs:
$M1$, $M2$, $M3$, $A5$ - systems with 1, 2, 3, 5 identical clouds correspondingly, arranged in a single row with constant cloud center separation of $4.0a_0$; $M14$ - system with 14 identical clouds with cloud center separation in a row equal to $4.0a_0$, separation between rows equal to $7.0a_0$; $M14_r$ - system with 14 clouds of random size in random positions.

time defining the duration of the initial compression phase, in our simulations $t_{CC} = 12t_{SC}$; χ is the cloud - unshocked ambient medium density contrast; γ is the unshocked cloud material specific heat ratio; F_{st} relates the unperturbed postshock pressure with the stagnation pressure $F_{st} \simeq 1 + 2.16/(1 + 10.7\{(\gamma + 1)\chi\}^{-1/2})$; $F_{c1} \approx 1.3$ relates the stagnation pressure with the shocked cloud pressure.

We should be able to determine, either from observations or from theoretical analysis, the thickness of the cloud layer $(\Delta x_N)_{max}$. This determines the class of the given cloud distribution, Λ_I or Λ_M. For distributions from the set Λ_I with average cloud separation $\langle \Delta y_N \rangle > d_{crit}$ evolution of the clouds during the compression, re-expansion, and destruction phases will proceed in the *non-interacting regime* and the formalism for a single cloud interaction with a shock wave (e.g. KMC) can be used to describe the system. On the other hand, if the cloud separation is less than critical, the clouds in the layer will merge into a single structure before their destruction is completed, and though the compression phase still can be considered independently for each cloud, evolution during the re-expansion and destruction phases proceeds in the *interacting regime*.

When the distribution belongs to the subset Λ_M it is necessary to determine the average cloud separation projected onto the direction of the flow $\langle \Delta x_N \rangle$ and compare it against L_{CD}: if $\langle \Delta x_N \rangle > L_{CD}$ evolution of the system can be roughly approximated as of a set of distributions from the subset Λ_I and the above "thin-layer case" analysis applies. If, on the other hand, $\langle \Delta x_N \rangle \leq L_{CD}$ (especially if $\langle \Delta y_N \rangle < d_{crit}$) the system evolution is dominated by cloud interactions and a thin layer formalism is inapplicable.

Finally we consider mass-loading. Here our principal conclusion is that mass-loading is not significant in the cases of strong shocks and supersonic winds interacting with inhomogeneities whose density contrast is in the range $10-1000$. In part this is due to short survival times of clouds as well as the very low mass loss rates of the clouds even during the times prior to their destruction. Therefore, mass-loading in such systems is not likely to have any appreciable effect on the overall dynamics of the global flow.

The major limitation of our current work is the purely hydrodynamic nature of our analysis that does not include any consideration of the magnetic fields. As was discussed by KMC, cold dense inhomogeneities (clouds) embedded in more tenuous hotter medium are inherently unstable against the dissipative action of diffusion and thermal conduction. Although weak magnetic fields, that are dynamically insignificant up to the moment of cloud destruction, can inhibit thermal conduction and diffusion and, therefore, stabilize the system of clouds, those magnetic fields may become dynamically important due to turbulent amplification during the mixing phase (see (Gregori et al. 2000) for a three-dimensional study of the wind interaction with a single magnetized cloud). We intend to provide a fully magnetohydrodynamic description of the interaction of a strong shock with a system of clouds in future work.

Acknowledgments. This work was supported in part by the NSF AST-0978765, NASA NAG5-8428, and DOE DE-FG02-00ER54600 grants.

The most recent results and animations of the numerical experiments, described above and not mentioned in the current paper, can be found at www.pas.rochester.edu/~wma.

References

Berger, M.J., & Colella, P. 1989, J. Comp. Phys., 82, 64

Berger, M.J., & Jameson, A. 1985, AIAA J., 23, 561

Berger, M.J., & LeVeque, R.J. 1998, SIAM J. Numer. Anal., 35, 2298

Gregori, G., Miniati, F., Ryu, D., & Jones, T.W. 2000, ApJ, 543, 775

Elvis, M. 2000, ApJ, 545, 63

Klein, R.I., McKee, C.F., & Colella, P. 1994, ApJ, 420, 213

Pietrini, P., & Torricelli-Ciamponi, G. 2000, A&A, 363, 455

Rees, M.J. 1987, MNRAS, 228, 47

Urry, C.M., & Padovani, P. 1995, PASP, 107, 803

Mass Outflow in Active Galactic Nuclei: New Perspectives
ASP Conference Series, Vol. 255, 2002
D.M. Crenshaw, S.B. Kraemer, and I.M. George

General Relativistic Effects on Emission from Outflows

Ran Sivron

Physics Department, Bucknell University, Lewisburg, PA, 17837

Sachiko Tsuruta

Physics Department, Montana State University, Bozeman, MT 59717

Abstract. Cool (not completely ionized) material exists in the "central engine" of active galactic nuclei (AGN). There is mounting evidence that in some cases this matter is in the form of an accretion disk. Recent observations indicate that some cool matter is also outflowing from some central engines. We argue that such cool matter may, in some cases, originate in the central engine. We calculate the line spectra from such outflows, and compare the results with the spectra from accretion disks. Some general relativistic effects are noticeable up to distances of $\sim 30 R_{Sch}$, due to a combination of lensing and gravitational and Doppler shifts. The skewed line profiles are reversed from the profiles of lines from accretion disks in the sense that the redshifted wing may be larger than the blueshifted wing. Future *XMM-Netwon (XMM)* and *Chandra X-ray Observatory (CXO)* observations, especially observations of high $\dot{m}$ sources, may reveal such line profiles.

1. Introduction

The X-ray spectra from AGN central engines and galactic black hole candidates is probably dominated by a power-law component from very hot electrons $((kT \sim m_e c^2)$, probably due to Comptonization of soft photons (Zdziarski 1988). Spectral features due to cool material in the central engine were predicted more than a decade ago (Guilbert & Rees 1988). Observations by various satellites showed those features, including asymmetric emission lines (Tanaka 1995; Nandra et al. 1997; Branduardi-Raymont et al. 2001) and Compton reflection features (e.g., Lee et al. 1999). Variations of these features on time scales of less than one day were also observed (Iwasawa 1996). It is now generally accepted that, at least the lines, originate within a few tens of R_{Sch} or less (Fabian 1998). Here $R_{Sch} = 2GM/c^2$ is the Schwarzchild radius, and M is the mass of the black hole.

Previous theorists assumed that the cool matter was in an accretion disk configuration (Fabian et al. 1989; Laor 1991; Reynolds & Begelman 1997). These authors considered general relativistic and Doppler effects on spectra emerging from a cold disk illuminated by a power-law source near a Schwarzchild or a Kerr black hole.

Although these theories agree with observations in the cases of Seyfert 1's MCG-6-30-15 and MKN 766 (Lee et al. 1999; Branduardi-Raymont et al. 2001) there are several reason to consider other configurations:

I. The discovery by Chandra of P-Cygni type outflows, and other outflows (Kaspi et al. 2000; Kaastra et al. 2000; Collinge et al. 2001). The Doppler shifts and broadening of lines are of order of thousands of km s^{-1}. Assuming that such effects are due to the black hole potential the matter must be emitting from $r \sim< 100$ even for these specific AGN (Sivron 2001, in preparation [hereafter S01]). Other AGN with higher $\dot{m}$ are expected to have faster outflows (S01).

II. Theoretically, significant outflows may dominate the spectrum. In cases in which the variability is non-linear, as may be the case in NLS1 (narrow line Seyfert 1's) and some QSOs, the outflows may then have a high covering factor, and may cool on short time scales (Sivron 1998; S01). Also, if the accretion rate is of order $\dot{m}$, or if there are significant magnetic stresses in the central engine cool filaments may form (Celotti & Rees 1998; Sivron & Tsuruta 1993). Here $\dot{m} = \dot{M}/\dot{M}_{Edd}$, $\dot{M}$ is the accretion rate, and $\dot{M}_{Edd} = 4\pi GMm_H/\Sigma_T c$ is the Eddington accretion rate. Even a small filling factor may lead to a large covering factor in such cases.

III. Although it makes sense (but, how did it get there?), one cannot assume a-priori that the cool matter already in the central engine has high angular momentum.

We therefore decided to calculate the line profile for a configuration most sensible other than disk: cool outflows from the vicinity of a Schwarzchild black hole. We present our method of obtaining spectra in section §2, our results in section §3 and a discussion of the implications of our results in section §4.

2. General assumptions and method of obtaining the spectrum

We are interested in the energy of a photon emerging from matter at r, θ in a Schwarzchild geometry with radial velocity v_r , and reaching $= \infty, \theta = \pi$ on a null congruence, and the flux of photons from r, θ (Misner, Thorne, & Wheeler 1972, hereafter MTW72; Schneider, Ehlers, & Falco 1992, hereafter SEF92; Fabian et al. 1989). Here $r = R/R_{Schwarzchild}$ is the distance from the black hole.

We included only photons coming from $r > r_{min} = 3$, assuming that there are no cool atoms *outflowing* from closer than the last stable orbit (this assumption can be relaxed; see Reynolds & Begelman [1997]). We also assumed $v_r \sim K\sqrt{1/r}$, where $K < 1$ is a constant parameter, and the speed of light is set to 1. To first order in v/c and $1/r$ the ratio of observed photon energy to emitted photon energy becomes

$$\frac{E_{obs}}{E_{em}} = \frac{(U_\alpha p^\alpha)_{em}}{(U_\alpha p^\alpha)_{obs}} \sim 1 - v_r \cos\theta - \frac{1}{2r}, \quad (1)$$

where $v_r > 0$ for outflow. Here U_α is the four velocity, p_α is the four momentum, and we follow the usual summation, using the Schwarzchild metric in spherical coordinates.

To find the flux from an extended outflow we assume that each point on the shell radiates isotropically. For small v_r, the magnification (ratio of the flux of the observed pixel to the flux of the emitting pixel) is the the determinant of small changes in vector image position to source position, of order

$$M \sim \frac{1}{\sin \Theta}, \tag{2}$$

(SEF92; Rauch & Blandford 1994; MTW72). Since every source generates two images in a Schwarzchild solution we considered the average of the two observed specific fluxes. The effect of the aberration of light (beaming) reduces the effects of lensing, but is generally small at $r > 3$, $K <\sim 1/3$. Although one does not expect black holes to lens stellar objects (MTW72), they seem to lens their own outflow or inflow just enough to skew a line profile (see section §3).

The total spectrum is determined by assuming a spatial distribution of emitters. We assumed an outflowing spherical shell. More complicated generalizations, such as distribution of clouds at different radii, absorption in the line of sight, and Compton reflection will be included in subsequent publications (S01).

3. Results

In this section we show the effects of outflows on a narrow line. For historical purposes we chose the Fe K_α line at 6.4 keV. Obviously, this treatment is the same for other emission lines.

The effects of a *stationary spherical shell* on a line spectrum are apparent in Figure 1. The change from 6.4 keV is due to gravitational redshift. The darkest line is for a shell at $r = 12$, the second darkest is at $r = 9$ and the light gray line is for $r = 3$.

In Figure 2, the spectrum is from a P-Cygni outflow with $K = 0.1$. The line with the largest red wing is for a shell at $r = 6$. The darkest line is for a shell at $r = 12$, and the last line is for a shell at $r = 18$. Note that the flux of the redshifted "wing" is always greater than the flux of the blue-shifted wing.

In Figure 3, we set $r = 10$ and changed the velocity. The darkest line is with $K = 0.3$, the second darkest line is with $K = 0.1$ and the third line is with $K = 0.01$. The various "sub-peaks" are the result of numerical integration.

In Figure 4, we show the effect of an obscuring disk on the shape of the line. The faint line is the line in the case that *except for a hole of size* $r = 3$ (the last stable orbit) half of the outflow is obscured by an accretion disk.

4. Discussion and Concluding Remarks

Although there are no observations that clearly support the existence of cool outflows from the central engines of AGN yet, such observations may be forthcoming. The resolution of *CXO* and, especially, *XMM* are sufficient for finding the general relativistic effects on outflows. We are especially hopeful that NLS1s may show such complex lines. In the cases of MCG-6-30-15 and MRK 776 (Figures 2 and 3 in Branduardi-Raymont et al. [2001]), we suggest that trying to fit the residuals after fitting a relativistic disk may imply radial flow. We are currently working on adding the effects of absorption in the line of sight, and the Kerr metric on emission from outflows.

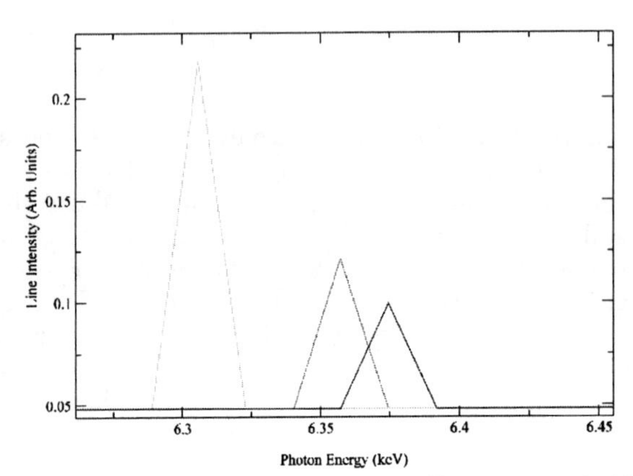

Figure 1. Stationary spherical shells at $r = 3, 9, 12$.

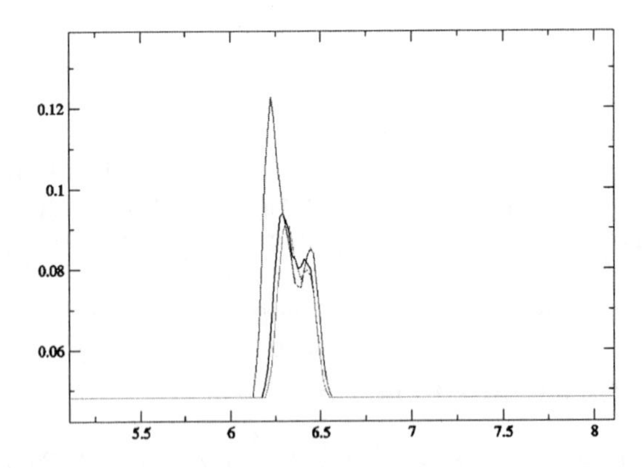

Figure 2. A P-Cygni outflow at $v_r = 0.1 v_{ff}$ at $r = 6, 12, 18$.

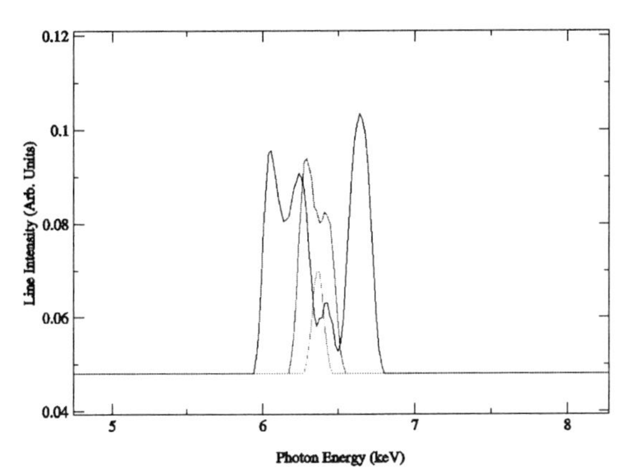

Figure 3. A P-Cygni outflow at $r = 10$ with $v_r = 0.01, 0.1, 0.3 v_{ff}$.

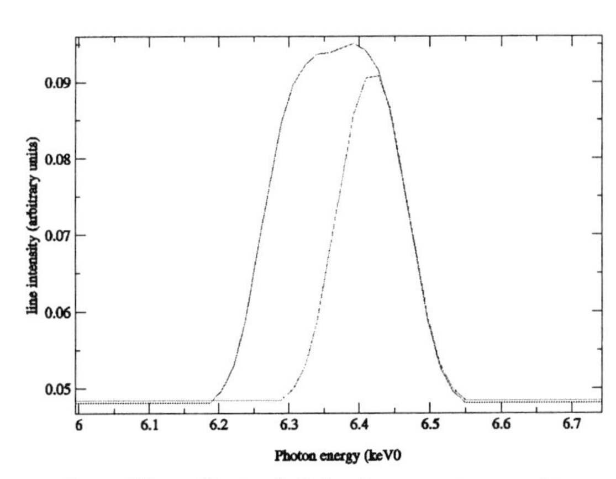

Figure 4. The effect of disk obscuration on line emission.

References

Branduardi-Raymont, G., et al. 2001, A & A, 365, L140

Collinge, M.J., et al. 2001, ApJ, in press (astro-ph/0104125)

Fabian, A.C. 1998, in Theory of Black Hole Accretion Disks, eds. Abramowicz M.A., Bjornsson, G. & Pringle J.E. (Cambridge University Press), 123

Fabian, A.C., Rees, M.J., Stella L., & White, N.E. 1989, MNRAS, 238, 729

Guilbert P.W, & Rees M.J. 1988, MNRAS, 233,475

Iwasawa, K., et al. 1996, MNRAS 282, 1038

Kaspi, S., et al. 2000, ApJ, 535, L17

Kaastra, J., et al. 2000, A&A 354, L83

Laor, A. 1991, ApJ, 376, L90

Lee, J.C., et al. 1999, MNRAS, 310, 973

Nandra, K., et al. 1997, ApJ, 477, 602

Pounds K.A., Nandra K., Stewart G.C., George, I.M., & Fabian A.C. 1990, Nature, 344, 132

Rauch K.P., & Blandford R.D. 1994, ApJ, 421, 46

Reynolds C.S., & Begelman, M.C. 1997, ApJ, 488, 109

Schneider P.M., Ehlers J., & Falco, E.E. 1993, Gravitational Lenses (Springer-verlag: New York)

Sivron, R. 1998, ApJ, 503, L57

Sivron, R., & Tsuruta, S. 1993, 402, 420

Zdziarski, A.A. 1988, MNRAS, 296, L51

Mass Outflow in Active Galactic Nuclei: New Perspectives
ASP Conference Series, Vol. 255, 2002
D.M. Crenshaw, S.B. Kraemer, and I.M. George

Are Quasars Fractal?

Mark C. Bottorff and Gary J. Ferland

Department of Physics and Astronomy, University of Kentucky, Lexington, KY 40506-0055

Abstract. In this paper we show that bright emission lines of quasars and other active galactic nuclei (AGNs) are reproduced if the gas distribution in the central parsec of these objects has a fractal geometry. While the detailed nature of AGN clouds is unknown recent models invoke magnetohydrodynamic (MHD) effects to explain their persistence. Numerical simulations of MHD turbulence have shown that a fractal cloud distribution is a result. Whatever the mechanism, fractal structure is observed in nature and may be present in AGNs as well. We therefore test a fractal distribution of gas in the broad line region (BLR) by calculating various emission line ratios it produces when illuminated by an AGN continuum and then comparing these to line ratios obtained from observations. We find eight out of eleven emission line ratios are reproduced. Additional successes of the model are that it explains the persistence of broad line region clouds without invoking a special confinement mechanism, it explains the number and column density of line of sight absorbers, and predicts the observed frequency of nuclear activity in galaxies.

1. The Role of Magnetic Fields

The fractal description of AGN gas represents the third part of our continuing study of large scale microturbulence in the BLR clouds of quasars and other AGN (see Bottorff & Ferland 2000; Bottorff et al. 2000; Bottorff & Ferland 2001). This introduction is used to review our previous efforts in order to set the context for the present topic.

1.1. Line widths

Rees (1987) suggested that BLR clouds are confined by magnetic fields in energy equipartition with the thermal motions in the cloud. The minimal confining magnetic field is $B \gtrsim \sqrt{8\pi nkT} \sim 0.6\sqrt{n_{10}T_4}$ G where $n_{10} \sim 1$ is the density in units of 10^{10} cm^{-3} and $T_4 \sim 1$ is the temperature in units of 10^4 K of 'typical' BLR clouds (Davison & Netzer 1979). In this case, of purely thermal motions, the velocity width of a single BLR cloud is $\sigma \sim 10$ km s^{-1}. While it may be the case that BLR clouds have only thermal widths it may also be that super thermal motions are present, possibly in the in the form of MHD waves. Such waves are observed in the interstellar medium where it has been found that the motions are turbulent and in energy equipartition with the magnetic field (Arons & Max 1974; Meyers & Goodman 1988a, 1988b). Thus

$$\frac{B^2}{8\pi} \approx \frac{1}{2}\rho\sigma_{\mathrm{B}}^2 \approx \frac{GM\rho}{R} \tag{1}$$

where $B^2/8\pi$ and $1/2\rho\sigma_{\mathrm{B}}^2$ are the magnetic pressure and MHD wave energy density respectively, ρ is the mass density and σ_{B} is the velocity width of the gas.

Broadening is considerably amplified however if the magnetic field is in equipartition with the gravitational energy density as indicated by the last term of equation 1 in which R is the radial distance of gas from a central black hole of mass M, and G is the universal constant of gravity. In this case, the line width will be greater than the virial width, the local velocity field will be highly supersonic and MHD broadening may actually account for the full width of an emission line (Bottorff & Ferland 2000).

1.2. Line transfer

No matter what its origin, the presence of microturbulent motion within a cloud has important effects on the emitted spectrum. Microturbulence affects the spectrum through changes in a line's intrinsic width. It is included in the Doppler width through

$$V = (V_{th}^2 + V_{turb}^2)^{1/2} \tag{2}$$

where V_{th} is the thermal width of the gas. Adding a velocity field in this way is appropriate when V_{turb} has a length scale that is shorter than the photon mean free path. The result is that the line spectrum grows stronger relative to the continuum as turbulence increases. This is because all lines more easily escape as the optical depth is spread out in velocity space, and permitted lines are selectively strengthened by continuum pumping. This helps to explain why emission lines appear too bright when they are assumed: (1) to have thermal width only and, (2) see the same continuum that we see (Bottorff et al. 2000; Dumont et al. 1998).

1.3. Why Fractal Clouds?

Our previous work suggests the BLR is a highly turbulent environment (Bottorff & Ferland 2000; Bottorff et al. 2000). It is found that high levels of turbulence exist elsewhere in nature (i.e.the interstellar medium) and have a fractal gas distribution (Elmegreen & Falagarone 1996). It is therefore natural to apply a fractal gas distribution to BLR clouds.

2. Fractal BLR Clouds

In this section we provide a brief outline of a simple fractal structure and how its four parameters can be normalized by observations of emission and absorption in AGN.

2.1. A Fractal Primer

A fractal is a self similar hierarchical clustering of structures. Each structure within a cluster is itself a cluster of smaller substructures and each of these is a cluster of even smaller subsubstructures. Three geometric parameters specify the relative size and spacing of structures in a fractal. The parameters are: (1) the multiplicity (N), defined as the number of substructures per structure, (2) the scale factor ($L > 1.0$), defined as the relative size of a structure compared to one of its N substructures and (3) the maximum hierarchy (H), defined as the number of sublevels ($h = 0, 1, 2, ..., H$) of structure in the fractal. A fourth parameter S sets the absolute scale by defining the size of the smallest structure (corresponding to $h = 0$) which we call a "cloud element". Thus size of a structure at level h is given by $R_h = SL^h$ and the number of cloud elements it contains is $N_h = N^h$. A fractal with maximum hierarchy H is therefore L^H times larger than a cloud element and consists of N^H of them. We note that the parameter N is usually written in terms of L and another parameter, D, called the fractal dimension. The relationship between N and D is defined by $N = L^D$.

A simple fractal is illustrated in Figure 1 (Figure 1a from Bottorff & Ferland [2001]). The Figure shows the two-dimensional projection onto a plane of a fractal structure in three-dimensional space. The fractal shown has $D = 2.3$, $L \approx 3.05$, and $H = 3$. In the figure the smallest spheres represent cloud elements. The value of L is chosen in this example to make the multiplicity an integer ($N = L^D \approx 13$). The spheres labeled 1,2, and 3 show substructures of h = 1,2, and 3 respectively. Note the number density of cloud elements decreases with increasing h (if $D \leq 3.0$). In a substructure with $h = 1$ individual cloud elements strongly overlap and individual cloud elements lack identity. For $h = 2$ and higher however the spacing is more open. We therefore define a BLR cloud to be any $h = 1$ structure (e.g., a cluster of cloud elements).

2.2. Normalizing the BLR Model

To describe a fractal BLR the parameters S, D, L, and H must be determined. The value $S \sim 10^{13}$ cm is inferred from the physical properties of 'typical' BLR clouds (Davidson & Netzer 1979). It is not possible determine D directly because a telescope capable of resolving detailed structure in the BLR is required. Instead $D = 2.3$ is adopted from the structural analysis of high resolution observations of the turbulent interstellar medium (Elmegreen & Falagarone 1996). The unknowns L, and H are determined simultaneously. One condition is obtained from $R_H/S = L^H$ where R_H is the size of the fractal. The other condition is from the global covering factor which can either be written as a scale free function $f = f(L, H)$ or alternatively as $f \propto R$ (Bottorff & Ferland 2001). The value of R_H is set equal to 5 times the size of the BLR which is in turn set by the dust sublimation radius (Laor & Netzer 1993). The choice of 5 reflects the fact that material beyond the BLR such as warm absorbers (Reynolds 1997; George et al. 1998) or UV absorbers (Crenshaw & Kraemer 1999) have covering fraction of ~ 0.5 and that of the BLR is ~ 0.1 (Peterson 1993). The resulting parameter values obtained from solving the two conditions are $L = 3.2$ and $H = 11$.

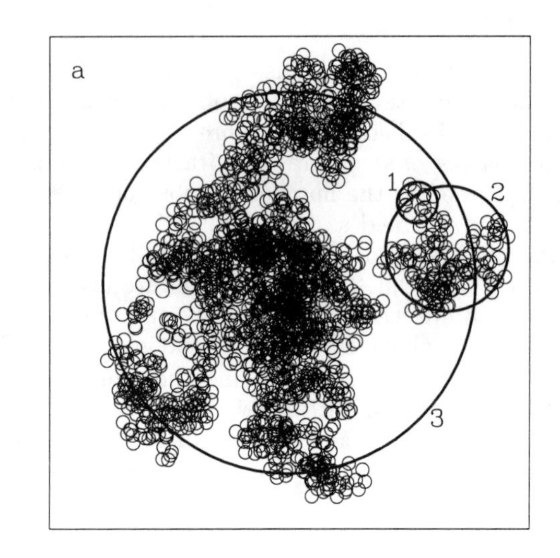

Figure 1. Simulated fractal of hierarchy 3 with $L \approx 3.05$ and $D = 2.3$. Note that the density of cloud elements decreases for larger structures.

3. Model Results

The four fractal parameters (as normalized above) may be combined to give additional (and more physical) properties of the fractal model (3.1). In addition photoionization modeling of the fractal gas distribution allows us to test whether or not the model is capable of reproducing observed emission line strengths (3.2).

3.1. Physical Properties

Of particular interest is the number of clouds, the number of line of sight clouds, and cloud column densities. A BLR cloud (given its definition as a $h = 1$ structure of cloud elements) has a column density of $\log(N_{cloud}[\mathrm{cm}^{-2}]) = 23.5 + \log(\mathrm{n}_{10})$ where n_{10} is the gas density of a cloud element in units of 10^{10} cm^{-3}. Since the clouds cluster together we expect that there will be strong cloud to cloud shielding. The result is there are about 6 BLR clouds along any given line of sight through the fractal, consistent with observations of Seyfert 1 galaxies (Crenshaw et al. 1999). In addition for every cloud directly exposed to the continuum about 60 are shielded. Since the total number of BLR clouds in the fractal model is $\sim 10^{9.3}$ the total illuminated is $\sim 10^{7.5}$. If the turbulence is in energy equipartition with gravity then only a small fraction of BLR clouds have large scale turbulence within them. The number with turbulence larger than 2500 km s^{-1} is $\sim 10^{4.0}$ and the number of these that are illuminated is only $\sim 10^{2.2}$. Thus relatively few clouds (less than 200) contribute to the wings of an emission line. The line wings should be smooth however due to the extremely broad contribution of each cloud to the total line profile. Rough estimates of the distribution of structures within the fractal show that the chance of finding material near the central black hole is 0.01 thus about 1 in 100 galaxies are expected to show nuclear activity (Bottorff & Ferland 2001). In addition we find that if material is already in the vicinity of the black hole the probability of there being additional material nearby is high so that quasi continuous feeding of the central black hole is possible.

3.2. Line Ratios Relative to Lyα

Table 1 shows the strengths of emission line blends relative to Lyα for the density ($n = 10^{10}$ cm^{-3}) that gives collectively the best fit to the data in Zheng et al. (1997) and Baldwin et al. (1995, 1996). The model is moderately successful in that 8 out of 11 ratios are reproduced by the model.

Table 1
Line Strength Relative to Lyα: Observation and Models

Emission Line Blend	Zheng	Baldwin	Strength/Lyα	2× ?
O III λ835+O II λ834	0.014		0.006	no
C III λ977+Lyγ λ973	0.009	0.007-0.20	0.036	no (high 4×)
N III λ990	0.011	0.013	0.007	yes
O VIλ1032+O VIλ1035+Lyβ λ1026	0.190	0.068-0.69	0.106	yes
N V λ1239+N V λ1243	0.110	0.069-0.99	0.030	no
Si IV λ1397+*O* IV] λ1402	0.075	0.022-0.50	0.076	yes
C IV λ1548+C IV λ1551	0.620	0.087-0.65	0.656	yes
He II λ1640+O III] λ1663+*Al* II λ1671	0.068	0.013-0.14	0.136	yes
C III] λ1909+Si III] λ1892 +Al III λ1859	0.163	0.076-0.74	0.153	yes
Mg II λ2796+Mg II λ2804	0.250	0.15-0.30[a]	0.144	yes
Hβ λ4861		0.07-0.20[a]	0.031	no (low 3×)
EW(Lyα)/1216	0.076	0.03-0.20	0.73	yes

[a]Ranges for observations are from Baldwin et al. 1995.

4. Conclusions

• A fractal BLR model is simultaneously consistent with observation deduced covering factor, column density, emission line strengths, and line ratios of BLRs in AGNs.

• Model absorption properties are consistent with line of sight column density as determined from X-ray observations.

• Rough estimates show that 1 in 100 of the galaxies that harbor a super massive black hole will show activity.

• Stochastic feeding of the central engine due to a fractal BLR may account for continuum variations and long term activity.

Acknowledgments. This work was supported by grants from NSF through grant AST-0071180 and NASA through its LTSA program.

References

Arav N., Barlow T.A., Laor A., & Blandford, R.D. 1997, MNRAS, 288, 1015

Arav N., Barlow T.A., Laor A., Sargent, & W.L.W, Blandford R.D. 1998, MNRAS, 297, 990

Arons J., & Max C.I. 1975, ApJ, 196, L77

Baldwin, J.A., Ferland, G.J., Korista, K.T., & Verner, D. 1995, ApJ, 455

Baldwin, J.A., et al. 1996, ApJ, 461, 644

Bottorff M.C., Korista K.T., Shlosman I., & Blandford R.D. 1997, ApJ 479, 200

Bottorff, M.C., & Ferland, G.J. 2000, MNRAS, 316,103

Bottorff, M.C., Ferland, G.J., Baldwin, J.A., & Korista, K.T. 2000, ApJ, 543

Bottorff, M.C., & Ferland, G.J. 2001, ApJ, 549,118

Crenshaw, D.M.,& Kraemer, S. B. 1999, ApJ, 521, 572

Crenshaw, D.M., et al. 1999, ApJ, 516, 750

Dietrich, M.,Wagner, S.J., T.J.-L. Courvoiser, Bock H., & North, P. 1999, 351, 31

Dumont, A.-M.,Collin-Souffrin, S., & Nazarova, L. 1998, A&A, 331,11

Elmegreen, B.G, & Falagarone, E. 1996, ApJ, 471, 816

Ferland, G.J, 1999, ASP Conference Series, Vol. 162, Quasars and Cosmology, ed. Ferland, G. J., & Baldwin J.A. (San Francisco: ASP), 147

Ferland, G. J., et al. 1996, ApJ 461, 683

George, I.M., et al 1998, ApJS, 114, 73

Hamann, F., et al. 1998 ApJ, 496, 761

Hamann, F., & Ferland, G.J. 1999, ARA&A 37, 487

Korista, K.T., Baldwin, J., Ferland, G., & Verner, D. 1997a, ApJS, 108, 401

Korista, K.T., Ferland, G., & Baldwin, J. 1997b, ApJ, 487, 555

Korista, K.T., Baldwin, J.A., & Ferland, G.J. 1998, ApJ, 507, 24

Myers P.C., & Goodman A.A., 1988a, ApJ, 326, L27

Myers P.C., & Goodman A.A., 1988b, ApJ, 329, 392

Netzer, H., et al. 1995, ApJ, 448, 27

Netzer H., & Laor, A. 1993, ApJ, 404, L51

Peterson, B.M. 1993, PASP, 105, 247

Rees, M.J. 1987, MNRAS, 228, 47P

Reynolds, C.S. 1997, MNRAS, 286, 513

Zheng, W., et al. 1997, ApJ, 475, 469

Mass Outflow in Active Galactic Nuclei: New Perspectives
ASP Conference Series, Vol. 255, 2002
D.M. Crenshaw, S.B. Kraemer, and I.M. George

Tests of a Structure for Quasars

Martin Elvis
Harvard-Smithsonian Center for Astrophysics, 60 Garden St., Cambridge MA02138 USA

Abstract. The model I recently proposed for the structure of quasars unifies all the emission, absorption, and reflection phenomenology of quasars and active galactic nuclei (AGN), and so is heavily overconstrained and readily tested.

Here I concentrate on how the model has performed against tests since publication - with many of the tests being reported at this meeting. I then begin to explore how these and future tests can discriminate between this wind model and 3 well-defined alternatives.

1. Introduction

Quasar research suffers from an overabundance of phenomenology. There has been a constant piling up of new details at all wavelengths with sadly little integration, let alone physical explanation. Nevertheless, stars were in a similar situation for 20 years (Lawrence 1987). There is hope that the complexities of quasars will resolve themselves the same way. I have proposed (Elvis 2000) a geometrical and kinematic model for quasars that does seem to subsume a great deal of the phenomenology of quasar emission and absorption lines into a simple scheme. Here I briefly outline the model and then concentrate on tests of the model, many reported at this meeting.

2. A Structure for Quasars

In Elvis (2000) I proposed that a flow of warm ($\sim 10^6$K) gas rises vertically from a *narrow range of radii* on an accretion disk. This flow then accelerates, angling outward (most likely under the influence of radiation pressure from the intense quasar continuum) until it forms thin conical wind moving radially (Figure 1). When the continuum source is viewed through this wind it shows narrow absorption lines (NALs) in both UV and X-ray (the X-ray "warm absorber"); when viewed down the radial flow the absorption is stronger and is seen over a large range of velocities down to $v(vertical)$, the "detachment velocity", so forming the Broad Absorption Line (BAL) quasars. Given the narrowness of the vertical flow ($\sim 0.1r$), the divergence of the continuum radiation at the turning point will be $\sim 6°$, giving 10% solid angle coverage, and so the correct fraction of BAL quasars. [The angle to the disk axis, 60°, is at present arbitrarily chosen to give the correct number of NAL and non-NAL quasars.]

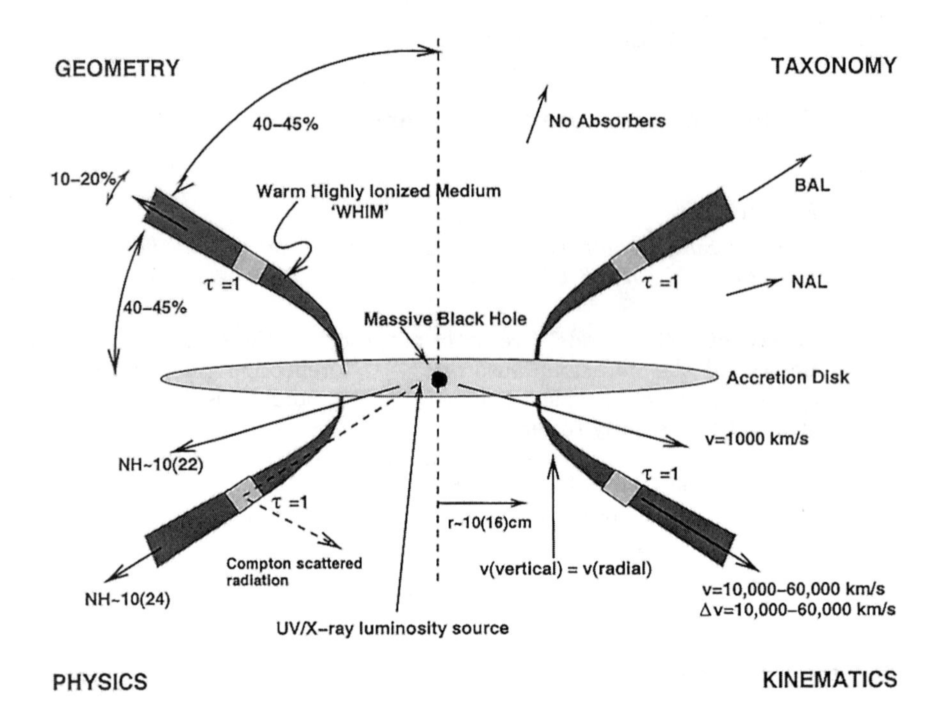

Figure 1. A Structure for Quasars

This 'Warm Highly Ionized Medium' (WHIM) has a cool phase (like the interstellar medium) with which it is in pressure equlibrium. This cool phase provides the clouds that emit the Broad Emission Lines (BELs). Since the BEL clouds move along with the WHIM they are not ripped apart by shear forces; and since the medium is only Compton thick along the radial flow direction rapid continuum variations are not smeared out. Both problems had long been a strong objections to pressure confined BEL clouds, but they are invalid in the proposed geometry.

The radial flow is Compton thick along the flow direction, and so will scatter all wavelengths passing along that direction. Since the flow is highly non-spherical the scattered radiation will be polarized. The solid angle covered by the radial flow is 10%-20%, so this fraction of all the continuum radiation will be scattered, leading to the filling in of the BAL troughs, and to an X-ray Compton hump in all AGN. Since the WHIM is only ionized to Fe XVII, there will be Fe-K fluorescence off the same structure at $\sim$100 eV equivalent width (EW). Some of the BEL radiation will also pass along the flow and will be scattered off the fast moving flow, producing the polarized, non-variable 'Very Broad Line Region'.

3. Tests of the Model

The model makes five main claims (each has been tested and passed):

Claim 1: The quasar wind is a warm, highly ionized medium (WHIM). The velocities of the UV absorption line systems must match X-ray absorbers. In NGC 3783 (Kaspi 2002) and NGC 4051 (Collinge et al., 2001) they do match. The line strengths must also be consistent across the X-ray and UV lines. In NGC 3783 (Kaspi 2002) they also match, if a particular EUV spectral slope is assumed. This slope is tightly constrained, which implies an accurate thermometer for the, otherwise hard to observe but energetically important, EUV continuum. Consistency with the NLR optical coronal lines is required. Accretion disk models predict specific EUV continuum shapes, which will be stringently tested if the X-ray/UV absorbers are the same.
Claim 2: BAL medium = NAL medium = WHIM. Since Compton scattering is wavelength independent the BAL covering factor in the soft X-ray band should be the same as in the optical. Partial covering of the right order is seen in the X-ray spectrum of one BAL quasar (Mathur et al. 2002). The low energy X-rays should be polarized like the optical BAL troughs, and have consistent fluxes.
Claim 3: WHIM = BEL region (BELR) Confining Medium. The high ionization BELs (P VII, Ne VIII) at ~770Å(rest) seen in some high redshift quasars (Hamann et al. 1995) could arise from the WHIM, hence should match its properties. In fact they do, with column densities of $\sim 10^{22}\mathrm{cm}^{-2}$, a temperature of 5×10^{5}K (if collisional), and a covering factor of 0.5 (and unlike the BELR for which 0.1 is a typical value). The radius of the NAL from the continuum source has to match the radius of the high ionization BELR. In NGC 5548 this works, but the NAL radius is poorly constrained (a factor 100). Short recombination times, which will determine the NAL density, are the key to a better constraint on the radius. Timescales of hours are likely, and require quite large continuum variations on this timescale or less to be measurable.
Claim 4: Narrow Range of radii: cylindrical WHIM/BELR. Arav et al.(2002) show that the covering factor or the NAL systems in NGC 5548 varies systematically with velocity in a way that is easily understood as a narrow flow of material across our line of sight. These constraints will determine many parameters of the flow (thickness, angles, density contrast, acceleration rate).

Cohen & Martel (2002) explain the rotation of polarization position angle of Hα in some radio galaxies by means of a rotating circularly symmetric structure, a disk or cylinder, which emits the BELs and scatters BEL photons off the opposite sides. A few percent of the Hα photons are polarized and red or blue shifted according to where they strike the opposite face. A WHIM shaped cylinder is hollow, making it simple for photons to cross to the other side. Similar PA rotations are implied for *all* edge-on AGN , i.e. those with NALs.

Laor & Brandt (2002) showed that weak EW([O III]), the presence of NALs, and broad C IV are correlated. If [O III] is isotropic and the continuum is from a disk then the EW([O III]) is an inclination indicator, so these effects are qualitatively predicted by the proposed structure.

Murray (comments at this meeting) notes that following a flare the BELs may become optically thin, revealing a double-peaked profile for a cylinder or disk geometry. He finds just such an effect in the NGC 5548 data base.
Claim 5: Reflection from Structure. Compton scattering off the radial part of the structure must produce a symmetric X-ray 6.4 keV Fe-K emission line with a width broader but comparable to the BELR widths. Since the WHIM is

optically thin to Fe-K, while BEL clouds are optically thick, the two profiles will be different, with Fe-K possibly showing a double peaked structure. This needs detailed calculations, since it can be measured with *Chandra X-ray Observatory* grating spectra.

The Fe-K emission line will respond to continuum changes with a smeared response determined by the size of the τ=1 scattering ring. Takashi, Inoue & Dotani (2001, in preparation) have found such an effect in NGC 4151, and derive an Fe-K scattering region size of 10^{17}cm. This is a few times larger than the CIV radius for NGC 4151 (9±2 light-days, 2×10^{16}cm, Kaspi et al. 1996), consistent with our structure, but is strongly inconsistent with an accretion disk or pc-scale torus origin

At any given angle there will be four dominant delay times relative to a central continuum flaring event, as the flare scatters off the near and far parts of the τ=1 rings above and below the disk plane. (The disk may obscure one or both parts of the lower ring.) These should show up in autocorrelation functions of the continuum at low amplitudes. Schild (1996) has found autocorrelation timescales in the gravitational lens Q0957+561 and shows them to be consistent with the double ring expected from this structure.

The X-ray 'Compton Hump' should show the same time smearing and delays as the Fe-K line. Moreover the Compton Hump should be polarized, just as the optical BAL troughs are polarized.

4. Alternative Wind Models

To reconcile the presence of both a fast (BAL) wind and a slow (NAL) wind in most AGN and quasars, we adopted a single wind with a specific geometry. It is worth considering alternative means of reconciling these, relaxing the constraint that they are the same flow. There are two main possibilities: either high luminosity objects have faster winds, or faster winds are emitted in some preferential direction , and slower winds in others. Here we assume that the models retain all the other features of our model: 1) they still try to combine the UV and X-ray absorbers in a single WHIM (our starting point), 2) they have the BEL clouds embedded in this wind, and 3) they try to explain reflection features via scattering off the fast wind.

4.1. Luminosity Dependent Velocities

In this hypothesis the fast BAL winds are emitted only by high luminosity quasars (into a ~10% solid angle), while NAL winds are found only at lower luminosities (for a ~50% solid angle). The wind no longer needs to arise from a narrow range of disk radii. There are a number of comments that can be made:
(1) In this scenario a continuum of widths should be found with a covering factor that decreases from ~50% at NGC 5548 luminosities ($L_X \sim 5\times10^{43}$erg s^{-1}, Elvis et al. 1978), to ~10% $L_X \sim 10^{46}$erg s^{-1} (PHL 5200, Mathur, Elvis, & Singh [1995]). There is a range of BAL velocities, but a line width vs. luminosity or covering factor analysis has not yet been performed.
(2) This model has a built-in explanation for the absence of low luminosity BALs; our model requires low luminosity AGN to be dustier.

(3) If the BELs are embedded in the fast BAL wind then they should have similar profiles, with no obvious physical cause for a 'detachment velocity'.
(4) Can high ionization BELs with large covering factor arise in this model?
(5) The scattering effects of the radial part of the wind (X-ray Fe-K emission line, Compton hump, optical polarized flux) will disappear in lower luminosity objects since the column density needed to produce scattering is much larger than the column density through the X-ray warm absorbers, and low luminosity AGN will have no fast wind out of the line of sight to produce the polarized, non-variable 'Very Broad Line Region'. The opposite is observed (Iwasawa & Taniguchi 1993).
(6) The BAL opening angle of 10% does not arise naturally from the narrow width of the wind origin site on the disk.
(7) Are there directions in which one looks through the wind? If so then NALs will be seen, the wind originates from a restricted range of radii and the picture reverts to something close to our model.

The absence of reflection effects in this version considerably weakens its unifying power.

4.2. Orientation Dependent Velocities

In this hypothesis the quasar wind has fast (BAL) velocities only in some directions (covering ~10% solid angle), and has slower (NAL) velocities in other directions (covering ~50% solid angle). There are two obvious preferential directions for the fast wind: *polar* and *equatorial.* Some comments are:
(1) Proga (2002) finds such a directional velocity stratification arising naturally from his simulations, with higher velocities toward the pole.
(2) Like our model, this hypothesis does not explain the absence of low luminosity BALs without invoking dust.
(3) Compton scattering features will arise naturally in all objects in this model, as in ours, if the fast wind has sufficient column density.
(4) The fast wind has to have a large column density of high velocity material when viewed end-on to reproduce the BAL observations. This arises naturally in our model, but is not obvious in this hypothesis. A large column density implies a large mass input rate at the wind base, decaying rapidly to larger radii (in the fast polar case; to smaller radii for the fast equatorial case).
(5) In the fast polar case a much (<1%) lower column density is required when viewed from other directions to avoid BALs being seen more often. A long thin BAL region is necessary, which would have to be non-divergent to maintain column density. This may make it difficult to cover a 10% solid angle. An equatorial fast wind avoids this problem.
(6) An equatorial fast wind has the faster material originating further from the continuum source, which seems unlikely since radiation pressure and Keplerian rotation both predict the opposite. A polar fast wind avoids this problem.

The difficulties in this version of the model lie primarily in theory: can the large column densities needed in the fast wind be produced, in an acceptable geometry? Neither the polar nor equatorial solution is fully appealing.

5. Conclusions

Because the model puts together so many aspects of quasar/AGN phenomenology it is highly overconstrained, and so readily tested. Our model has passed quite a number of tests already, but they are not yet as stringent as one would like. Quite simple extensions of the model (e.g., to type 2 objects, Risaliti, Elvis, & Nicastro [2001]) suggest that much more of the quasar/AGN phenomenology can be incorporated with only a handful of extra variables. With luck then, quasars will now enter a period of rapid development of their physics, allowing their physical evolution to be understood, and placing them constructively within cosmology.

This work was supported in part by NASA grant NAG5-6078.

References

Arav, N., et al. 2002, in Mass Outflow in Active Galactic Nueclei: New Perspectives, eds. D.M. Crenshaw, S.B. Kraemer, & I.M. George (San Francisco: ASP), p. 179

Elvis, M. 2000, ApJ, 545, 63

Elvis, M., et al. 1978, MNRAS, 183, 129

Cohen, M., & Martel A. 2002, in Mass Outflow in Active Galactic Nueclei: New Perspectives, eds. D.M. Crenshaw, S.B. Kraemer, & I.M. George (San Francisco: ASP), p. 255

Collinge, M.J., et al., 2001, ApJ in press.

Hamann, F., Shields, J., Ferland, G., & Korista, K. 1995, ApJ, 454, 688

Iwasawa, K., & Taniguchi, Y. 1993, ApJL, 413, L15

Kaspi, S. 2002, in Mass Outflow in Active Galactic Nueclei: New Perspectives, eds. D.M. Crenshaw, S.B. Kraemer, & I.M. George (San Francisco: ASP), p.7

Kaspi, S., et al. 1996, ApJ, 470, 336

Lawrence, A. 1987, PASP, 99, 309

Laor, A. & Brant, W.N. 2002, in Mass Outflow in Active Galactic Nueclei: New Perspectives, eds. D.M. Crenshaw, S.B. Kraemer, & I.M. George (San Francisco: ASP), p. 99

Mathur, S., et al. 2002, in Mass Outflow in Active Galactic Nueclei: New Perspectives, eds. D.M. Crenshaw, S.B. Kraemer, & I.M. George (San Francisco: ASP), p. 151

Mathur, S., Elvis, M., & Wilkes, B.J. 1995, ApJ, 452, 230

Mathur, S., Elvis, M., & Singh, K.P. 1995, ApJL, 455, L9

Proga, D. 2002, in Mass Outflow in Active Galactic Nueclei: New Perspectives, eds. D.M. Crenshaw, S.B. Kraemer, & I.M. George (San Francisco: ASP), p. 309

Schild, R. 1996, ApJ, 464, 125

Risaliti, G., Elvis, M., & Nicastro F. 2001, ApJ, submitted

Mass Outflow in Active Galactic Nuclei: New Perspectives
ASP Conference Series, Vol. 255, 2002
D.M. Crenshaw, S.B. Kraemer, and I.M. George

Dynamics of Line-Driven Disk Winds

Daniel Proga

NASA Goddard Space Flight Center, Laboratory for High Energy Astrophysics, Code 662, Greenbelt, MD 20771

Abstract. I review the main results from recent 2-D, time-dependent hydrodynamic models of radiation-driven winds from accretion disks in active galactic nuclei (AGN). I also discuss the physical conditions needed for a disk wind to be shielded from the strong X-rays and to be accelerated to hypersonic velocities. I conclude with a few remarks on winds in hot stars, low mass young stellar objects, cataclysmic variables, low mass X-ray binaries, and galactic black holes and future work.

1. Introduction

A key constraint on any model for the origin of AGN outflows is the ionization balance. On one hand we observe very high luminosities in X-rays and the UV and on the other hand we observe spectral lines from moderately and highly ionized species. One wonders then how the gas avoids full photoionization and we see any spectral lines at all. Two mechanisms have been proposed to resolve the so-called overionization problem: (i) the AGN outflows have filling factors less than one and consist of dense clouds and (ii) the filling factors equal one but the outflows are shielded from the powerful radiation by some material located between the central engine and the outflow (e.g., Krolik 1999). Only limited citations will be possible due to space limitations; my apologies in advance.

One plausible scenario for AGN outflows is that a wind is driven from an accretion disk around a black hole. Radiation pressure due to spectral lines is one of the forces that have been suggested to accelerate outflows in AGNs.

Our understanding of how line-driving produces powerful high velocity winds is based on the studies of winds in hot stars (e.g., Castor, Abbott, Klein 1975, hereafter CAK). The key element of the CAK model is that the momentum is extracted most efficiently from the radiation field via line opacity. CAK showed that the radiation force due to lines, $F^{rad,l}$ can be stronger than the radiation force due to electron-scattering, $F^{rad,e}$ by up to several orders of magnitude (i.e., $F^{rad,l}/F^{rad,e} < M_{max} \approx 2000$). Thus even a star that radiates at around 0.05% (i.e., $1/M_{max}$) of its Eddington limit, L_E, can have a strong wind.

To apply line-driven stellar wind models to AGN we have to take into account at least two important differences: (i) the difference in geometry – stellar winds are to a good approximation spherically symmetric, whereas the wind in AGN likely arises from a disk and is therefore axisymmetric; and (ii) the difference in the spectral energy distribution – hot stars radiate mostly the UV (the UV luminosity, L_{UV}, accounts for most of the total luminosity, L), whereas

AGN radiate strongly both in the UV and X-rays (L_{UV} and L_X are comparable). The latter difference has two important consequences on UV line driving: not all AGN radiation contributes to driving, and even worse, the X-rays that to do not contribute to UV line driving can ionize the gas and reduce the number of transition that can scatter the UV photons (in the case of a fully ionized gas, $F^{rad,l} = 0$ and by definition $M_{max} = 0!$).

The consequences of the difference in geometry have been recently studied using 2-D axisymmetric numerical hydrodynamical simulations (e.g., Proga, Stone & Drew 1998, hereafter PSD98). These simulations were focused on cataclysmic variables (CVs), which, as do hot stars, radiate mostly in the UV. In particular, PSD98 explored the impact upon the mass-loss rate, $\dot{M}_w$ and outflow geometry caused by varying the system luminosity and the radiation field geometry. A striking outcome was that winds driven from, and illuminated solely by, an accretion disk yield complex, unsteady outflow. In this case, time-independent quantities can be determined only after averaging over several flow timescales. On the other hand, if winds are illuminated by radiation mainly from the central object, then the disk yields steady outflow. PSD98 also found that $\dot{M}_w$ is a strong function of the total luminosity, while the outflow geometry is determined by the geometry of the radiation field. For high system luminosities, the disk mass-loss rate scales with the luminosity in a way similar to stellar mass loss. As the system luminosity decreases below a critical value (about twice the effective Eddington limit, L_E/M_{max}) the mass-loss rate decreases quickly to zero. Matter is fed into the fast stream from within a few central object radii. In other words, the mass-loss rate per unit area decreases sharply with radius. The terminal velocity of the stream is similar to that of the terminal velocity of a corresponding spherical stellar wind, i.e., $v_\infty \sim$ a few v_{esc}, where v_{esc}, is the escape velocity from the photosphere. Thus the difference in geometry changes the wind geometry and time behavior but has less effect on $\dot{M}_w$ and v_∞.

Proga, Stone & Kallman (2000, hereafter PSK) made another step in studying line-driven winds from accretion disks. To assess how winds can be driven from a disk in the presence of very strong ionizing radiation (as in AGN) they adopted the approach from PSD98 with three major modifications: 1) calculation of the parameters of the line force based on the wind properties, 2) inclusion of optical depth effects on the continuum photons, and 3) inclusion of radiative heating and cooling of the gas. In next section I will review results from PSK and from related calculations by Proga & Kallman (2001, PK hereafter).

2. Hydrodynamical Simulations For AGN

In PSK, we calculated a few disk wind models for the mass of the non-rotating black hole, $M_{BH} = 10^8$ M$_\odot$. To determine the radiation field from the disk, we assumed the mass accretion rate $\dot{M}_a = 1.8$ M$_\odot$ yr^{-1}. These system parameters yield the disk Eddington number, $L_D/L_E \equiv \Gamma_D = 0.5$ and the disk inner radius, $r_* = 8.8 \times 10^{13}$ cm. For the radiation field from the central engine, we assumed that one half to the central engine luminosity is radiated in the UV and the other half in X-rays. The total central engine luminosity was assumed equal to the accretion disk luminosity. To calculate the gas temperature, we assumed the temperature of the X-ray radiation, $T_X = 10$ KeV (see PSK for more details).

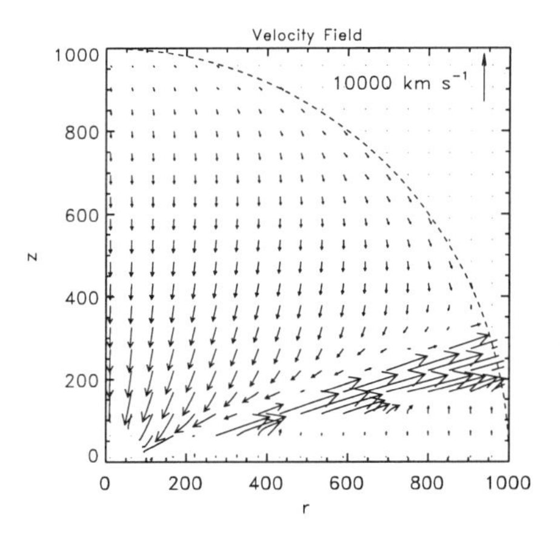

Figure 1. A map of the velocity field (the poloidal component only) for the line-driven wind from a disk accreting on a 10^8 $M_\odot$ black hole (PSK's Fig. 2). The rotation axis of the disk is along the left hand vertical frame, while the mid-plane of the disk is along the lower horizontal frame. The position on the figure is expressed in units of the disk inner radius (e.g., 100 $r_* = 8.8 \times 10^{15}$ cm).

For a fixed disk atmosphere and central radiation source, the most important parameter of the PSK model is the wind X-ray opacity κ_X that determines the optical depth for the X-rays from the central object, τ_X.

Figure 1 presents the poloidal velocity of the PSK wind model for the X-ray opacity $\kappa_X = 40$ g^{-1} cm^2 for the photoionization parameter $\xi \equiv 4\pi F_{\rm X}/n < 10^5$ and $\kappa_X = 0.4$ g^{-1} cm^2 otherwise. Here $F_{\rm X}$ is the local X-ray flux and n is the number density of the gas. The UV wind opacity, κ_{UV} was assumed 0.4 g^{-1} cm^2 for all ξ. The arrows in Figure 1 show that the gas streamlines are perpendicular to the disk over some height that increases with radius. The streamlines then bend away from the central object and converge. The region where the flow is moving almost radially outward is associated with a high-velocity, high density stream. PSK's calculation follows (i) a hot, low density flow with negative radial velocity in the polar region (ii) a dense, warm and fast equatorial outflow from the disk, (iii) a transitional zone in which the disk outflow is hot and struggles to escape the system.

PSK found that the local disk radiation can launch a wind from the disk despite strong ionizing radiation from the central object. The central radiation may overionize the supersonic portion of the flow and severely reduce the wind velocity. To produce a fast disk wind the X-ray opacity must be higher than the UV opacity by a factor $\gtrsim 100$ for the photoionization parameter $\xi < 10^5$. For lower relative X-ray opacity (i.e., $\kappa_X/\kappa_{UV} < 100$), the gas can be still launched from the disk by the UV disk radiation but the gas velocity never exceeds the escape velocity. To radially accelerate the gas lifted by the local disk radiation, the column density, $N_{\rm H}$ must be large enough to reduce the X-ray radiation but not too large to reduce the UV flux from the central engine. In other words, this requires $\tau_{\rm X} > 1$ and at the same time, $\tau_{UV} \ll 1$.

Our disk wind model can explain many aspects of AGN outflows. For example, the fast stream with the terminal velocity of ~ 15000 km s^{-1} can be identified as a BAL region in QSOs. The stream density and column density are comparable to those observed in BAL QSOs. Because the stream is narrow, BALs should be seen only for a narrow range of inclination angles. Synthetic line profiles calculated based of this model confirm that the model can explain the observations (Proga 2001, in preparation). In particular, the line profiles strongly vary with the inclination angle. The fast stream can produces a strong, broad and blue-shifted absorption component, and even multi-trough structure.

Additionally, our model illustrates that terminal velocity of line-driven winds does not have to been coupled to v_{esc}, it can be much smaller as in the transient zone in which the disk outflow is overionized and loses support from the line force. It is possible that this slow inner flow, launched and driven by radiation, can explain warm absorbers and outflows in the Seyfert I galaxies.

Let us return to the problem of launching a wind in the presence of X-rays: when can gas be lifted from the disk by the local UV radiation instead of being heated and overionized by the X-ray radiation? This question can be qualitatively answered in terms of the CAK model developed to study stars; results from PSD98 for disks showed that this approach is reasonable. For a given L_E, the stronger the UV radiation from the disk, the higher the wind mass-loss rate and subsequently the wind density. If we assume fixed X-ray radiation and no X-ray attenuation, we will find that if the UV radiation is strong enough, it can launch a wind of so high density that the X-rays will be unable to ionize the wind. The density of ionizing photons is simply too low compared to the gas density, in other words the photoionization parameter will be low. The X-ray attenuation becomes essential when gas accelerates because the gas density decreases and eventually the X-ray may overionize the gas.

The line force is not always strong enough to push gas of sufficiently high density. To examine in more detail what happens then, it is helpful to consider low mass X-ray binaries (LMXBs). LMXBs resemble AGN in a few respects, for example, in their X-ray/bolometric luminosity ratios. Recent calculations by PK found that in the case of LMXBs, the local disk radiation cannot launch a wind from the disk because of strong ionizing radiation from the central object. Unphysically high X-ray opacities ($\kappa_X \geq 10^5$ g^{-1} cm^2) are required to shield the UV emitting disk and to allow the line force to drive a disk wind. However the same X-ray radiation that inhibits line driving heats the disk and can produce a hot bipolar wind or corona above the disk. PK's results are consistent with the UV observations of LMXB which show no obvious spectral features associated with strong and fast disk winds.

3. A Few Remarks

An important aspect of studying AGN outflows is to understand why some AGN of the same type have outflows and some do not (e.g., QSOs with and without broad absorption lines, BALs), and why different types of AGN have outflows of different appearance (e.g., narrow UV absorption lines in the Seyfert I galaxies and BALs in QSOs). The talks and posters of this workshop have provided many wonderful examples of the variety of outflows in AGN. To address the problem of AGN outflows in general, I will build upon theoretical results for line-driven winds I have discussed so far. Fortunately, the 2-D hydrodynamical models can be understood fairly well using concepts from the original CAK model and the 2-D results can be approximated by analytic formulae.

For example, it is possible to derive analytic formulae to estimate the photoionization parameter at the base of a line-driven disk wind (PK). PK compared results of line-driven disk wind models for accretion disks in LMXBs and AGN. They found that the key parameter determining the role of the line force is not merely the presence of the luminous UV zone in the disk and the presence of

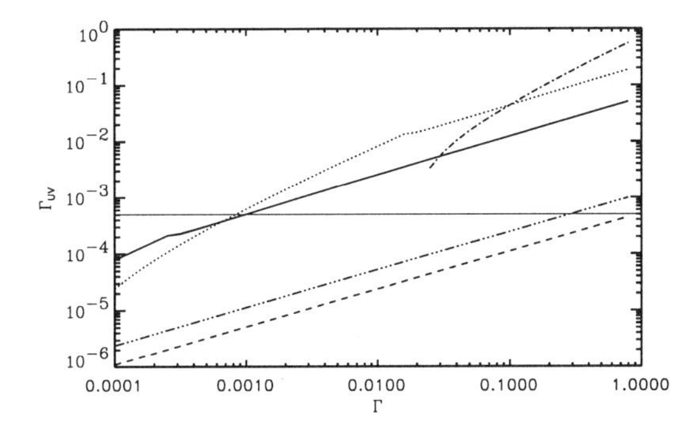

Figure 2. The UV Eddington number as a function of the total Eddington number for AGN, CVs, LMXBs, GBHs and FU Ori stars. See the text for the explanation of the lines.

X-rays, but also the distance of this zone from the center. This result is not surprising because the closer the UV zone to the center, the higher the UV contribution to the total luminosity. That in turn implies a stronger line force and subsequently a denser disk wind launched by the line force. As I already mentioned, the density of the disk wind critically determines whether the wind will stay in a lower ionization state in the presence of X-ray radiation and be further accelerated by the line force to supersonic velocities. Therefore in a general case, we ought to consider not only the total luminosity but also the UV luminosity, L_{UV} as well as the Eddington number of the UV emitting part of the disk, Γ_{UV}. The UV disk radiation is the one that is to drive a wind, the remaining disk radiation has either little impact on the wind (i.e., the optical and infrared radiation of cold disk) or can reduce the line driving (i.e., X-rays of a hot disk). Simply put, the difference, $(1-\Gamma_{UV})L$ is the luminosity that can 'damage' a wind emerging from the UV disk.

Figure 2 shows the UV Eddington number, Γ_{UV} as a function of the total Eddington number, Γ for various accreting objects: massive black hole ($M_{BH} = 10^8$ M$_\odot$, the dotted line), stellar black hole ($M_{BH} = 10$ M$_\odot$, the triplet dot-dashed line), neutron star ($M_{NS} = 1.4$ M$_\odot$ the dashed line), white dwarf ($M_{WD} = 0.6$ M$_\odot$ the solid line), and low mass young stellar object ($M_{YSO} = 0.2$ M$_\odot$ the dot-dashed line). Note that for fixed mass and radius of the accreting object, Γ is proportional to the mass accretion rate. For simplicity, I estimate L_{UV} by integrating the disk intensity over the disk surface of the effective temperature between 8,000 K and 50,000 K. Detailed photoionization calculations are needed to determine what is a contribution to the line force from radiation at the temperatures beyond this range. To calculate the disk intensity and temperature I used the standard Shakura & Sunyaev disk model (Shakura & Sunyaev 1973). I define the UV Eddington number as the ratio between L_{UV} and L_E. The solid horizontal line marks the UV Eddington number above which line driving can drive a wind if there are no X-rays in a system, $\Gamma_{UV}(M_{max}+1) < 1$. Inclusion of X-rays will move this line up because the X-rays will reduce M_{max}.

Figure 2 allows an easy identification of objects capable of driving winds by lines. They are accretion disks around: massive black holes and white dwarfs

(AGN and CVs with $\Gamma \gtrsim 0.001$) and low mass young stellar object (FU Ori stars with $\Gamma \gtrsim$ a few $\times$ 0.01 !!!). The systems that have too low Γ_{UV} to drive wind are accretion disks around low mass compact objects (LMXB and galactic black holes, GBH). My classification of objects is consistent with the observations, in the sense that the objects I identified as capable of driving winds have been observed to have winds, whereas the objects with too low Γ_{UV} do not exhibit strong spectral features associated with winds. Additionally my conclusions based on this simple figure are consistent with detailed numerical calculations (those for FU Ori and GBH have not been published). I do not claim here that UV driving can fully explain the observed outflows but simply make a point that UV driving can drive some winds.

The lesson from the above exercise is that if we repeat it for AGN of different masses of the black hole and different luminosities (subdivided to the UV and X-rays) we can gain some insight into the nature of all AGN outflows.

Future work in hydrodynamical simulations of mass outflows can only benefit from simple analyses such as this. In fact, there are a few difficult problems that one encounters while making theoretical studies. For example, it is hard to make detailed photoionization calculations in connection with multidimensional time dependent calculations. However such photoionization calculations are crucial to confirm if the UV and X-ray opacities are such that the corresponding optical depths are as required. Future work on hydrodynamical models should include modeling the disk internal structure while modeling disk winds. An important limitation of the PSK approach is that it is valid for the gas pressure dominated disk. PSK did not include in their calculations the whole UV disk because its inner part is likely radiation dominated. Nevertheless they found that the mass-loss rate is about 25% of the assumed mass accretion rate. If we assume that the whole UV disk is gas pressure dominated, the mass loss rate will exceed the mass accretion rate by a factor of 10 or more. Does this result mean that the line force can drive such a strong wind that it can significantly change accretion on a black hole? If this is true then it would be very interesting because the models for line-driven disk winds in CVs predict $\dot{M}_w$ that is too low to completely explain the observations.

Acknowledgments. The work presented in this paper was performed while I held a NRC Research Associateship at NASA/GSFC. I thank J.E. Drew, T.R. Kallman, D. Kazanas, S.J. Kenyon, and J.M. Stone for useful discussions. I am grateful to S.J.K. for bringing to my attention outflows in FU Ori stars.

References

Castor, J.I., Abbott, D.C., Klein, R.I. 1975, ApJ, 195, 157 (CAK)

Krolik, J.H. 1999, Active galactic nuclei: from the central black hole to the galactic environment, Princeton, N.J.: Princeton University Press

Proga, D., Kallman, T.R. 2001, ApJ, in press (astro-ph/0109064, PK)

Proga, D., Stone J.M., & Drew J.E. 1998, MNRAS, 295, 595 (PSD)

Proga, D., Stone J.M., & Kallman T.R. 2000, ApJ, 543, 686 (PSK)

Shakura N.I., & Sunyaev R.A. 1973 A&A, 24, 337

Mass Outflow in Active Galactic Nuclei: New Perspectives
ASP Conference Series, Vol. 255, 2002
D.M. Crenshaw, S.B. Kraemer, and I.M. George

Acceleration of Highly Ionized gas by X-ray Radiation Pressure in AGN

Doron Chelouche & Hagai Netzer

School of Physics and Astronomy and the Wise Observatory, The Beverly and Raymond Sackler Faculty of Exact Sciences, Tel Aviv University, Tel Aviv 69978, Israel

Abstract. We model the structure, dynamics and spectral features of the highly ionized X-ray gas observed in low-luminosity, type 1 active galactic nuclei (AGN). We suggest a physical connection between the component seen by X-ray absorption and the outflow suggested by the UV lines. The observed properties of such flows are difficult to reconcile with current radiation pressure driven flow models. We discuss different scenarios in which radiation pressure force, mainly due to bound-free absorption, accelerates the gas and show that these agree well with present X-ray observations.

1. Introduction

Recent studies of low-luminosity type 1 AGN suggest that some 70% of all sources show absorption features in their X-ray spectra (George et al. 1998). These features are attributed to O VII and O VIII absorption edges and imply the presence of highly ionized gas (HIG) along the line of sight with column densities of $\sim 10^{22}\text{cm}^{-2}$. High resolution X-ray observations show narrow, blueshifted absorption lines (NAL) implying outflows with ~ 1000 km s^{-1} velocities (Kaastra et al. 2000; Kaspi et al. 2000). Narrow UV absorption lines, such as C IV λ1549 and Lyα (Crenshaw et al. 1999), are also seen in many objects, indicating outflows with similar velocities. High resolution UV spectroscopy of these lines show several distinct components with velocity dispersions of ~ 100 km s^{-1}.

In this contribution we compare the NAL phenomenon to the more studied, broad absorption line (BAL) phenomenon in BALQSOs, which is explained, at least qualitatively, by radiation pressure driven flows. We discuss the applicability of such models to X-ray flows and study two extreme scenarios in which the HIG is either in the form of geometrically thin clouds or in the form of a geometrically thick, continuous flow. We also discuss the dynamics and the spectrum predicted by both models and suggest possible ways to distinguish between them.

2. BAL models and NAL flows

To quantify the acceleration due to radiation pressure force, we use the force multiplier formalism adapted by Chelouche & Netzer (2001a). The force multi-

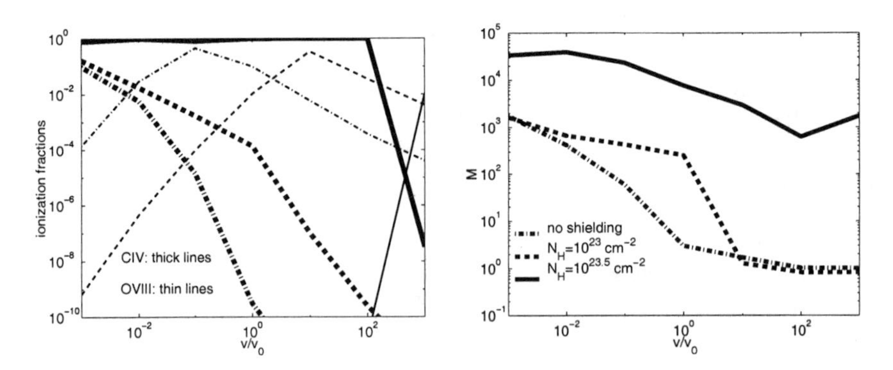

Figure 1. Left (a): O VII and C IV abundances as a function of the normalized velocity for an optically thin, continuous flow for various columns of the shielding gas. Note the rapid decline of the C IV abundance with v when no shielding applies. Right (b): The normalized force multiplier, M. Note the more effective radiation pressure force over a large velocity range for the case of large X-ray shielding (legend is common to both graphs).

plier, M, is the ratio of the radiation pressure force acting on the gas due to all absorption and scattering processes (Compton, bound-bound, bound-free, and free-free) to the radiation pressure force due to Compton scattering in highly ionized, Compton thin gas.

To study the effects of the shape of the X-ray ionizing continuum on the physics of radiation pressure driven, optically thin flows, we consider a stationary cloud of highly ionized gas (O VII and O VIII ionization fractions are of order unity at the illuminated surface of the cloud) which is situated **between the flow and the continuum source**. The cloud acts as an effective soft X-ray shield and is characterized by its column density, N_H. The flow obeys the continuity equation $\rho r^2 v =$ const, where ρ is the gas density and r, v are the distance and velocity respectively. In this case the ionization parameter U_x (defined over the 0.1 – 10 keV range), which determines the ionization level of the flow, is proportional to v.

Figure 1 shows the ionization structure and the force multiplier of such flows under various columns of the shielding gas. When exposed to a typical, unattenuated, AGN continuum, the flow becomes highly ionized at low velocities. The effective opacity is at higher energies where the incident flux is lower and hence M decreases rapidly with v. X-ray shielding of the flow can prevent it from reaching high ionization states and maintain a large radiation pressure force over a wide velocity range. Effective shielding requires a very low soft X-ray flux and therefore large absorbing columns ($N_H \gtrsim 10^{23}$ cm^{-2}, Figure 1) for a typical AGN continuum. Such columns are inferred from recent X-ray observations of BALQSOs (e.g., Gallagher et al. 2001, Green et al. 2001) but are not consistent with the soft X-ray spectrum of Sy1s (George et al. 1998). Therefore, BAL flow models which require shielding (e.g., Murray et al. 1995; Proga, Stone, & Kallman 2000) may not be applicable to NAL flows in Sy1s.

3. Proposed dynamical models for the HIG

We consider two extreme models of radiation pressure driven, photoionized flows: the cloud model which assumes that the HIG flow is made of clouds that are confined by magnetic pressure, and the wind model which assumes a continuous steady state flow originating from a disk. Both models required photoionization calculations that were preformed by ION2000 (Netzer 1996) and radiation pressure calculations that take into account all absorption and scattering processes. In both models we assume the following, standard equation of motion

$$v\frac{dv}{dr} = \frac{1}{\rho}\left[\frac{n_e \sigma_T L}{4\pi r^2 c}\left(M - \frac{L_{\rm Edd}}{L}\right) - \frac{dP}{dr}\right]. \quad (1)$$

where the gravitational term is expressed by the Eddington ratio $L/L_{\rm Edd}$. The following is a brief description of both models.

3.1. The cloud model

In this model (Chelouche & Netzer 2001a), geometrically thin HIG clouds obtain a quasi-hydrostatic density structure (e.g., Blumenthal & Mathews 1979) by adjusting to the outer magnetic pressure, P, on timescales shorter than dynamical timescales. The equation of motion can therefore be reduced to a ballistic equation, where we define an average force multiplier, $\langle M \rangle$ which is proportional to the cloud's bulk acceleration due to radiation pressure force. For typical HIG column densities, most of the radiation pressure force is due to bound-free absorption. Lines contribute only ~20% due to large optical depths at line center and the lack of appreciable velocity gradients within the cloud.

Figure 2a shows $\langle M \rangle$ as a function of the cloud's column density, N_H, and the ionization parameter U_x. For large values of U_x, the gas becomes more ionized and radiation pressure force is less effective. As N_H increases, radiation pressure force across the cloud drops due to optical depth effects and $\langle M \rangle$ decreases. Beyond $N_H \simeq 10^{22}$ cm^{-2}, the ionization structure changes and the gas becomes more opaque to the UV radiation. Since the UV part of the spectrum of typical AGN has higher flux, this results in a larger $\langle M \rangle$. For a cloud which maintains a constant column density and constant ionization parameter along its path (i.e., $P \propto r^{-2}$), the final velocity, v_∞, assuming $v = 0$ at the launching point (r_0), is

$$v_\infty = 1350 \left(\frac{L_{45}}{r_{17}}\right)^{1/2} \left(\langle M \rangle - \frac{L_{\rm edd}}{L}\right)^{1/2} \text{ km s}^{-1}, \quad (2)$$

where L_{45} and r_{17} are the source luminosity and launching radius, in units of 10^{45} erg s^{-1} and 10^{17} cm, respectively. Figure 2b shows several velocity profiles for different magnetic pressure profiles. Pressure profiles resulting in a decreasing ionization parameter with distance are most effective in accelerating the clouds.

A common feature to most geometrically thin clouds in quasi-hydrostatic equilibrium is a positive density gradient towards the leading (non-illuminated) surface. Such clouds are less ionized compared with uniform density clouds and produce different spectral imprints (more pronounced absorption in the soft X-ray and UV part of the spectrum, see Figure 3a). For a hydrostatic HIG cloud

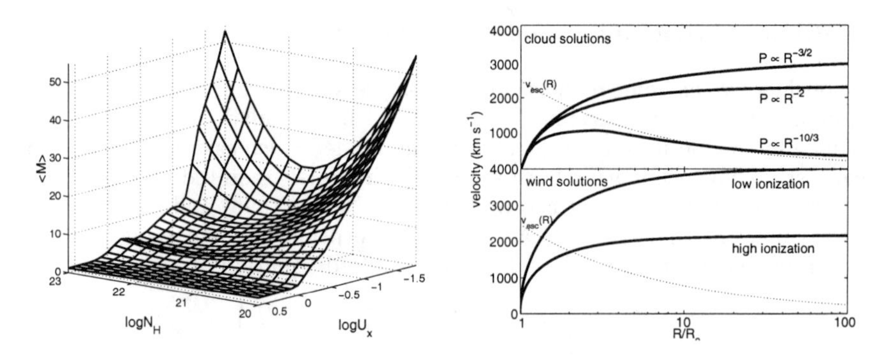

Figure 2. Left (a): The average force multiplier, $\langle M \rangle$, as a function of the ionization parameter, U_x and the column density, N_H. Note that for typical HIG parameters ($U_x \sim 0.1, N_H = 10^{22}\mathrm{cm}^{-2}$), radiative acceleration is effective for $L/L_{\mathrm{edd}} \gtrsim 0.1$ (see Eq. 1). Right (b): Velocity profiles for cloud and wind models. In all models $L = 10^{45}$ erg s^{-1}, $R_0 = 10^{17}$cm and $L/L_{\mathrm{edd}} = 0.5$. Note that winds have larger velocity gradients at their launching point and reach their final velocity on shorter distances compared to clouds (winds are assumed to start subsonically).

with typical warm absorption properties ($U_x = 0.15$, $N_H = 10^{22}$ cm^{-2}), a C IV column density of 10^{14} cm^{-2} is obtained. This is an order of magnitude larger than the C IV columns obtained for uniform density cloud models with the same U_x and N_H, and in better agreement with observations of C IV absorption lines.

3.2. The wind model

The wind models presented here (Chelouche & Netzer 2001b, in preparation) pertain to a freely expanding disk wind which is exposed to the primary ionizing continuum of the AGN (i.e., no shielding except for self-shielding). The continuum source is assumed to be point-like. We compute self-consistent ionization, density, pressure, and dynamical structure via iterations between the radiative transfer equations (for lines and continuum), the equation of motion, the continuity equation and the photoionization/thermal equations. Only steady state solutions of the dynamics and photoionization equations are considered.

Unlike the case of BAL flows, HIG winds become highly ionized due to expansion via the continuity condition since the X-ray continuum suffers only mild absorption (the column density of the shielding gas is $\ll 10^{23}$ cm^{-2}, Figure 1a). A practical problem is the fact that the HIG is not optically thin to continuum and line absorption and the validity of the commonly used Sobolev approximation is questionable. For these reasons, and in order to calculate accurate absorption line profiles, we have developed a code which calculates the radiation pressure force in $\sim$ 1000 of the most important resonance lines by integrating over line profiles across the flow.

Figure 2b depicts the wind velocity profiles for two extreme cases. The lower velocity flow corresponds to a more ionized wind with a higher U_x ($\sim 10^{-2.5}$)

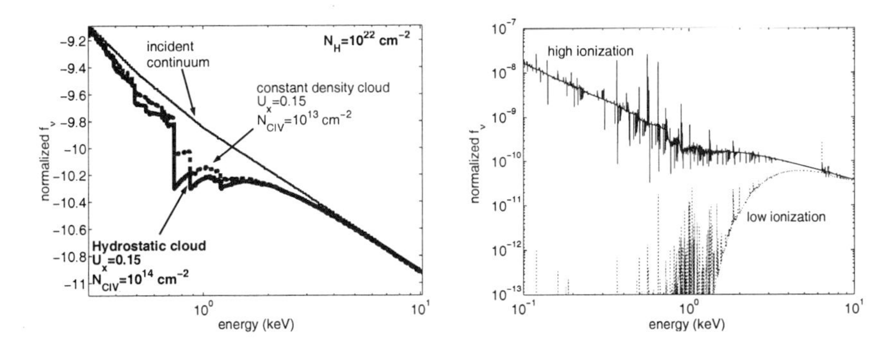

Figure 3. Left (a): Absorption features of hydrostatic, accelerating clouds (solid line) compared to uniform density clouds (dashed lines). Note the change in the C IV column density (see text). Right (b): X-ray spectra of wind models. Note the warm absorption features of high ionization winds and the heavily absorbed X-ray flux in low ionization winds.

at its base. The X-ray spectrum of this model shows typical warm absorption features (Figure 3b). The rapid expansion of the wind results in narrow UV lines (e.g., C IV $\lambda1549$) that are centered around low velocities (< 100 km s^{-1}; Figure 4). Higher ionization lines, such as O VIII $\lambda18.97$, are broader and blueshifted to larger velocities (< 1000 km s^{-1}, Figure 4). The second model has $U_x = 10^{-3.5}$ at its base. The lower ionization state and the somewhat higher gas columns at the base of the flow, result in effective self-shielding of the X-ray flux (Figure 3b). This effects the gas farther out in two ways. First, the gas is less ionized compared to the unshielded case (Figure 1a) and, therefore has a higher opacity which contributes to self-shielding. Second, the radiation pressure force increases (Figure 1b) which drives the flow to higher velocities. These processes result in a highly absorbed X-ray continuum and broader, more blueshifted, UV lines in low ionization flows (Figure 4).

4. Distinguishing wind from cloud models

In order to distinguish between the above two scenarios, we need (a) simultaneous multi-wavelength observations and (b) variability studies of absorption lines. Spectral predictions of the cloud and wind model are different since, for the HIG wind model, we expect to see spectral features associated with a large range of ionization stages on accounts of their stratified density structure, while the HIG cloud model is characterized by a more uniform ionization structure. Variablity studies of lines and edges will put constraints on the gas density and distance. The effect of continuum variability on the flow dynamics of NALs may prove to be a formidable task since typical dynamical timescales (given the HIG estimated distance) are of the order of 100 years. Time dependent modeling of the structure and dynamics of flows should be a prime objective of future theoretical research.

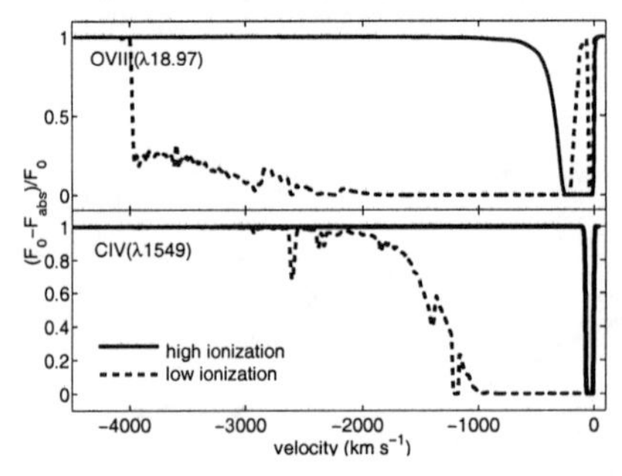

Figure 4. C IV λ1549 and O VIII λ18.97 line profiles in wind models (see text; doublets appear as a single component after summing their oscillator strengths). Note that C IV and O VIII are saturated in both models. C IV is formed close to the base of the flow and has a lower velocity compared to O VIII. The bumps in the line profiles of the low ionization flow are due to fluctuations in the ionization structure.

Acknowledgments. This research is supported by the Israel Science Foundation.

References

Blumenthal, G. R., & Mathews, W. G. 1989 ,ApJ, 233, 479

Chelouche, D., & Netzer, H. 2001a, MNRAS, 326, 916

Crenshaw, D. M., Kraemer, S. B., Boggess, A., Maran, S. P., Mushotzky, R. F., & Wu, C. 1999, ApJ, 516, 750

Gallagher, S. C., Brandt, W. N., Chartas, G., & Garmire, G. P. 2001, in Mass Outflow in Active Galactic Nuclei: New Perspectives, eds. D.M. Crenshaw, S. B. Kraemer, & I.M. George (San Francisco: ASP), p. 25

George, I. M., Turner, T. J., Netzer, H., Nandra, K., Mushotzky, R. F., & Yaqoob, T. 1998, ApJS, 114, 73

Green, P. J., Aldcroft, T. L., Mathur, S., Wilkes, B. J., & Elvis, M. 2001, ApJ, 558, 109

Kaastra, J. S., Mewe, R., Liedahl, D. A., Komossa, S., & Brinkman, A. C. 2000, A&A, 354, L83

Kaspi, S., Brandt, W. N., Netzer, H., Sambruna, R., Chartas, G., Garmire, G. P., & Nousek, J. A. 2000, ApJ, 535, L17

Murray, N., Chiang, J., Grossman, S. A., & Voit, G. M. 1995, ApJ, 451, 496

Netzer, H. 1996, ApJ, 473, 781

Proga, D., Stone, J. M., & Kallman, T. R. 2000, ApJ, 543, 686

Mass Outflow in Active Galactic Nuclei: New Perspectives
ASP Conference Series, Vol. 255, 2002
D.M. Crenshaw, S.B. Kraemer, and I.M. George

Molecular Gas in the Nucleus of NGC 1068: Observations of the H_2 2.12μm and Br_γ Emission Lines.

E. Galliano & D. Alloin

European Southern Observatory, Casilla 19001, Santiago, Chile

Abstract. Results about the distribution and kinematics of the molecular material around the nucleus of the Seyfert 2 galaxy NGC 1068 are reported. Spectroscopic data were obtained using the *Very Large Telescope* (VLT) *Infrared Spectrometer And Array Camera* (ISAAC) at the *Paranal Observatory* of the *European Southern Observatory*. The H_2 emission line at 2.12 μm and the Br_γ emission line at 2.56 μm were observed simultaneously at high spatial ($0''.5$) and spectral (35 km s^{-1}) resolutions. The H_2 emitting gas is found to be distributed along the East-West direction in two emission knots located on both sides of the nucleus at a distance of 70 pc. The Eastern knot is much brighter that the Western one. A velocity difference of 140 km s^{-1} between these two knots is measured. We believe that this velocity difference is the rotational signature of the outer part of the dusty/molecular obscuring material invoked in the "unification" theory of Active Galactic Nuclei (AGN). No Br_γ emission is detected in these knots, whereas a strong Br_γ emitting knot is seen, North of the central engine. It is spatially coincident with the radio knot C, where the small scale radio jet is redirected, probably after collision with a molecular cloud.

1. Introduction

NGC 1068 is a well studied Seyfert 2 galaxy located at a distance of 14.4 Mpc (Bland-Hawthorn et al. 1997). This corresponds to an observational scale of 70 pc per arcsec. The predicted size of the molecular torus around the AGN ranges from 1 to 100 pc. Consequently, NGC 1068 is an ideal target for the direct search of such a torus. Adaptive optics high resolution K, L and M band images of the nuclear region of this galaxy have unveiled the presence and structure of hot to warm dust (Rouan et al. 1998; Marco & Alloin 2000) within the central arcsec scale region. The source extension found along PA=102° was interpreted as the edge-on signature of the molecular/dusty obscuring torus. This interpretation was supported by the fact that this PA is perpendicular to the axis of the ionization cone. Another important piece of information to probe the existence of the torus is of course its kinematical status. Such information concerning the cold molecular gas has been derived from CO(2-1) line interferometric observations (Schinnerer et al. 2000). Here, we chose to probe instead the warm molecular gas component which can be observed through the 2.12 μm H_2 emission line. The warm H_2 emitting mass derived

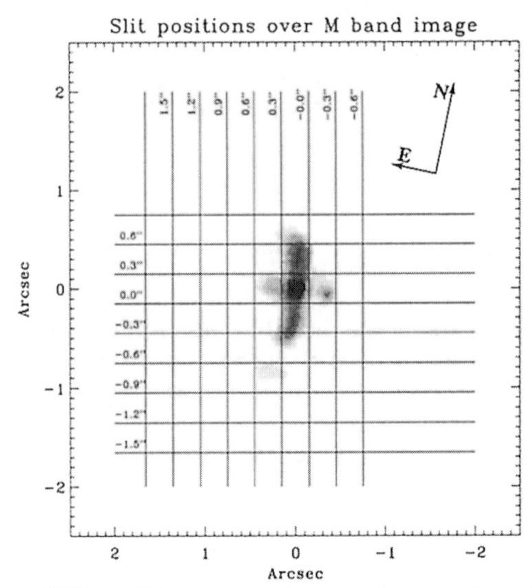

Figure 1. Slit positions superimposed over the adaptive optics M Band image of the nuclear region in NGC 1068 (Marco & Alloin 2000)

from H_2 line flux measurements is 4 orders of magnitude smaller than the mass of cold molecular gas derived from CO emission line studies. The warm gas is certainly heated by some interaction with the active nucleus, like shocks between an outflow and molecular clouds, or by nuclear X-rays. This warm component may hence have a different kinematical signature than the cold one. Therefore, to be complete in the study of the molecular environment of the AGN, both cold and warm gas should be considered. High spatial and spectral resolution data are necessary to detect a rotation of the order of 100 km s^{-1} on a spatial scale of 1 to 2 arcsec: VLT/ISAAC is the appropriate instrument.

2. Observations

The observations were performed using the SWS1 mode of ISAAC installed at the Nasmith focus of VLT/ANTU telescope. They were carried out under very good seeing conditions in 1999 August and 2000 December. The FWHM on the images was $0''.5$ in the K band, and the use of a $0''.3$ slit led to a spectral resolution of 8000 (35 km s^{-1} at 2.1 μm). Two sets of observations were acquired: one with the slit positionned along PA=102° (perpendicularly to the axis of the ionization cone) and one along PA=12° (along the ionization cone axis). The quality of the data was probed by the consistency of these two data sets. The location of the central engine (radio source S1 in Muxlow et al. 1996; see Figure 5 below) is coincident with the unresolved K emission peak (Marco, Alloin, & Beuzit 1997). The adopted observing technique was the following: after taking a spectrum with the slit centered on the core, the slit position was offset by $0''.3$ several times on both sides of the nucleus. We extracted a series of spectra, each three pixels high ($0''.45$), along the slit with a sliding step of one

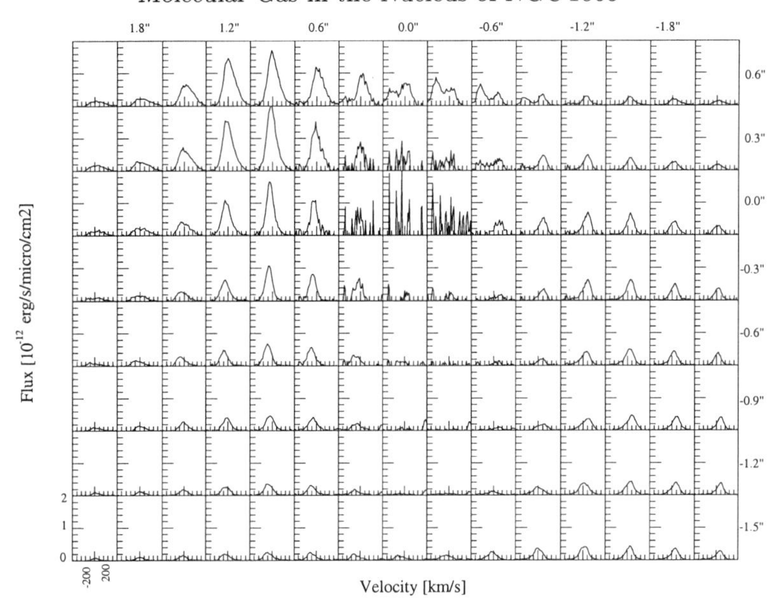

Figure 2. Flux-calibrated H_2 emission-line profiles after subtracting the continuum. Each panel corresponds to $0''.3 \times 0''.45$. The central engine (K continuum maximum) is located at $[0''.,0''.0]$. The intense noise due to high continuum level, as well as the low H_2 line intensity prevent measurements of the line in the central area. The x-axis corresponds to PA=102° and the y-axis to PA=12°. The x-offset is positive to the East and the y-offset is positive to the North.

pixel. The spectra presented here were continuum substracted. They therefore show a high noise level at the location of the continuum maximum. The total covered area is shown in Figure 1. For a complete description of the observations and data reduction, please refer to Galliano & Alloin (2001).

3. Results and Discussion

The main results of this study are presented in a Letter (Alloin et al. 2001), while a detailed discussion is provided in Galliano & Alloin (2001). The 2.12 μm H_2 emission line profiles over the whole central region are shown in Figure 2. PA=102° corresponds to the horizontal axis and each panel shows the emission from a $0''.3$ (slit width) $\times$ $0''.45$ (3 pixels wide aperture along the slit) area. Two-dimensional maps of the emission in H_2 and Br_γ were reconstructed by measuring the line fluxes on each spectrum. These maps are shown on Figures 3, 4 and 5.

3.1. The Ionized Component: A Witness of the Impact of the Small-scale Radio Jet on a Molecular Cloud

The Br_γ reconstructed map shows only one bright emission knot located at $[0''.45$ N $\pm 0''.15$ and $0''.1$ W $\pm 0''.15]$ from the nucleus (Figure 5). This corresponds to

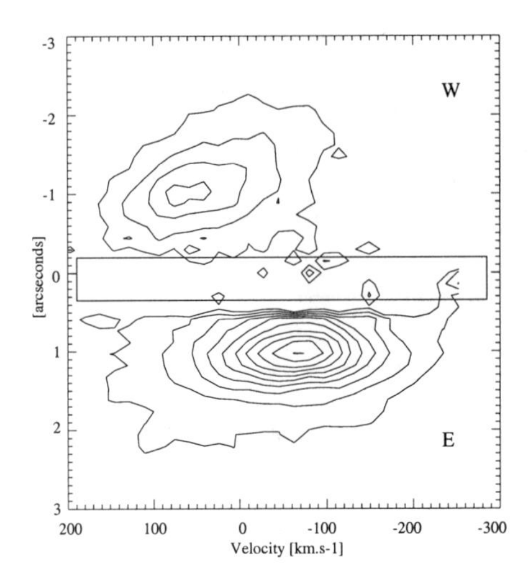

Figure 3. Position-velocity diagram of the central H_2 emission line spectrum at PA=102°.

the change in direction of the radio jet (knot C) seen in the *MERLIN* image (Muxlow et al. 1996; see Figure 5). The collision between the jet and a large molecular cloud may be the cause of the ionized hydrogen Br_γ emission. Whithin error-bars, this Br_γ emission knot coincides with the brightest [O III] λ5007 cloud (B) on the *Hubble Space Telescope* map by Macchetto et al. (1994). This allows to position the O III map with respect to the radio map, and thus with respect to the central engine. We conclude indeed that the Br_γ emission knot *is* radio knot C *and is* [O III] cloud B.

3.2. The Molecular Component: A Rotating Disk and a Radial Outflow

The H_2 emission is distributed in an extended component, on which are superimposed a bright emission knot to the East, and a much fainter emission knot to the West. The location of the central engine corresponds to the minimum of H_2 emission (Figure 4). The Eastern and Western knots, regardless of their large intensity difference, can be identified as the same kinematical component: molecular material rotating around the central engine. The velocity difference between these knots is 140 km s^{-1}. This result is enlightened on the position-velocity diagram shown in Figure 3. An enclosed mass of 10^8 $M_\odot$ inside a 70 pc radius around the nucleus can be derived assuming Keplerian rotation up to that distance. This value is comparable to that found by Schinnerer et al. (2000) from CO data. A central mass of 1.5×10^7 $M_\odot$ was derived from observations of the H_2O maser on a 1 pc scale, where the rotation is actually Keplerian. Thanks to the high spectral resolution, the profiles could be resolved and are found to be asymmetrical. The low velocity wings of the profiles (red wing for the blue, Eastern profiles and blue wing for the red, Western profiles) are almost always

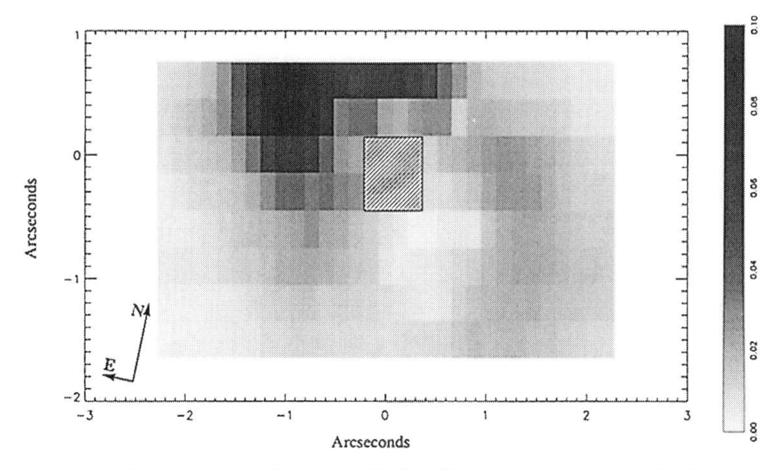

Figure 4. Reconstructed map of the H_2 emission around the nucleus. The restored "pixel" size is $0''.15\times0''.3$. The value of each pixel is the total integrated line flux of the corresponding $0''.45\times0''.3$ area divided by a factor of three to keep the total flux. The scale is in 10^{-14} erg s^{-1} cm^{-2} μm^{-1}. The central region was hidden because the line flux could not be measured.

extended. Simulations in the frame of a rotating inclined disk were performed in order to probe if this asymmetry could be a pure consequence of the seeing effects (Galliano & Alloin 2001). We find that the observed profile asymmetries cannot be explained entirely by seeing effects and we propose a kinematical model with two components: a rotating disk together with a radial outflow in the plane of this disk. This allows to reproduce very well the profile shapes and their spatial evolution. The heating source of the H_2 emitting material remains an unclear issue: X-rays and shocks are the two most probable mechanisms of excitation of H_2 emission in this context. An attractive idea would be that the shock between the outflow and the rotating material is the cause of the heating, but this does not seem to be energetically sustainable. Direct exposition of the rotating material to the nuclear X-rays is also a likely cause of excitation. The material in our line of sight is opaque even at X-ray wavelengths (Matt et al. 1997), but the surrounding obscuring clouds do not necessarily have an axisymmetric distribution. The fact that the H_2 line intensities are very different on the Eastern and Western sides of the central engine strongly supports this idea. We intend to observe other transitions of H_2 in order to clarify the issue of the excitation mechanism.

References

Alloin, D., Galliano, E., Cuby, J. G., Marco, O., Rouan, D., Clénet, Y., Granato, G. L., & Franceschini, A. 2001, A&A, 369, L33

Bland-Hawthorn, J., Gallimore, J., Tacconi, L. Brinks, E., Baum, S.A., Antonucci, R. R. J., & Cecil, G. N. 1997, Ap&SS, 248, 9

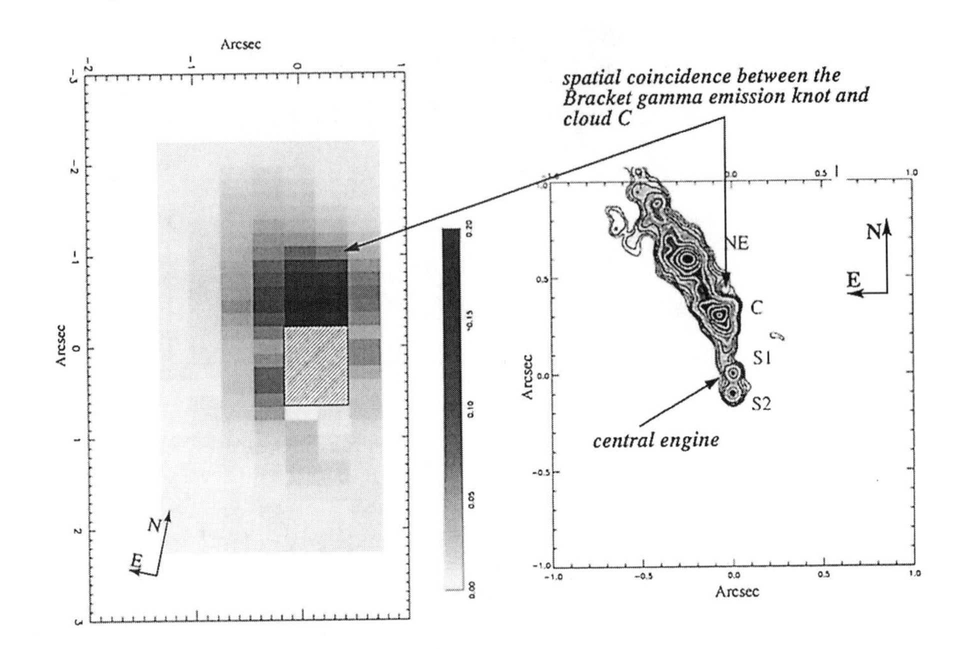

Figure 5. Comparison between our Br_γ map and the radio jet *MERLIN* image by Muxlow et al. (1996). The Br_γ was reconstructed from the data set at PA=102° following the same technique for the H_2 image. The intensity scale unit is 10^{-14} erg s^{-1} cm^{-2} μm^{-1}. Within the positional error bars, the Br_γ knot coincides with the radio source where the small scale radio jet is redirected.

Galliano, E., & Alloin, D. 2001, A&A, submitted

Machetto, F., Capetti, A., Sparks, W., Axon, D.J. & Boksenberg, A. 1994, ApJ, 435, L15

Marco, O., & Alloin, D. 2000, A&A, 353, 465

Marco, O., Alloin, D., & Beuzit, J. L. 1997, A&A, 320, 399

Matt, G., Guainazzi, M., Frontera, F., Bassani, L., Brandt, W. N., Fabian, A. C., Fiore, F., Haardt, F., Iwasawa, K., Maiolino, R., Malaguti, G., Marconi, A., Matteuzzi, A., Molendi, S., Perola, G. C., Piraino, S., & Piro, L. 1997, A&A, 325, L13

Muxlow, T. W., Pedlar, A., Holloway, A., Gallimore, J. F., & Antonucci, R. R. J. 1996, MNRAS, 278, 854

Rouan, D., Rigaut, F., Alloin, D., Doyon, R., Lai, O., Crampton, D., Gendron, E., & Arsenault, R. 1998, A&A, 339, 687

Schinnerer, E., Eckart, A., Tacconi, L. J., Genzel, R., & Downes, D. 2000, ApJ, 533, 850

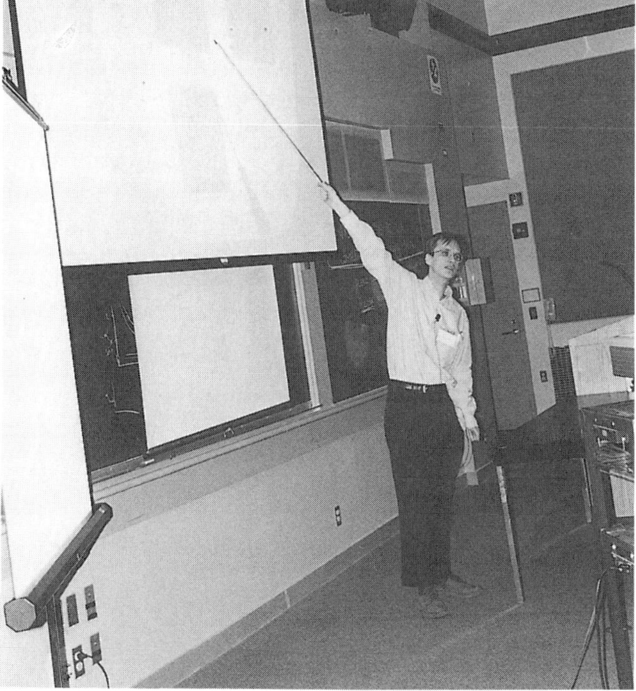

Mass Outflow in Active Galactic Nuclei: New Perspectives
ASP Conference Series, Vol. 255, 2002
D.M. Crenshaw, S.B. Kraemer, and I.M. George

Workshop Summary

Ray J. Weymann

Carnegie Observatories, 813 Santa Barbara Street, Pasadena, CA 91107, USA.

Abstract. The present level of our understanding of the AGN outflow phenomenon is reviewed in the context of the work presented at this workshop. We compare the present level of understanding with that when a similar workshop was held four years ago in Pasadena. Enormous progress has been made, especially in the quality and quantity of X-ray data. Nevertheless, several fundamental questions remain unanswered, including the determination of the distance of the material from the AGN nucleus, whether the observed spectroscopic features should be thought of as ballistic motion or "patterns", and the primary mechanism giving rise to inhomogeneities in the flows. Neither do we yet understand the relative importance played by viewing angle compared to the age of the AGN along a presumed evolutionary sequence, nor the role played by radio power. Some directions for further progress are suggested, among them being a far more detailed and intensive effort at modeling the flows, along the lines discussed at the workshop.

1. Introduction

Four years ago a workshop on this same topic was held in Pasadena. At that time I reviewed the state of our understanding of the Active Galactic Nuclei (AGN) outflow phenomenon from the point of view of an optical spectroscopist (Weymann 1997), and a comprehensive summary of that workshop was also given by Norman and Arav (1997). These two papers list a number of questions which I posed at the beginning of that workshop, together with an assessment at its close of what progress had been made during it. Those articles serve as a convenient point of departure to discuss the very significant progress which has been made in the intervening four years, along with a discussion of the problems which seem to me still in need of further work and clarification.

Needless to say, the amount of material covered in the oral and poster presentations during this workshop was so large and so wide ranging that it would be impossible for me to do it all justice, even if I had the expertise to do so. Thus, I must give only the most cursory mention of, or omit altogether, much important work presented during the workshop, and concentrate instead on those areas of special interest to me. I have thus broken this summary discussion into the following topics, although they are all inter-related: 1) Column Densities and Abundances; 2) The X-ray Data; 3) The Role of Radio Power; 4) Dynamical Models: Challenges and Promise; 5) The Cosmic Connection.

Though I will not attempt point-by-point acknowledgments, I have greatly benefited from numerous conversations with several workshop participants, both during and subsequent to the workshop, for which I thank them. However, any errors of fact or interpretation are solely my responsibility.

2. Column Densities and Abundances

The excellent data set and very careful analysis of the line profiles in Q0946+301 by Arav et al. (2002; see also Arav et al. 2001) represents the standard which future analyses of this type must meet. This work firmly establishes the importance of taking into account the heavy saturation of many of the transitions and the serious errors in the column density estimates of various ions if this is not taken into account. Therefore an implicit caveat associated with all column density estimates must be understood unless there is clear evidence that the transitions in question are not saturated.

This work also points out the fact that the local covering factor[1] is often distinctly smaller than unity and moreover may vary strongly across an absorption trough. This work also should lay to rest the likelihood of super-metallicity, though it is still intriguing that there may be some very interesting abundance anomalies (e.g. Phosphorus), though definitive results must await a more detailed photoionization treatment including possible absorption of high energy photons by underlying gas.

3. The X-ray Data

The most striking advance since the Pasadena workshop is the amazing growth in both the quantity and quality of X-ray observations bearing on the AGN Outflow phenomenon. Several years ago the "very high" ionization species O VI was known to exist, and it was a very uncertain speculation whether even higher ionization markers (e.g. Ne VIII) would ever be securely established. Now it is clear that representatives of the Lithium, Helium and Hydrogen isoelectronic lines or edges march all the way up the periodic table until Fe XXVI! There is a fairly general consensus that photoionization dominates, testimony to the powerful X-ray flux illuminating the outflowing gas.

Among the various X-ray issues touched upon, I was particularly struck by:

---

[1] I plead that workers in this field distinguish between: (i) The *local* covering factor (hereafter LCF)—the fraction of the effective continuum source covered by a particular ionic species at a given velocity along the line of sight. (ii) The *global* covering factor (hereafter GCF) —the fraction of 4π steradians at the source which are covered by any particular system–e.g. the gas producing the totality of emission or absorption lines of a given species, or material which is thick to the Lyman limit, etc. It should be pointed out however, that the LCF as usually determined is really an upper limit to the geometrical LCF for two reasons: a) Scattered radiation almost certainly contributes to the residual intensity (as indicated by polarization data); b) In some disk wind models, the broad absorption line (BAL) region creates a significant amount of emission.

1) Green et al.'s result (2002; see also Green et al. 2001) that the composite spectrum of a sample of HiBALs[2] shows intrinsic X-ray power and a spectral index essentially the same as for RQ-NrmQSOs, but with X-ray absorbing column densities $\sim 6 \times 10^{22}$ cm^{-2}, and some "leakage" or scattering of these X-rays as well. Mathur et al. (2002; see also Mathur et al. 2001) presented the highest quality individual X-ray spectrum of a BALQSO of PHL5200, which may be considered to be a very mild LoBAL. In this case a powerful intrinsic X-ray luminosity is also present but is attenuated by an even higher column density and with the additional intriguing property of a steeper hard X-ray spectrum similar to those found in the narrow-line Seyfert 1 objects, thus strengthening the relation between the high-luminosity (or high redshift) LoBALs and low luminosity/low redshift NLS1s. Additionally, these observations make it self-evident that objects exhibiting strong outflows are precisely those which seem to be absorbing a signifcant fraction of the incident X-ray power, as discussed below in connection with the models.

2) Although there are exceptions, it appears that more often than not there are X-ray absorption systems which share very similar velocity structure with that defined by the traditional UV lines O VI, N V and C IV, but generally the derived X-ray column densities are substantially higher than those for the optical lines. While 2–, 3–, or n–zone models can be constructed, it seems likely that a continuum of density fluctuations is present in a given velocity range, and that stratified models in which X-ray absorbing gas resides *only* in front of the UV absorbing gas are too simple. The consistently higher column densities inferred from the X-ray data also means that the mass ejection rates are at least an order of magnitude larger than older estimates.

3) The exciting possibility of observing a wider range of phenomena which can be unambiguously attributed to the relativistic portions of accretion disks was presented by Branduardi-Raymont et al. (2002; see also Branduardi-Raymont et al. 2001) undoubtedly drove the passion with which this model—as opposed to the "dusty X-ray absorber" model (Lee et al. 2001)—was debated. I do not have the expertise to make an informed judgment on this issue, and indeed to do so would be futile at this stage: Much higher quality X-ray spectra are in the offing, and, with the admonition of Netzer (comments at this meeting) that they must be modeled using all the available information on the wide range of atomic transitions which may contribute, I would think that this issue will be settled shortly. It does seem clear, however, that absorbing gas and dust is present, so the question is whether accretion disk emission line signatures will survive. Whether or not bound-free X-ray absorption is measurable as well as bound-bound is relevant but not the central issue. It is regrettable that reddening in MCG−6-30-15 only permits low resolution *HST*/STIS spectra to be obtained, since *FUSE* and high resolution STIS spectra to study N V and O VI would be of great interest.

[2] I adopt the acronyms: HiBALs—BALQSOs showing no strong absorption from Al III, Mg II or Fe II; LoBALs—BALQSOs showing these three species, but in addition always very strong absorption from the high ionization species as well; NALQSOs—QSOs showing much narrower absorption features than the typical BALQSO, but whose variability, LCF, or complex profiles strongly indicate ejection. Finally, NrmQSOs— normal QSOs showing no absorption clearly intrinsic in origin. RQ and RL are appended to indicate weak or strong radio sources as needed.

4. The Role of Radio Power

Brotherton's summary (Brotherton 2002) of the *FIRST* collaboration (Becker et al. 2000) drove home four facts: 1) BALQSOs of both the LoBAL and HiBAL type are present at moderate levels of radio power. 2) At least among this sample and up to a certain level of radio power, the LoBAL fraction is strikingly high. 3) The BALQSO phenomenon occurs among objects with both flat and steep spectra, whose radio properties suggest that at least some are *not* being viewed near the equatorial plane. (See also the discussion below of Mrk231 in connection with polarization data.)

It would be nice to be able to quantify what the allowed range of viewing angles is. It is also not clear what the implication of these observations are for models: Do they indicate that the outflow may emerge at angles much larger than the opening angle of the putative accretion disk, or do they indicate that the disk simply has a very large opening angle? 4) The radio properties associated with the RL-BALQSOs seem to be GHz-peaked, compact sources.

As one who was partly responsible for emphasizing in the past the "radio-avoidance" of BALQSOs, it should be remarked that nature is rarely, if ever discontinuous. Thus, I think that too much has been made of whether a given object should, or should not, be officially classified as "radio loud" or not or whether an object should or should not be included as a BALQSO if its "balnicity index" (BI) is zero. Rather, we need to look at trends to understand what connection there is between radio power and morphology and the character of outflows. In particular: (1) The BI was defined when it was believed that the NALQSOs near zero outflow velocity represented a phenomenon which was different from the BALQSO phenomenon. While in some cases this may be true, it is now clear that in most cases the NALQSOs represent a milder manifestation of outflow from the nucleus. Thus, I believe that while some definition of the strength of outflowing absorption is useful, the BI definition ought to be modified to include outflowing gas represented by the NAL systems. (2) It is still true that, to the best of my knowledge, among the *most powerful* radio sources—e.g. a large, magnitude–limited but still unpublished sample of 3C, 4C and PKS sources, BALQSOs are strikingly absent, though NALs near the emission line redshift are common. Murray (comments at this meeting) suggested that the intrinsic difference in X-ray properties is a key factor in governing the intensity of the outflow because of the amount of shielding possible and this needs to be quantitatively explored over the full range of radio power. In my opinion, though, we still do not understand the role of radio power in the outflow phenomenon.

5. Dynamical Models: Challenges and Promise

5.1. Some Relevant Observational Challenges

Two issues strike me as posing interesting challenges for the models: These are: i) The evidence of significant density contrasts within the outflow and the correspondingly small characteristic sizes of absorbing regions, ii) The striking stability of the overall structure in the absorption profiles, but with occasional absorption features appearing and disappearing.

Density Fluctuations and Filling Factors From the point of view of the basic physics and models for the outflow, the main question seems to me to be whether we are dealing with filling factors for the various ionization and density regimes of a factor of a few or a few orders of magnitude. If the latter is the case then the nagging "confinement mechanism" problem, which I had hoped four years ago might by now have been laid to rest, is still with us.

The most cogent observations for refining these estimates seem to me to be observations of, or limits on, the strength of absorption arising from excited states. De Kool et al. (2002) provided a summary of electron density estimates and the corresponding distance estimates based upon the ionization parameter for several objects. Other examples of "clouds" were provided by Everett, Konigl, & Arav (2002) and also by Hamann et al. (2002; see also Hamann et al. 2001). Several examples of inferring electron densities and inferred distances from excited state transitions were reviewed by Kraemer et al. (2002) in NGC 4151. The limits on, or estimates of, the electron density from excited state transitions seem persuasive and indicate a huge range in densities and distances. The inferred distances (from ionization parameter arguments) together with the inferred total column densities *for these particular absorbing species* imply filling factors which are $<< 1$. Although the inferred distances may be subject to uncertainties due to X-ray shielding it is difficult to believe that in some of the cases they are so wildly off as to overcome these problems. In "system A" the inferred thickness of $\sim 10^{16}$ cm implies a sound crossing time of only a few hundred years, a tiny fraction of the travel time if this cloud originated in the nucleus–and it is hard to see how it could otherwise acquire this velocity.

What I find especially interesting about the NGC 4151 example is the presence of very high densities, ($> 10^{9.5}$) as inferred from the C III excited state transition, with similar conclusions: For derived column densities associated with the gas whose ionization produces the C III, the inferred longitudinal dimensions are so small that the sound (or Alfven) crossing times seem very short compared to the travel time lifetimes. Thus, this issue must be addressed on all scales of the flow. In this last example, the column densities are more difficult to estimate because of saturated lines, but one would have to be off by one or two orders of magnitude to relieve the problem. In some objects one finds that the absorbing material covers a significant fraction of the gas giving rise to the broad emission, but in other cases the outflowing gas does not even cover the continuum source, so that very small lateral dimensions for the absorbing gas are required.

Dynamical Stability It is striking that, while there are obvious changes in the intensity of the absorbing systems (whether due to changes in covering factor or ionization state is not entirely clear), the general structure of the absorption troughs is remarkably stable. In some cases though, features appear and disappear. The limits on the acceleration of particular absorption features have been quantified and leads to extremely low limits on the acceleration of particular parcels of gas, both in luminous BALQSOs (Foltz et al. 1987) and in low luminosity AGNs (Weymann et al. 1997). Two possibilities have been suggested: (i) The gas is sufficiently distant at the time of observation that it has long ago been accelerated (ii) We are not viewing individual parcels of gas ("clouds"), but rather stable patterns.

5.2. Models

The simple model consisting of "clouds" being accelerated faces the well-known confinement problem in yielding the small apparent filling factors. Whether a two-phase medium can really arise and comoving cooler denser clouds survive I believe is still an open question. At a given distance from the continuum source the different ionization levels will produce significantly different radiative forces and differential velocities. Can such clouds then persist (and what forms them in the first place)?

The "shielded accretion disk wind" (SADW) model pioneered by Murray and Chiang and their collaborators (see Murray & Chiang 1998) and extended into a more detailed two-dimensional model as described by Proga (2002) at the workshop (see also Proga, Stone, & Kallmann 2000) has a great deal to recommend it, and accounts for many of the essential features for the AGN outflow in a fairly natural way. More heuristic, phenomenological orientation-dependent "unified" models attempting to tie together the wide range of phenomena seen in AGNs were discussed by Ganguly (2002; see also Ganguly et al. 2001) and even more ambitiously by Elvis (2002; see also Elvis 2000) employing similar geometries, though differing in details.

The X-ray shielding in the SADW model allows the flow to be moved in to small distances making the scale of the flow smaller and mitigating the confinement problem. As noted above, the presence of substantial X-ray absorption required by this model now seems well established.

A variant on the SADW model was proposed by deKool & Begelman (1995) who invoked magnetic confinement and differential radiation forces to compress and confine the winds to thin sheets above the equatorial plane. In their model the material is at significantly larger distances than in the Proga model, responding to the concerns that, since the absorbing gas must be beyond most of the gas responsible for the broad emission lines, and since reverberation analyses seemed to require larger characteristic sizes than the SADW model, larger launching distances must be involved. They also worried that the transit time for features in the flow would be prohibitively short for the typical SADW models. In the SADW model, no separate "Broad Emission Line Region" (BELR) is invoked. A crucial factor in the SADW emission line analysis is the anisotropic emission due to the high optical depths of, e.g. C IV, and this apparently avoids conflict with the reverberation results (Chiang & Murray 1996). A related question is whether the SADW models can avoid the double peaked emission line profiles one expects. As described by Murray at the workshop (comments at this meeting; see also Murray & Chiang 1997), this same effect also leads to the prediction of a single peak from the optically thin C III] line, since this is formed further out where the kinematic properties of the gas are different. What is the prediction for the (weaker) He IIλ1640 line, or better yet the (still weaker) He IIλ4686 line which is probably formed in the same region as C IV, (and this latter line is surely optically thin)?

Whether the highly structured absorption features seen in so many of the outflows can be produced in the standard SADW models remains to be seen. Proga et al. (2000) note that Kelvin–Helmholtz "knots" with quite large density enhancements propagate through the flow, but whether these can be responsible for the high density condensations noted above in NGC 4151, while at the same time avoiding the detection of changes in the velocity of the material remains to be seen, and as Proga (2002) pointed out, the behavior of these knots in the real 3D world may be different than in the 2D calculations. Can this be investigated in a 3D numerical calculation, but using the 2D kinematics as a platform for the 3D work?

The virtue of the SADW models is that, given enough work and computer time, many more detailed comparisons with the observations in terms of emission and absorption line strengths and profiles (and polarization too?!) can be carried out and it will be interesting to see if they can meet these tests without "extra" physics (e.g. magnetic fields).

If the complex velocity structure seen in the outflow cannot naturally be produced in the standard SADW models, then some variant of the deKool & Begelman model seems to me compelling. A particular model which appeals to me involves non-axisymmetric flux tubes emerging from the rotating disk, which produce inherently small filling factors, with perhaps "beads" along the tubes responsible for further density contrasts. It was objected that these would rotate in and out of the line of sight on too short a time scale. Whether this is true depends crucially on how far these tubes really are from the black hole when they cross the line of sight: For a central mass of $10^8\ M\odot$ and a distance of 1 pc a "ribbon" of width 2×10^{15} cm takes one year to cross the line of sight. Smaller distances clearly require thicker ribbons. Moreover there are clear instances where absorption systems *do* come and go, which are most plausibly attributed to transverse motion, one instance of which was described by Crenshaw, Kraemer, & Gabel (2002). Another striking instance was observed by Boroson et al. (1991) in Mrk231. Such instances also occur among high luminosity objects, generally at the higher ejection velocities. Monitoring on a shorter time scale to determine the time scale for the appearance and disappearance of such systems would be very interesting.

It was also objected that the lack of success with magnetic confinement in attempts at controlled fusion is *a priori* evidence against magnetic confinement operating in nature. Anyone who doesn't believe in magnetic confinement and the ability to produce such beaded ribbons (a picture which greatly appeals to me for an object like Q1303+308 whose spectrum I showed at the workshop) is invited to look at the spectacular movies of the sun taken with the *Transition Region and Coronal Explorer* (*TRACE*) spacecraft and available on their web site.

While it is simplistic to look at pictures of phenomena taking place on such a wildly different scale and infer their applicability to AGN scales, nevertheless the *TRACE* images offer an existence proof for such structures, in which magnetic confinement clearly plays a critical role.

None of these models address (and are not really intended to address) what appears to be the very small filling factors at very large distances (e.g. $\sim$ kpc scale for NGC 4151 and the 10's of kpc deduced for 3C 191 by Hamann et al. (2001). High velocity knots at very large distances compared to the BELR are directly shown in the images presented by Cecil (2002) in NGC 1068. These knots, moving at speeds much higher than can be produced by gravitational forces, and at distances much larger than those envisaged in the SADW-type models, pose a separate challenge.

6. The Cosmic Connection

I believe it is fair to say that at present the study of this fascinating phenomenon of outflows from AGNs is regarded by most astronomers who are not workers in this area as a rather esoteric sub–discipline with little connection to "fundamental problems". I believe this is an understandable but incorrect view, and I believe there are at least two aspects of outflow from AGNs which bear directly on the more "cosmic" and large scale questions which warrant further attention.

6.1. Evolution or Orientation?

As described above, attempts to describe the range of AGN phenomena as primarily arising from orientation effects are very popular, and the models of Ganguly (2002) and Elvis (2002) provide very appealing examples of such attempts. There are substantial objections to this approach however, noted above in connection with the radio-loud BALQSO discussion. The problem seems most acute in connection with the LoBALS, where it was first pointed out by Boroson and Meyers (1992) that there were substantial differences between NrmQSOs and LoBALS involving observables which presumably are orientation-independent, namely the far infrared luminosities and the strength of the [O III] forbidden line. These, and other considerations led to the "evolution" picture in which an AGN commences life as a nearly fully shrouded object with large far IR luminosity, evolves into a LoBal with large covering factor to ionizing radiation (hence low [O III] emission), then, perhaps into a HiBAL as the column density and dust content in the ejecta are emitted, and finally to a NrmQSO.

The usual lesson in astrophysics is that nature is not that simple and it would astonish me if viewing angle did not play a critical factor in what we see, but I would be equally astonished if AGNs did not evolve in fundamental ways over time scales associated with QSO lifetimes (whose ejection phase must last at least $\sim 10^6$ to $\sim 10^8$ years simply to account for the transit time implied by the narrow, distant components in NGC 4151 and 3C 191.) Neither of the arguments for the two limiting views seem to me at present compelling. I would like to see a far better and more extensive data set selected from, e.g. the Sloan data base of QSOs, so that boxes in portions of redshift and luminosity space can be studied, and which include NrmQSOs, HiBALs and LoBALs. The result of Weymann et al. (1991) has been cited to indicate that there are no significant differences in the emission line properties of NrmQSOs and BALQSOs. I would be yet again astonished if much higher S/N data of a larger and more carefully selected sample did not reveal differences in profile and strength, and these should surely provide clues regarding this issue. The region from He IIλ4686 through Hβ and [O III]λ5007 is especially interesting in this regard, and the rapid improvement in near IR spectroscopic instrumentation on large telescopes needs to be exploited. For objects showing somewhat lower levels of ionization in their outflows, very high quality data for the He Iλ3889 and λ10830 transitions would be worth obtaining, since they arise from a density-sensitive metastable level and the atomic physics of this ion are well understood (Osterbrock 1989).

The two observational areas during the workshop which seemed to me under-represented were IR spectroscopy (the rationale for which I have just mentioned) and spectropolarimetry. The latter was represented in some poster papers, but did not arouse much discussion during the workshop. Spectropolarimetry, while very photon-demanding, seems to me to have the promise to make further important contributions to our understanding of the orientation/evolution question and I hope its practioners will strongly pursue it. As an example of this, I mention the work of Cohen & Martel (2002) in their written contribution to these workshop proceedings, and in particular their interpretation of the polarization properties of Hα in Mrk 231. In this object, the polarization position angle and percentage polarization of Hα changes dramatically across the line profile. Cohen & Martel interpret this in terms of a model involving an obscured rotating Hα emitting annulus whose emission is reflected and partially polarized by the unobscured portion of a dusty torus of larger radius. The model requires that the angle with respect to the axis be quite small–not much exceeding 20 degrees–to explain the polarization properties. Although Cohen & Martel addressed only the polarization properties of Hα in the context of Mrk 231 as a Seyfert 1 galaxy, it has, as alluded to above, and discussed in detail by Smith et al. (1995), all the attributes of a LoBAL. Thus, if the geometric model of Cohen & Martel is correct, this indicates that outflows of several thousand km s^{-1}occur even for viewing angles which are nearly pole-on. However, other geometrical models have been proposed (e.g. Smith et al. 1995). The well-studied spectroscopic and polarization properties of this object, the appearance of a new absorption component, and the remarkable occurance of absorption of He I from a highly excited metastable state in the same velocity system as strong Na I absorption all suggest that this object is worthy of much more intensive efforts at modeling and ongoing observations (extending, if at all possible, down to C IV.)

6.2. Influence of AGN Outflows on Galactic Properties and Large-scale Evolution

There was also very little discussion concerning the influence of AGN outflows on the large scale evolution of the host galaxy. The very large matter, momentum, and kinetic energy deposition represented by the outflows in the luminous BALQSOs must surely have a significant impact on the ISM, hence star formation history, in the host galaxy. Several authors, e.g. Fabian (1999) have given semi-quantitative discussions of the whole scenario of black hole formation and galaxy evolution taking into account the AGN outflows, and many years ago Schiano (1985) studied the hydrodynamics of the outflow in a disk galaxy environment. It would be useful to pursue such detailed investigations further. With the advent of adaptive optics, larger space telescopes, and the future large aperture and high spatial resolution ground-based telescopes being discussed, perhaps future workshops will have direct images revealing in some detail the impact of the more powerful AGN outflows on the host galaxy.

Acknowledgments. I am especially grateful to Nahum Arav, Rajib Ganguly, Paul Green, Steve Kraemer, Kirk Korista, Smita Mathur, and Patrick Ogle for extended phone conversations following the workshop, as well as to Roger Blandford, Mark Giampapa and Paul Bellan for comments concerning magnetic confinement and solar phenomena.

Finally, I thank Mike Crenshaw, Steve Kraemer, and their colleagues on the organizing committee for putting together a very stimulating workshop as well as for their care in the planning which went into it, and I thank them for inviting me to participate.

References

Arav, N., et al. 2002, in Mass Outflow in Active Galactic Nuclei: New Perspectives, eds. D.M. Crenshaw, S.B. Kraemer, & I.M. George (San Francisco: ASP), p. 179

Arav, N., deKool, M., Korista, K., Crenshaw, M., van Breugel, W., Brotherton, M., Green, R., Pettini, M., Wills, B., Wills, D., deVries, W., Becker, B., Brandt, W., Cohen, M., Foltz, C., Green, P., Junkkarinen, V., Koratkar, A., Laor, A., Laurent-Muehleisen, S., Mathur, S., Murray, N., & Ogle, P. 2001, submitted to ApJ

Becker, R., White, R. Gregg, M., Brotherton, M., Laurent-Muehleisen, S., & Arav, N. 2000, ApJ, 538, 72

Boroson, T., Meyers, K., Morris, S., & Persson, S. 1991, ApJ370, L19

Boroson, T.A., & Meyers, K.A. 1992, ApJ, 397, 442

Branduardi-Raymont, G., et al. 2002, in Mass Outflow in Active Galactic Nuclei: New Perspectives, eds. D.M. Crenshaw, S.B. Kraemer, & I.M. George (San Francisco: ASP), p. 31

Branduardi-Raymont, G., Sako,M.,Kahn, S., Brinkman, A., Kaastra, J., & Page, M.J. 2001, A&A, 365, L140

Brotherton, M., et al. 2002, in Mass Outflow in Active Galactic Nuclei: New Perspectives, eds. D.M. Crenshaw, S.B. Kraemer, & I.M. George (San Francisco: ASP), p. 155

Cecil, G. 2002, in Mass Outflow in Active Galactic Nuclei: New Perspectives, eds. D.M. Crenshaw, S.B. Kraemer, & I.M. George (San Francisco: ASP), p. 227

Chiang, J., & Murray, N. 1996, ApJ, 466, 704

Crenshaw, D.M., Kraemer, S.B., & Gabel, J.R. 2002, in Mass Outflow in Active Galactic Nuclei: New Perspectives, eds. D.M. Crenshaw, S.B. Kraemer, & I.M. George (San Francisco: ASP), p. 87.

de Kool, M., et al. 2002, in Mass Outflow in Active Galactic Nuclei: New Perspectives, eds. D.M. Crenshaw, S.B. Kraemer, & I.M. George (San Francisco: ASP), p. 183

de Kool, M., & Begelman, M. 1995, ApJ, 455, 448

Elvis, M. 2002, in Mass Outflow in Active Galactic Nuclei: New Perspectives, eds. D.M. Crenshaw, S.B. Kraemer, & I.M. George (San Francisco: ASP), p. 303

Elvis, M. 2000, ApJ, 545, 63

Everett, J., Konigle, A., & Arav, N. 2002, in Mass Outflow in Active Galactic Nuclei: New Perspectives, eds. D.M. Crenshaw, S.B. Kraemer, & I.M. George (San Francisco: ASP), p. 189

Fabian, A. 1999, MNRAS, 308, L39

Foltz, C., Weymann, R., Morris, S., & Turnshek, D. 1987, ApJ, 317, 450

Ganguly, R. 2002, in Mass Outflow in Active Galactic Nuclei: New Perspectives, eds. D.M. Crenshaw, S.B. Kraemer, & I.M. George (San Francisco: ASP), p. 109

Ganguly, R., Bond, N., Charlton, J., Eracleous, M., Brandt, W., & Churchill, C. 2001, ApJin press; astro-ph 0010192

Green, P., et al. 2002, in Mass Outflow in Active Galactic Nuclei: New Perspectives, eds. D.M. Crenshaw, S.B. Kraemer, & I.M. George (San Francisco: ASP), p. 19

Green, P., Aldcroft, T., Mathur, S., Wilkes, B., & Elvis, M. 2001, in preparation

Hamann, F., et al. 2002, in Mass Outflow in Active Galactic Nuclei: New Perspectives, eds. D.M. Crenshaw, S.B. Kraemer, & I.M. George (San Francisco: ASP), p. 137

Hamann, F., Barlow, T., Chaffee, F., Foltz, C., & Weymann, R. 2001, ApJ, 550, 142

Kraemer, S.B., et al. 2002, in Mass Outflow in Active Galactic Nuclei: New Perspectives, eds. D.M. Crenshaw, S.B. Kraemer, & I.M. George (San Francisco: ASP), p. 93

Lee, J., Ogle, P., Canizares, C., Marshall, H., Shultz, N., Morales, R., Fabian, A., & Kazushi, I. 2001, astro-ph 0101065

Mathur, S., et al. 2002, in Mass Outflow in Active Galactic Nuclei: New Perspectives, eds. D.M. Crenshaw, S.B. Kraemer, & I.M. George (San Francisco: ASP), p. 151

Mathur, S., Matt, G., Green, P., Elvis, M., & Singh, K.P. 2001, ApJ551, L13

Murray, N., & Chiang, J. 1997, ApJ, 474, 91

Murray, N., & Chiang, J. 1998, ApJ, 494, 125 and references therein

Norman, C., & Arav, N. 1997, in ASP Conf. Ser. Vol. 128, Mass Ejection from AGN, ed. N. Arav, I. Shlosman & R. Weymann (San Francisco: ASP), 291

Osterbrock, D. E. 1989, Astrophysics of Gaseous Nebula and Active Galactic Nuclei (Mill Valley California: University Science Books)

Proga, D. 2002, in Mass Outflow in Active Galactic Nuclei: New Perspectives, eds. D.M. Crenshaw, S.B. Kraemer, & I.M. George (San Francisco: ASP), p. 309

Proga, D., Stone, J., & Kallman, T. 2000, ApJ, 543, 686

Schiano, A. 1985, ApJ, 299, 24

Smith, P., Schmidt, G., Allen, R. & Angel, J.R.P. 1995, ApJ, 444, 146

Weymann, R., Morris, S., Foltz, C., & Hewett, P. 1991, ApJ, 373, 23

Weymann, R. 1997, in ASP Conf. Ser. Vol. 128, Mass Ejection from AGN, ed. N. Arav, I. Shlosman & R. Weymann (San Francisco: ASP), 3

Weymann, R., Morris, S., Gray, & Hutchings, J. 1997, ApJ, 483, 717

Author Index

First-author contributions in **bold**.

Object Index

Contributions in which object included in the title indicated in **bold**.

ASTRONOMICAL SOCIETY OF THE PACIFIC
CONFERENCE SERIES

and

INTERNATIONAL ASTRONOMICAL UNION
VOLUMES

Published
by

The Astronomical Society of the Pacific
(ASP)

ASP CONFERENCE SERIES VOLUMES

Published by the Astronomical Society of the Pacific

PUBLISHED: 1988 (* asterisk means OUT OF STOCK)

Vol. CS -1 PROGRESS AND OPPORTUNITIES IN SOUTHERN HEMISPHERE OPTICAL ASTRONOMY: CTIO 25TH Anniversary Symposium
eds. V. M. Blanco and M. M. Phillips
ISBN 0-937707-18-X

Vol. CS-2 PROCEEDINGS OF A WORKSHOP ON OPTICAL SURVEYS FOR QUASARS
eds. Patrick S. Osmer, Alain C. Porter, Richard F. Green, and Craig B. Foltz
ISBN 0-937707-19-8

Vol. CS-3 FIBER OPTICS IN ASTRONOMY
ed. Samuel C. Barden
ISBN 0-937707-20-1

Vol. CS-4 THE EXTRAGALACTIC DISTANCE SCALE:
Proceedings of the ASP 100th Anniversary Symposium
eds. Sidney van den Bergh and Christopher J. Pritchet
ISBN 0-937707-21-X

Vol. CS-5 THE MINNESOTA LECTURES ON CLUSTERS OF GALAXIES AND LARGE-SCALE STRUCTURE
ed. John M. Dickey
ISBN 0-937707-22-8

PUBLISHED: 1989

Vol. CS-6 SYNTHESIS IMAGING IN RADIO ASTRONOMY: A Collection of Lectures from the Third NRAO Synthesis Imaging Summer School
eds. Richard A. Perley, Frederic R. Schwab, and Alan H. Bridle
ISBN 0-937707-23-6

PUBLISHED: 1990

Vol. CS-7 PROPERTIES OF HOT LUMINOUS STARS: Boulder-Munich Workshop
ed. Catharine D. Garmany
ISBN 0-937707-24-4

Vol. CS-8* CCDs IN ASTRONOMY
ed. George H. Jacoby
ISBN 0-937707-25-2

Vol. CS-9 COOL STARS, STELLAR SYSTEMS, AND THE SUN: Sixth Cambridge Workshop
ed. George Wallerstein
ISBN 0-937707-27-9

Vol. CS-10* EVOLUTION OF THE UNIVERSE OF GALAXIES:
Edwin Hubble Centennial Symposium
ed. Richard G. Kron
ISBN 0-937707-28-7

Vol. CS-11 CONFRONTATION BETWEEN STELLAR PULSATION AND EVOLUTION
eds. Carla Cacciari and Gisella Clementini
ISBN 0-937707-30-9

Vol. CS-12 THE EVOLUTION OF THE INTERSTELLAR MEDIUM
ed. Leo Blitz
ISBN 0-937707-31-7

PUBLISHED: 1991

Vol. CS-13 THE FORMATION AND EVOLUTION OF STAR CLUSTERS
ed. Kenneth Janes
ISBN 0-937707-32-5

ASP CONFERENCE SERIES VOLUMES

Published by the Astronomical Society of the Pacific

PUBLISHED: 1991 (* asterisk means OUT OF STOCK)

Vol. CS-14 ASTROPHYSICS WITH INFRARED ARRAYS
ed. Richard Elston
ISBN 0-937707-33-3

Vol. CS-15 LARGE-SCALE STRUCTURES AND PECULIAR MOTIONS IN THE UNIVERSE
eds. David W. Latham and L. A. Nicolaci da Costa
ISBN 0-937707-34-1

Vol. CS-16 Proceedings of the 3rd Haystack Observatory Conference on ATOMS, IONS, AND MOLECULES: NEW RESULTS IN SPECTRAL LINE ASTROPHYSICS
eds. Aubrey D. Haschick and Paul T. P. Ho
ISBN 0-937707-35-X

Vol. CS-17 LIGHT POLLUTION, RADIO INTERFERENCE, AND SPACE DEBRIS
ed. David L. Crawford
ISBN 0-937707-36-8

Vol. CS-18 THE INTERPRETATION OF MODERN SYNTHESIS OBSERVATIONS OF SPIRAL GALAXIES
eds. Nebojsa Duric and Patrick C. Crane
ISBN 0-937707-37-6

Vol. CS-19 RADIO INTERFEROMETRY: THEORY, TECHNIQUES, AND APPLICATIONS, IAU Colloquium 131
eds. T. J. Cornwell and R. A. Perley
ISBN 0-937707-38-4

Vol. CS-20 FRONTIERS OF STELLAR EVOLUTION:
50th Anniversary McDonald Observatory (1939-1989)
ed. David L. Lambert
ISBN 0-937707-39-2

Vol. CS-21 THE SPACE DISTRIBUTION OF QUASARS
ed . David Crampton
ISBN 0-937707-40-6

PUBLISHED: 1992

Vol. CS-22 NONISOTROPIC AND VARIABLE OUTFLOWS FROM STARS
eds. Laurent Drissen, Claus Leitherer, and Antonella Nota
ISBN 0-937707-41-4

Vol CS-23 ASTRONOMICAL CCD OBSERVING AND REDUCTION TECHNIQUES
ed. Steve B. Howell
ISBN 0-937707-42-4

Vol. CS-24 COSMOLOGY AND LARGE-SCALE STRUCTURE IN THE UNIVERSE
ed. Reinaldo R. de Carvalho
ISBN 0-937707-43-0

Vol. CS-25 ASTRONOMICAL DATA ANALYSIS, SOFTWARE AND SYSTEMS I - (ADASS I)
eds. Diana M. Worrall, Chris Biemesderfer, and Jeannette Barnes
ISBN 0-937707-44-9

Vol. CS-26 COOL STARS, STELLAR SYSTEMS, AND THE SUN:
Seventh Cambridge Workshop
eds. Mark S. Giampapa and Jay A. Bookbinder
ISBN 0-937707-45-7

Vol. CS-27 THE SOLAR CYCLE: Proceedings of the
National Solar Observatory/Sacramento Peak 12th Summer Workshop
ed. Karen L. Harvey
ISBN 0-937707-46-5

ASP CONFERENCE SERIES VOLUMES

Published by the Astronomical Society of the Pacific

PUBLISHED: 1992 (asterisk means OUT OF STOCK)

Vol. CS-28 AUTOMATED TELESCOPES FOR PHOTOMETRY AND IMAGING
eds. Saul J. Adelman, Robert J. Dukes, Jr., and Carol J. Adelman
ISBN 0-937707-47-3

Vol. CS-29 Viña del Mar Workshop on CATACLYSMIC VARIABLE STARS
ed. Nikolaus Vogt
ISBN 0-937707-48-1

Vol. CS-30 VARIABLE STARS AND GALAXIES
ed. Brian Warner
ISBN 0-937707-49-X

Vol. CS-31 RELATIONSHIPS BETWEEN ACTIVE GALACTIC NUCLEI AND STARBURST GALAXIES
ed. Alexei V. Filippenko
ISBN 0-937707-50-3

Vol. CS-32 COMPLEMENTARY APPROACHES TO DOUBLE AND MULTIPLE STAR RESEARCH, IAU Colloquium 135
eds. Harold A. McAlister and William I. Hartkopf
ISBN 0-937707-51-1

Vol. CS-33 RESEARCH AMATEUR ASTRONOMY
ed. Stephen J. Edberg
ISBN 0-937707-52-X

Vol. CS-34 ROBOTIC TELESCOPES IN THE 1990's
ed. Alexei V. Filippenko
ISBN 0-937707-53-8

PUBLISHED: 1993

Vol. CS-35* MASSIVE STARS: THEIR LIVES IN THE INTERSTELLAR MEDIUM
eds. Joseph P. Cassinelli and Edward B. Churchwell
ISBN 0-937707-54-6

Vol. CS-36 PLANETS AROUND PULSARS
ed. J. A. Phillips, S. E. Thorsett, and S. R. Kulkarni
ISBN 0-937707-55-4

Vol. CS-37 FIBER OPTICS IN ASTRONOMY II
ed. Peter M. Gray
ISBN 0-937707-56-2

Vol. CS-38 NEW FRONTIERS IN BINARY STAR RESEARCH: Pacific Rim Colloquium
eds. K. C. Leung and I.-S. Nha
ISBN 0-937707-57-0

Vol. CS-39 THE MINNESOTA LECTURES ON THE STRUCTURE AND DYNAMICS OF THE MILKY WAY
ed. Roberta M. Humphreys
ISBN 0-937707-58-9

Vol. CS-40 INSIDE THE STARS, IAU Colloquium 137
eds. Werner W. Weiss and Annie Baglin
ISBN 0-937707-59-7

Vol. CS-41 ASTRONOMICAL INFRARED SPECTROSCOPY: FUTURE OBSERVATIONAL DIRECTIONS
ed. Sun Kwok
ISBN 0-937707-60-0

ASP CONFERENCE SERIES VOLUMES

Published by the Astronomical Society of the Pacific

PUBLISHED: 1993 (* asterisk means OUT OF STOCK)

Vol. CS-42 GONG 1992: SEISMIC INVESTIGATION OF THE SUN AND STARS
ed. Timothy M. Brown
ISBN 0-937707-61-9

Vol. CS-43 SKY SURVEYS: PROTOSTARS TO PROTOGALAXIES
ed. B. T. Soifer
ISBN 0-937707-62-7

Vol. CS-44 PECULIAR VERSUS NORMAL PHENOMENA IN A-TYPE AND RELATED STARS, IAU Colloquium 138
eds. M. M. Dworetsky, F. Castelli, and R. Faraggiana
ISBN 0-937707-63-5

Vol. CS-45 LUMINOUS HIGH-LATITUDE STARS
ed. Dimitar D. Sasselov
ISBN 0-937707-64-3

Vol. CS-46 THE MAGNETIC AND VELOCITY FIELDS OF SOLAR ACTIVE REGIONS, IAU Colloquium 141
eds. Harold Zirin, Guoxiang Ai, and Haimin Wang
ISBN 0-937707-65-1

Vol. CS-47 THIRD DECENNIAL US-USSR CONFERENCE ON SETI -- Santa Cruz, California, USA
ed. G. Seth Shostak
ISBN 0-937707-66-X

Vol. CS-48 THE GLOBULAR CLUSTER-GALAXY CONNECTION
eds. Graeme H. Smith and Jean P. Brodie
ISBN 0-937707-67-8

Vol. CS-49 GALAXY EVOLUTION: THE MILKY WAY PERSPECTIVE
ed. Steven R. Majewski
ISBN 0-937707-68-6

Vol. CS-50 STRUCTURE AND DYNAMICS OF GLOBULAR CLUSTERS
eds. S. G. Djorgovski and G. Meylan
ISBN 0-937707-69-4

Vol. CS-51 OBSERVATIONAL COSMOLOGY
eds. Guido Chincarini, Angela Iovino, Tommaso Maccacaro, and Dario Maccagni
ISBN 0-937707-70-8

Vol. CS-52 ASTRONOMICAL DATA ANALYSIS SOFTWARE AND SYSTEMS II - (ADASS II)
eds. R. J. Hanisch, R. J. V. Brissenden, and Jeannette Barnes
ISBN 0-937707-71-6

Vol. CS-53 BLUE STRAGGLERS
ed. Rex A. Saffer
ISBN 0-937707-72-4

PUBLISHED: 1994

Vol. CS-54* THE FIRST STROMLO SYMPOSIUM: THE PHYSICS OF ACTIVE GALAXIES
eds. Geoffrey V. Bicknell, Michael A. Dopita, and Peter J. Quinn
ISBN 0-937707-73-2

Vol. CS-55 OPTICAL ASTRONOMY FROM THE EARTH AND MOON
eds. Diane M. Pyper and Ronald J. Angione
ISBN 0-937707-74-0

Vol. CS-56 INTERACTING BINARY STARS
ed. Allen W. Shafter
ISBN 0-937707-75-9

ASP CONFERENCE SERIES VOLUMES

Published by the Astronomical Society of the Pacific

PUBLISHED: 1994 (* asterisk means OUT OF STOCK)

Vol. CS-57 STELLAR AND CIRCUMSTELLAR ASTROPHYSICS
eds. George Wallerstein and Alberto Noriega-Crespo
ISBN 0-937707-76-7

Vol. CS-58* THE FIRST SYMPOSIUM ON THE INFRARED CIRRUS AND DIFFUSE INTERSTELLAR CLOUDS
eds. Roc M. Cutri and William B. Latter
ISBN 0-937707-77-5

Vol. CS-59 ASTRONOMY WITH MILLIMETER AND SUBMILLIMETER WAVE INTERFEROMETRY,
IAU Colloquium 140
eds. M. Ishiguro and Wm. J. Welch
ISBN 0-937707-78-3

Vol. CS-60 THE MK PROCESS AT 50 YEARS: A POWERFUL TOOL FOR ASTROPHYSICAL INSIGHT, A Workshop of the Vatican Observatory --Tucson, Arizona, USA
eds. C. J. Corbally, R. O. Gray, and R. F. Garrison
ISBN 0-937707-79-1

Vol. CS-61 ASTRONOMICAL DATA ANALYSIS SOFTWARE AND SYSTEMS III - (ADASS III)
eds. Dennis R. Crabtree, R. J. Hanisch, and Jeannette Barnes
ISBN 0-937707-80-5

Vol. CS-62 THE NATURE AND EVOLUTIONARY STATUS OF HERBIG Ae/Be STARS
eds. Pik Sin Thé, Mario R. Pérez, and Ed P. J. van den Heuvel
ISBN 0-9837707-81-3

Vol. CS-63 SEVENTY-FIVE YEARS OF HIRAYAMA ASTEROID FAMILIES: THE ROLE OF COLLISIONS IN THE SOLAR SYSTEM HISTORY
eds. Yoshihide Kozai, Richard P. Binzel, and Tomohiro Hirayama
ISBN 0-937707-82-1

Vol. CS-64* COOL STARS, STELLAR SYSTEMS, AND THE SUN:
Eighth Cambridge Workshop
ed. Jean-Pierre Caillault
ISBN 0-937707-83-X

Vol. CS-65* CLOUDS, CORES, AND LOW MASS STARS:
The Fourth Haystack Observatory Conference
eds. Dan P. Clemens and Richard Barvainis
ISBN 0-937707-84-8

Vol. CS-66* PHYSICS OF THE GASEOUS AND STELLAR DISKS OF THE GALAXY
ed. Ivan R. King
ISBN 0-937707-85-6

Vol. CS-67 UNVEILING LARGE-SCALE STRUCTURES BEHIND THE MILKY WAY
eds. C. Balkowski and R. C. Kraan-Korteweg
ISBN 0-937707-86-4

Vol. CS-68* SOLAR ACTIVE REGION EVOLUTION: COMPARING MODELS WITH OBSERVATIONS
eds. K. S. Balasubramaniam and George W. Simon
ISBN 0-937707-87-2

Vol. CS-69 REVERBERATION MAPPING OF THE BROAD-LINE REGION IN ACTIVE GALACTIC NUCLEI
eds. P. M. Gondhalekar, K. Horne, and B. M. Peterson
ISBN 0-937707-88-0

Vol. CS-70* GROUPS OF GALAXIES
eds. Otto-G. Richter and Kirk Borne
ISBN 0-937707-89-9

ASP CONFERENCE SERIES VOLUMES

Published by the Astronomical Society of the Pacific

PUBLISHED: 1995 (* asterisk means OUT OF STOCK)

Vol. CS-71 TRIDIMENSIONAL OPTICAL SPECTROSCOPIC METHODS IN ASTROPHYSICS, IAU Colloquium 149
eds. Georges Comte and Michel Marcelin
ISBN 0-937707-90-2

Vol. CS-72 MILLISECOND PULSARS: A DECADE OF SURPRISE
eds. A. S Fruchter, M. Tavani, and D. C. Backer
ISBN 0-937707-91-0

Vol. CS-73 AIRBORNE ASTRONOMY SYMPOSIUM ON THE GALACTIC ECOSYSTEM: FROM GAS TO STARS TO DUST
eds. Michael R. Haas, Jacqueline A. Davidson, and Edwin F. Erickson
ISBN 0-937707-92-9

Vol. CS-74 PROGRESS IN THE SEARCH FOR EXTRATERRESTRIAL LIFE: 1993 Bioastronomy Symposium
ed. G. Seth Shostak
ISBN 0-937707-93-7

Vol. CS-75 MULTI-FEED SYSTEMS FOR RADIO TELESCOPES
eds. Darrel T. Emerson and John M. Payne
ISBN 0-937707-94-5

Vol. CS-76 GONG '94: HELIO- AND ASTERO-SEISMOLOGY FROM THE EARTH AND SPACE
eds. Roger K. Ulrich, Edward J. Rhodes, Jr., and Werner Däppen
ISBN 0-937707-95-3

Vol. CS-77 ASTRONOMICAL DATA ANALYSIS SOFTWARE AND SYSTEMS IV - (ADASS IV)
eds. R. A. Shaw, H. E. Payne, and J. J. E. Hayes
ISBN 0-937707-96-1

Vol. CS-78 ASTROPHYSICAL APPLICATIONS OF POWERFUL NEW DATABASES: Joint Discussion No. 16 of the 22nd General Assembly of the IAU
eds. S. J. Adelman and W. L. Wiese
ISBN 0-937707-97-X

Vol. CS-79* ROBOTIC TELESCOPES: CURRENT CAPABILITIES, PRESENT DEVELOPMENTS, AND FUTURE PROSPECTS FOR AUTOMATED ASTRONOMY
eds. Gregory W. Henry and Joel A. Eaton
ISBN 0-937707-98-8

Vol. CS-80* THE PHYSICS OF THE INTERSTELLAR MEDIUM AND INTERGALACTIC MEDIUM
eds. A. Ferrara, C. F. McKee, C. Heiles, and P. R. Shapiro
ISBN 0-937707-99-6

Vol. CS-81 LABORATORY AND ASTRONOMICAL HIGH RESOLUTION SPECTRA
eds. A. J. Sauval, R. Blomme, and N. Grevesse
ISBN 1-886733-01-5

Vol. CS-82* VERY LONG BASELINE INTERFEROMETRY AND THE VLBA
eds. J. A. Zensus, P. J. Diamond, and P. J. Napier
ISBN 1-886733-02-3

Vol. CS-83* ASTROPHYSICAL APPLICATIONS OF STELLAR PULSATION, IAU Colloquium 155
eds. R. S. Stobie and P. A. Whitelock
ISBN 1-886733-03-1

ATLAS INFRARED ATLAS OF THE ARCTURUS SPECTRUM, 0.9 - 5.3 μm
eds. Kenneth Hinkle, Lloyd Wallace, and William Livingston
ISBN: 1-886733-04-X

ASP CONFERENCE SERIES VOLUMES

Published by the Astronomical Society of the Pacific

PUBLISHED: 1995 (* asterisk means OUT OF STOCK)

Vol. CS-84 THE FUTURE UTILIZATION OF SCHMIDT TELESCOPES, IAU Colloquium 148
eds. Jessica Chapman, Russell Cannon, Sandra Harrison, and Bambang Hidayat
ISBN 1-886733-05-8

Vol. CS-85* CAPE WORKSHOP ON MAGNETIC CATACLYSMIC VARIABLES
eds. D. A. H. Buckley and B. Warner
ISBN 1-886733-06-6

Vol. CS-86 FRESH VIEWS OF ELLIPTICAL GALAXIES
eds. Alberto Buzzoni, Alvio Renzini, and Alfonso Serrano
ISBN 1-886733-07-4

PUBLISHED: 1996

Vol. CS-87 NEW OBSERVING MODES FOR THE NEXT CENTURY
eds. Todd Boroson, John Davies, and Ian Robson
ISBN 1-886733-08-2

Vol. CS-88* CLUSTERS, LENSING, AND THE FUTURE OF THE UNIVERSE
eds. Virginia Trimble and Andreas Reisenegger
ISBN 1-886733-09-0

Vol. CS-89 ASTRONOMY EDUCATION: CURRENT DEVELOPMENTS, FUTURE COORDINATION
ed. John R. Percy
ISBN 1-886733-10-4

Vol. CS-90 THE ORIGINS, EVOLUTION, AND DESTINIES OF BINARY STARS IN CLUSTERS
eds. E. F. Milone and J. -C. Mermilliod
ISBN 1-886733-11-2

Vol. CS-91 BARRED GALAXIES, IAU Colloquium 157
eds. R. Buta, D. A. Crocker, and B. G. Elmegreen
ISBN 1-886733-12-0

Vol. CS-92* FORMATION OF THE GALACTIC HALO .... INSIDE AND OUT
eds. Heather L. Morrison and Ata Sarajedini
ISBN 1-886733-13-9

Vol. CS-93 RADIO EMISSION FROM THE STARS AND THE SUN
eds. A. R. Taylor and J. M. Paredes
ISBN 1-886733-14-7

Vol. CS-94 MAPPING, MEASURING, AND MODELING THE UNIVERSE
eds. Peter Coles, Vicent J. Martinez, and Maria-Jesus Pons-Borderia
ISBN 1-886733-15-5

Vol. CS-95 SOLAR DRIVERS OF INTERPLANETARY AND TERRESTRIAL DISTURBANCES: Proceedings of 16th International Workshop National Solar Observatory/Sacramento Peak
eds. K. S. Balasubramaniam, Stephen L. Keil, and Raymond N. Smartt
ISBN 1-886733-16-3

Vol. CS-96 HYDROGEN-DEFICIENT STARS
eds. C. S. Jeffery and U. Heber
ISBN 1-886733-17-1

Vol. CS-97 POLARIMETRY OF THE INTERSTELLAR MEDIUM
eds. W. G. Roberge and D. C. B. Whittet
ISBN 1-886733-18-X

ASP CONFERENCE SERIES VOLUMES

Published by the Astronomical Society of the Pacific

PUBLISHED: 1996 (* asterisk means OUT OF STOCK)

Vol. CS-98 FROM STARS TO GALAXIES: THE IMPACT OF STELLAR PHYSICS ON GALAXY EVOLUTION
eds. Claus Leitherer, Uta Fritze-von Alvensleben, and John Huchra
ISBN 1-886733-19-8

Vol. CS-99 COSMIC ABUNDANCES:
Proceedings of the 6th Annual October Astrophysics Conference
eds. Stephen S. Holt and George Sonneborn
ISBN 1-886733-20-1

Vol. CS-100 ENERGY TRANSPORT IN RADIO GALAXIES AND QUASARS
eds. P. E. Hardee, A. H. Bridle, and J. A. Zensus
ISBN 1-886733-21-X

Vol. CS-101 ASTRONOMICAL DATA ANALYSIS SOFTWARE AND SYSTEMS V – (ADASS V)
eds. George H. Jacoby and Jeannette Barnes
ISBN 1080-7926

Vol. CS-102 THE GALACTIC CENTER, 4th ESO/CTIO Workshop
ed. Roland Gredel
ISBN 1-886733-22-8

Vol. CS-103 THE PHYSICS OF LINERS IN VIEW OF RECENT OBSERVATIONS
eds. M. Eracleous, A. Koratkar, C. Leitherer, and L. Ho
ISBN 1-886733-23-6

Vol. CS-104 PHYSICS, CHEMISTRY, AND DYNAMICS OF INTERPLANETARY DUST, IAU Colloquium 150
eds. Bo Å. S. Gustafson and Martha S. Hanner
ISBN 1-886733-24-4

Vol. CS-105 PULSARS: PROBLEMS AND PROGRESS, IAU Colloquium 160
ed. S. Johnston, M. A. Walker, and M. Bailes
ISBN 1-886733-25-2

Vol. CS-106 THE MINNESOTA LECTURES ON EXTRAGALACTIC NEUTRAL HYDROGEN
ed. Evan D. Skillman
ISBN 1-886733-26-0

Vol. CS-107 COMPLETING THE INVENTORY OF THE SOLAR SYSTEM:
A Symposium held in conjunction with the 106th Annual Meeting of the ASP
eds. Terrence W. Rettig and Joseph M. Hahn
ISBN 1-886733-27-9

Vol. CS-108 M.A.S.S. -- MODEL ATMOSPHERES AND SPECTRUM SYNTHESIS:
5th Vienna - Workshop
eds. Saul J. Adelman, Friedrich Kupka, and Werner W. Weiss
ISBN 1-886733-28-7

Vol. CS-109 COOL STARS, STELLAR SYSTEMS, AND THE SUN: Ninth Cambridge Workshop
eds. Roberto Pallavicini and Andrea K. Dupree
ISBN 1-886733-29-5

Vol. CS-110 BLAZAR CONTINUUM VARIABILITY
eds. H. R. Miller, J. R. Webb, and J. C. Noble
ISBN 1-886733-30-9

Vol. CS-111 MAGNETIC RECONNECTION IN THE SOLAR ATMOSPHERE:
Proceedings of a Yohkoh Conference
eds. R. D. Bentley and J. T. Mariska
ISBN 1-886733-31-7

ASP CONFERENCE SERIES VOLUMES

Published by the Astronomical Society of the Pacific

PUBLISHED: 1996 (* asterisk means OUT OF STOCK)

Vol. CS-112 THE HISTORY OF THE MILKY WAY AND ITS SATELLITE SYSTEM
eds. Andreas Burkert, Dieter H. Hartmann, and Steven R. Majewski
ISBN 1-886733-32-5

PUBLISHED: 1997

Vol. CS-113 EMISSION LINES IN ACTIVE GALAXIES: NEW METHODS AND TECHNIQUES, IAU Colloquium 159
eds. B. M. Peterson, F.-Z. Cheng, and A. S. Wilson
ISBN 1-886733-33-3

Vol. CS-114 YOUNG GALAXIES AND QSO ABSORPTION-LINE SYSTEMS
eds. Sueli M. Viegas, Ruth Gruenwald, and Reinaldo R. de Carvalho
ISBN 1-886733-34-1

Vol. CS-115 GALACTIC CLUSTER COOLING FLOWS
ed. Noam Soker
ISBN 1-886733-35-X

Vol. CS-116 THE SECOND STROMLO SYMPOSIUM: THE NATURE OF ELLIPTICAL GALAXIES
eds. M. Arnaboldi, G. S. Da Costa, and P. Saha
ISBN 1-886733-36-8

Vol. CS-117 DARK AND VISIBLE MATTER IN GALAXIES
eds. Massimo Persic and Paolo Salucci
ISBN-1-886733-37-6

Vol. CS-118 FIRST ADVANCES IN SOLAR PHYSICS EUROCONFERENCE: ADVANCES IN THE PHYSICS OF SUNSPOTS
eds. B. Schmieder. J. C. del Toro Iniesta, and M. Vázquez
ISBN 1-886733-38-4

Vol. CS-119 PLANETS BEYOND THE SOLAR SYSTEM AND THE NEXT GENERATION OF SPACE MISSIONS
ed. David R. Soderblom
ISBN 1-886733-39-2

Vol. CS-120 LUMINOUS BLUE VARIABLES: MASSIVE STARS IN TRANSITION
eds. Antonella Nota and Henny J. G. L. M. Lamers
ISBN 1-886733-40-6

Vol. CS-121 ACCRETION PHENOMENA AND RELATED OUTFLOWS, IAU Colloquium 163
eds. D. T. Wickramasinghe, G. V. Bicknell, and L. Ferrario
ISBN 1-886733-41-4

Vol. CS-122 FROM STARDUST TO PLANETESIMALS: Symposium held as part of the 108th Annual Meeting of the ASP
eds. Yvonne J. Pendleton and A. G. G. M. Tielens
ISBN 1-886733-42-2

Vol. CS-123 THE 12th 'KINGSTON MEETING': COMPUTATIONAL ASTROPHYSICS
eds. David A. Clarke and Michael J. West
ISBN 1-886733-43-0

Vol. CS-124 DIFFUSE INFRARED RADIATION AND THE IRTS
eds. Haruyuki Okuda, Toshio Matsumoto, and Thomas Roellig
ISBN 1-886733-44-9

Vol. CS-125 ASTRONOMICAL DATA ANALYSIS SOFTWARE AND SYSTEMS VI
eds. Gareth Hunt and H. E. Payne
ISBN 1-886733-45-7

ASP CONFERENCE SERIES VOLUMES

Published by the Astronomical Society of the Pacific

PUBLISHED: 1997 (* asterisk means OUT OF STOCK)

Vol. CS-126 FROM QUANTUM FLUCTUATIONS TO COSMOLOGICAL STRUCTURES
eds. David Valls-Gabaud, Martin A. Hendry, Paolo Molaro, and Khalil Chamcham
ISBN 1-886733-46-5

Vol. CS-127 PROPER MOTIONS AND GALACTIC ASTRONOMY
ed. Roberta M. Humphreys
ISBN 1-886733-47-3

Vol. CS-128 MASS EJECTION FROM AGN (Active Galactic Nuclei)
eds. N. Arav, I. Shlosman, and R. J. Weymann
ISBN 1-886733-48-1

Vol. CS-129 THE GEORGE GAMOW SYMPOSIUM
eds. E. Harper, W. C. Parke, and G. D. Anderson
ISBN 1-886733-49-X

Vol. CS-130 THE THIRD PACIFIC RIM CONFERENCE ON RECENT DEVELOPMENT ON BINARY STAR RESEARCH
eds. Kam-Ching Leung
ISBN 1-886733-50-3

PUBLISHED: 1998

Vol. CS-131 BOULDER-MUNICH II: PROPERTIES OF HOT, LUMINOUS STARS
ed. Ian D. Howarth
ISBN 1-886733-51-1

Vol. CS-132 STAR FORMATION WITH THE INFRARED SPACE OBSERVATORY (ISO)
eds. João L. Yun and René Liseau
ISBN 1-886733-52-X

Vol. CS-133 SCIENCE WITH THE NGST (Next Generation Space Telescope)
eds. Eric P. Smith and Anuradha Koratkar
ISBN 1-886733-53-8

Vol. CS-134 BROWN DWARFS AND EXTRASOLAR PLANETS
eds. Rafael Rebolo, Eduardo L. Martin, and Maria Rosa Zapatero Osorio
ISBN 1-886733-54-6

Vol. CS-135 A HALF CENTURY OF STELLAR PULSATION INTERPRETATIONS: A TRIBUTE TO ARTHUR N. COX
eds. P. A. Bradley and J. A. Guzik
ISBN 1-886733-55-4

Vol. CS-136 GALACTIC HALOS: A UC SANTA CRUZ WORKSHOP
ed. Dennis Zaritsky
ISBN 1-886733-56-2

Vol. CS-137 WILD STARS IN THE OLD WEST: PROCEEDINGS OF THE 13th NORTH AMERICAN WORKSHOP ON CATACLYSMIC VARIABLES AND RELATED OBJECTS
eds. S. Howell, E. Kuulkers, and C. Woodward
ISBN 1-886733-57-0

Vol. CS-138 1997 PACIFIC RIM CONFERENCE ON STELLAR ASTROPHYSICS
eds. Kwing Lam Chan, K. S. Cheng, and H. P. Singh
ISBN 1-886733-58-9

Vol. CS-139 PRESERVING THE ASTRONOMICAL WINDOWS: Proceedings of Joint Discussion No. 5 of the 23rd General Assembly of the IAU
eds. Syuzo Isobe and Tomohiro Hirayama
ISBN 1-886733-59-7

ASP CONFERENCE SERIES VOLUMES

Published by the Astronomical Society of the Pacific

PUBLISHED: 1998 (* asterisk means OUT OF STOCK)

Vol. CS-140 SYNOPTIC SOLAR PHYSICS --18th NSO/Sacramento Peak Summer Workshop
eds. K. S. Balasubramaniam, J. W. Harvey, and D. M. Rabin
ISBN 1-886733-60-0

Vol. CS-141 ASTROPHYSICS FROM ANTARCTICA:
A Symposium held as a part of the 109th Annual Meeting of the ASP
eds. Giles Novak and Randall H. Landsberg
ISBN 1-886733-61-9

Vol. CS-142 THE STELLAR INITIAL MASS FUNCTION: 38th Herstmonceux Conference
eds. Gerry Gilmore and Debbie Howell
ISBN 1-886733-62-7

Vol. CS-143* THE SCIENTIFIC IMPACT OF THE GODDARD HIGH RESOLUTION SPECTROGRAPH (GHRS)
eds. John C. Brandt, Thomas B. Ake III, and Carolyn Collins Petersen
ISBN 1-886733-63-5

Vol. CS-144 RADIO EMISSION FROM GALACTIC AND EXTRAGALACTIC COMPACT SOURCES, IAU Colloquium 164
eds. J. Anton Zensus, G. B. Taylor, and J. M. Wrobel
ISBN 1-886733-64-3

Vol. CS-145 ASTRONOMICAL DATA ANALYSIS SOFTWARE AND SYSTEMS VII – (ADASS VII)
eds. Rudolf Albrecht, Richard N. Hook, and Howard A. Bushouse
ISBN 1-886733-65-1

Vol. CS-146 THE YOUNG UNIVERSE GALAXY FORMATION AND EVOLUTION AT INTERMEDIATE AND HIGH REDSHIFT
eds. S. D'Odorico, A. Fontana, and E. Giallongo
ISBN 1-886733-66-X

Vol. CS-147 ABUNDANCE PROFILES: DIAGNOSTIC TOOLS FOR GALAXY HISTORY
eds. Daniel Friedli, Mike Edmunds, Carmelle Robert, and Laurent Drissen
ISBN 1-886733-67-8

Vol. CS-148 ORIGINS
eds. Charles E. Woodward, J. Michael Shull, and Harley A. Thronson, Jr.
ISBN 1-886733-68-6

Vol. CS-149 SOLAR SYSTEM FORMATION AND EVOLUTION
eds. D. Lazzaro, R. Vieira Martins, S. Ferraz-Mello, J. Fernández, and C. Beaugé
ISBN 1-886733-69-4

Vol. CS-150 NEW PERSPECTIVES ON SOLAR PROMINENCES, IAU Colloquium 167
eds. David Webb, David Rust, and Brigitte Schmieder
ISBN 1-886733-70-8

Vol. CS-151 COSMIC MICROWAVE BACKGROUND AND LARGE SCALE STRUCTURES OF THE UNIVERSE
eds. Yong-Ik Byun and Kin-Wang Ng
ISBN 1-886733-71-6

Vol. CS-152 FIBER OPTICS IN ASTRONOMY III
eds. S. Arribas, E. Mediavilla, and F. Watson
ISBN 1-886733-72-4

Vol. CS-153 LIBRARY AND INFORMATION SERVICES IN ASTRONOMY III -- (LISA III)
eds. Uta Grothkopf, Heinz Andernach, Sarah Stevens-Rayburn, and Monique Gomez
ISBN 1-886733-73-2

ASP CONFERENCE SERIES VOLUMES

Published by the Astronomical Society of the Pacific

PUBLISHED: 1998 (* asterisk means OUT OF STOCK)

Vol. CS-154 COOL STARS, STELLAR SYSTEMS AND THE SUN: Tenth Cambridge Workshop
eds. Robert A. Donahue and Jay A. Bookbinder
ISBN 1-886733-74-0

Vol. CS-155 SECOND ADVANCES IN SOLAR PHYSICS EUROCONFERENCE: THREE-DIMENSIONAL STRUCTURE OF SOLAR ACTIVE REGIONS
eds. Costas E. Alissandrakis and Brigitte Schmieder
ISBN 1-886733-75-9

PUBLISHED: 1999

Vol. CS-156 HIGHLY REDSHIFTED RADIO LINES
eds. C. L. Carilli, S. J. E. Radford, K. M. Menten, and G. I. Langston
ISBN 1-886733-76-7

Vol. CS-157 ANNAPOLIS WORKSHOP ON MAGNETIC CATACLYSMIC VARIABLES
eds. Coel Hellier and Koji Mukai
ISBN 1-886733-77-5

Vol. CS-158 SOLAR AND STELLAR ACTIVITY: SIMILARITIES AND DIFFERENCES
eds. C. J. Butler and J. G. Doyle
ISBN 1-886733-78-3

Vol. CS-159 BL LAC PHENOMENON
eds. Leo O. Takalo and Aimo Sillanpää
ISBN 1-886733-79-1

Vol. CS-160 ASTROPHYSICAL DISCS: An EC Summer School
eds. J. A. Sellwood and Jeremy Goodman
ISBN 1-886733-80-5

Vol. CS-161 HIGH ENERGY PROCESSES IN ACCRETING BLACK HOLES
eds. Juri Poutanen and Roland Svensson
ISBN 1-886733-81-3

Vol. CS-162 QUASARS AND COSMOLOGY
eds. Gary Ferland and Jack Baldwin
ISBN 1-886733-83-X

Vol. CS-163 STAR FORMATION IN EARLY-TYPE GALAXIES
eds. Jordi Cepa and Patricia Carral
ISBN 1-886733-84-8

Vol. CS-164 ULTRAVIOLET–OPTICAL SPACE ASTRONOMY BEYOND HST
eds. Jon A. Morse, J. Michael Shull, and Anne L. Kinney
ISBN 1-886733-85-6

Vol. CS-165 THE THIRD STROMLO SYMPOSIUM: THE GALACTIC HALO
eds. Brad K. Gibson, Tim S. Axelrod, and Mary E. Putman
ISBN 1-886733-86-4

Vol. CS-166 STROMLO WORKSHOP ON HIGH-VELOCITY CLOUDS
eds. Brad K. Gibson and Mary E. Putman
ISBN 1-886733-87-2

Vol. CS-167 HARMONIZING COSMIC DISTANCE SCALES IN A POST-HIPPARCOS ERA
eds. Daniel Egret and André Heck
ISBN 1-886733-88-0

Vol. CS-168 NEW PERSPECTIVES ON THE INTERSTELLAR MEDIUM
eds. A. R. Taylor, T. L. Landecker, and G. Joncas
ISBN 1-886733-89-9

ASP CONFERENCE SERIES VOLUMES

Published by the Astronomical Society of the Pacific

PUBLISHED: 1999 (* asterisk means OUT OF STOCK)

Vol. CS-169 11th EUROPEAN WORKSHOP ON WHITE DWARFS
eds. J.-E. Solheim and E. G. Meištas
ISBN 1-886733-91-0

Vol. CS-170 THE LOW SURFACE BRIGHTNESS UNIVERSE, IAU Colloquium 171
eds. J. I. Davies, C. Impey, and S. Phillipps
ISBN 1-886733-92-9

Vol. CS-171 LiBeB, COSMIC RAYS, AND RELATED X- AND GAMMA-RAYS
eds. Reuven Ramaty, Elisabeth Vangioni-Flam, Michel Cassé, and Keith Olive
ISBN 1-886733-93-7

Vol. CS-172 ASTRONOMICAL DATA ANALYSIS SOFTWARE AND SYSTEMS VIII
eds. David M. Mehringer, Raymond L. Plante, and Douglas A. Roberts
ISBN 1-886733-94-5

Vol. CS-173 THEORY AND TESTS OF CONVECTION IN STELLAR STRUCTURE: First Granada Workshop
ed. Álvaro Giménez, Edward F. Guinan, and Benjamín Montesinos
ISBN 1-886733-95-3

Vol. CS-174 CATCHING THE PERFECT WAVE: ADAPTIVE OPTICS AND INTERFEROMETRY IN THE 21st CENTURY,
A Symposium held as a part of the 110th Annual Meeting of the ASP
eds. Sergio R. Restaino, William Junor, and Nebojsa Duric
ISBN 1-886733-96-1

Vol. CS-175 STRUCTURE AND KINEMATICS OF QUASAR BROAD LINE REGIONS
eds. C. M. Gaskell, W. N. Brandt, M. Dietrich, D. Dultzin-Hacyan, and M. Eracleous
ISBN 1-886733-97-X

Vol. CS-176 OBSERVATIONAL COSMOLOGY: THE DEVELOPMENT OF GALAXY SYSTEMS
eds. Giuliano Giuricin, Marino Mezzetti, and Paolo Salucci
ISBN 1-58381-000-5

Vol. CS-177 ASTROPHYSICS WITH INFRARED SURVEYS: A Prelude to SIRTF
eds. Michael D. Bicay, Chas A. Beichman, Roc M. Cutri, and Barry F. Madore
ISBN 1-58381-001-3

Vol. CS-178 STELLAR DYNAMOS: NONLINEARITY AND CHAOTIC FLOWS
eds. Manuel Núñez and Antonio Ferriz-Mas
ISBN 1-58381-002-1

Vol. CS-179 ETA CARINAE AT THE MILLENNIUM
eds. Jon A. Morse, Roberta M. Humphreys, and Augusto Damineli
ISBN 1-58381-003-X

Vol. CS-180 SYNTHESIS IMAGING IN RADIO ASTRONOMY II
eds. G. B. Taylor, C. L. Carilli, and R. A. Perley
ISBN 1-58381-005-6

Vol. CS-181 MICROWAVE FOREGROUNDS
eds. Angelica de Oliveira-Costa and Max Tegmark
ISBN 1-58381-006-4

Vol. CS-182 GALAXY DYNAMICS: A Rutgers Symposium
eds. David Merritt, J. A. Sellwood, and Monica Valluri
ISBN 1-58381-007-2

Vol. CS-183 HIGH RESOLUTION SOLAR PHYSICS: THEORY, OBSERVATIONS, AND TECHNIQUES
eds. T. R. Rimmele, K. S. Balasubramaniam, and R. R. Radick
ISBN 1-58381-009-9

ASP CONFERENCE SERIES VOLUMES

Published by the Astronomical Society of the Pacific

PUBLISHED: 1999 (* asterisk means OUT OF STOCK)

Vol. CS-184 THIRD ADVANCES IN SOLAR PHYSICS EUROCONFERENCE: MAGNETIC FIELDS AND OSCILLATIONS
eds. B. Schmieder, A. Hofmann, and J. Staude
ISBN 1-58381-010-2

Vol. CS-185 PRECISE STELLAR RADIAL VELOCITIES, IAU Colloquium 170
eds. J. B. Hearnshaw and C. D. Scarfe
ISBN 1-58381-011-0

Vol. CS-186 THE CENTRAL PARSECS OF THE GALAXY
eds. Heino Falcke, Angela Cotera, Wolfgang J. Duschl, Fulvio Melia, and Marcia J. Rieke
ISBN 1-58381-012-9

Vol. CS-187 THE EVOLUTION OF GALAXIES ON COSMOLOGICAL TIMESCALES
eds. J. E. Beckman and T. J. Mahoney
ISBN 1-58381-013-7

Vol. CS-188 OPTICAL AND INFRARED SPECTROSCOPY OF CIRCUMSTELLAR MATTER
eds. Eike W. Guenther, Bringfried Stecklum, and Sylvio Klose
ISBN 1-58381-014-5

Vol. CS-189 CCD PRECISION PHOTOMETRY WORKSHOP
eds. Eric R. Craine, Roy A. Tucker, and Jeannette Barnes
ISBN 1-58381-015-3

Vol. CS-190 GAMMA-RAY BURSTS: THE FIRST THREE MINUTES
eds. Juri Poutanen and Roland Svensson
ISBN 1-58381-016-1

Vol. CS-191 PHOTOMETRIC REDSHIFTS AND HIGH REDSHIFT GALAXIES
eds. Ray J. Weymann, Lisa J. Storrie-Lombardi, Marcin Sawicki, and Robert J. Brunner
ISBN 1-58381-017-X

Vol. CS-192 SPECTROPHOTOMETRIC DATING OF STARS AND GALAXIES
ed. I. Hubeny, S. R. Heap, and R. H. Cornett
ISBN 1-58381-018-8

Vol. CS-193 THE HY-REDSHIFT UNIVERSE: GALAXY FORMATION AND EVOLUTION AT HIGH REDSHIFT
eds. Andrew J. Bunker and Wil J. M. van Breugel
ISBN 1-58381-019-6

Vol. CS-194 WORKING ON THE FRINGE: OPTICAL AND IR INTERFEROMETRY FROM GROUND AND SPACE
eds. Stephen Unwin and Robert Stachnik
ISBN 1-58381-020-X

PUBLISHED: 2000

Vol. CS-195 IMAGING THE UNIVERSE IN THREE DIMENSIONS: Astrophysics with Advanced Multi-Wavelength Imaging Devices
eds. W. van Breugel and J. Bland-Hawthorn
ISBN 1-58381-022-6

Vol. CS-196 THERMAL EMISSION SPECTROSCOPY AND ANALYSIS OF DUST, DISKS, AND REGOLITHS
eds. Michael L. Sitko, Ann L. Sprague, and David K. Lynch
ISBN: 1-58381-023-4

Vol. CS-197 XVth IAP MEETING DYNAMICS OF GALAXIES: FROM THE EARLY UNIVERSE TO THE PRESENT
eds. F. Combes, G. A. Mamon, and V. Charmandaris
ISBN: 1-58381-24-2

ASP CONFERENCE SERIES VOLUMES

Published by the Astronomical Society of the Pacific

PUBLISHED: 2000 (* asterisk means OUT OF STOCK)

Vol. CS-198 EUROCONFERENCE ON "STELLAR CLUSTERS AND ASSOCIATIONS: CONVECTION, ROTATION, AND DYNAMOS"
eds. R. Pallavicini, G. Micela, and S. Sciortino
ISBN: 1-58381-25-0

Vol. CS-199 ASYMMETRICAL PLANETARY NEBULAE II: FROM ORIGINS TO MICROSTRUCTURES
eds. J. H. Kastner, N. Soker, and S. Rappaport
ISBN: 1-58381-026-9

Vol. CS-200 CLUSTERING AT HIGH REDSHIFT
eds. A. Mazure, O. Le Fèvre, and V. Le Brun
ISBN: 1-58381-027-7

Vol. CS-201 COSMIC FLOWS 1999: TOWARDS AN UNDERSTANDING OF LARGE-SCALE STRUCTURES
eds. Stéphane Courteau, Michael A. Strauss, and Jeffrey A. Willick
ISBN: 1-58381-028-5

Vol. CS-202* PULSAR ASTRONOMY – 2000 AND BEYOND, IAU Colloquium 177
eds. M. Kramer, N. Wex, and R. Wielebinski
ISBN: 1-58381-029-3

Vol. CS-203 THE IMPACT OF LARGE-SCALE SURVEYS ON PULSATING STAR RESEARCH, IAU Colloquium 176
eds. L. Szabados and D. W. Kurtz
ISBN: 1-58381-030-7

Vol. CS-204 THERMAL AND IONIZATION ASPECTS OF FLOWS FROM HOT STARS: OBSERVATIONS AND THEORY
eds. Henny J. G. L. M. Lamers and Arved Sapar
ISBN: 1-58381-031-5

Vol. CS-205 THE LAST TOTAL SOLAR ECLIPSE OF THE MILLENNIUM IN TURKEY
eds. W. C. Livingston and A. Özgüç
ISBN: 1-58381-032-3

Vol. CS-206 HIGH ENERGY SOLAR PHYSICS – *ANTICIPATING HESSI*
eds. Reuven Ramaty and Natalie Mandzhavidze
ISBN: 1-58381-033-1

Vol. CS-207 NGST SCIENCE AND TECHNOLOGY EXPOSITION
eds. Eric P. Smith and Knox S. Long
ISBN: 1-58381-036-6

ATLAS VISIBLE AND NEAR INFRARED ATLAS OF THE ARCTURUS SPECTRUM 3727-9300 Å
eds. Kenneth Hinkle, Lloyd Wallace, Jeff Valenti, and Dianne Harmer
ISBN: 1-58381-037-4

Vol. CS-208 POLAR MOTION: HISTORICAL AND SCIENTIFIC PROBLEMS, IAU Colloquium 178
eds. Steven Dick, Dennis McCarthy, and Brian Luzum
ISBN: 1-58381-039-0

Vol. CS-209 SMALL GALAXY GROUPS, IAU Colloquium 174
eds. Mauri J. Valtonen and Chris Flynn
ISBN: 1-58381-040-4

Vol. CS-210 DELTA SCUTI AND RELATED STARS: Reference Handbook and Proceedings of the 6th Vienna Workshop in Astrophysics
eds. Michel Breger and Michael Houston Montgomery
ISBN: 1-58381-043-9

ASP CONFERENCE SERIES VOLUMES

Published by the Astronomical Society of the Pacific

PUBLISHED: 2000 (* asterisk means OUT OF STOCK)

Vol. CS-211 MASSIVE STELLAR CLUSTERS
eds. Ariane Lançon and Christian M. Boily
ISBN: 1-58381-042-0

Vol. CS-212 FROM GIANT PLANETS TO COOL STARS
eds. Caitlin A. Griffith and Mark S. Marley
ISBN: 1-58381-041-2

Vol. CS-213 BIOASTRONOMY `99: A NEW ERA IN BIOASTRONOMY
eds. Guillermo A. Lemarchand and Karen J. Meech
ISBN: 1-58381-044-7

Vol. CS-214 THE Be PHENOMENON IN EARLY-TYPE STARS, IAU Colloquium 175
eds. Myron A. Smith, Huib F. Henrichs and Juan Fabregat
ISBN: 1-58381-045-5

Vol. CS-215 COSMIC EVOLUTION AND GALAXY FORMATION:
STRUCTURE, INTERACTIONS AND FEEDBACK
The 3rd Guillermo Haro Astrophysics Conference
eds. José Franco, Elena Terlevich, Omar López-Cruz, and Itziar Aretxaga
ISBN: 1-58381-046-3

Vol. CS-216 ASTRONOMICAL DATA ANALYSIS SOFTWARE AND SYSTEMS IX
eds. Nadine Manset, Christian Veillet, and Dennis Crabtree
ISBN: 1-58381-047-1 ISSN: 1080-7926

Vol. CS-217 IMAGING AT RADIO THROUGH SUBMILLIMETER WAVELENGTHS
eds. Jeffrey G. Mangum and Simon J. E. Radford
ISBN: 1-58381-049-8

Vol. CS-218 MAPPING THE HIDDEN UNIVERSE: THE UNIVERSE BEHIND THE MILKY WAY
THE UNIVERSE IN HI
eds. Renée C. Kraan-Korteweg, Patricia A. Henning, and Heinz Andernach
ISBN: 1-58381-050-1

Vol. CS-219 DISKS, PLANETESIMALS, AND PLANETS
eds. F. Garzón, C. Eiroa, D. de Winter, and T. J. Mahoney
ISBN: 1-58381-051-X

Vol. CS-220 AMATEUR - PROFESSIONAL PARTNERSHIPS IN ASTRONOMY:
The 111th Annual Meeting of the ASP
eds. John R. Percy and Joseph B. Wilson
ISBN: 1-58381-052-8

Vol. CS-221 STARS, GAS AND DUST IN GALAXIES: EXPLORING THE LINKS
eds. Danielle Alloin, Knut Olsen, and Gaspar Galaz
ISBN: 1-58381-053-6

PUBLISHED: 2001

Vol. CS-222 THE PHYSICS OF GALAXY FORMATION
eds. M. Umemura and H. Susa
ISBN: 1-58381-054-4

Vol. CS-223 COOL STARS, STELLAR SYSTEMS AND THE SUN:
Eleventh Cambridge Workshop
eds. Ramón J. García López, Rafael Rebolo, and María Zapatero Osorio
ISBN: 1-58381-056-0

Vol. CS-224 PROBING THE PHYSICS OF ACTIVE GALACTIC NUCLEI
BY MULTIWAVELENGTH MONITORING
eds. Bradley M. Peterson, Ronald S. Polidan, and Richard W. Pogge
ISBN: 1-58381-055-2

ASP CONFERENCE SERIES VOLUMES

Published by the Astronomical Society of the Pacific

PUBLISHED: 2001 (* asterisk means OUT OF STOCK)

Vol. CS-225 VIRTUAL OBSERVATORIES OF THE FUTURE
eds. Robert J. Brunner, S. George Djorgovski, and Alex S. Szalay
ISBN: 1-58381-057-9

Vol. CS-226 12th EUROPEAN CONFERENCE ON WHITE DWARFS
eds. J. L. Provencal, H. L. Shipman, J. MacDonald, and S. Goodchild
ISBN: 1-58381-058-7

Vol. CS-227 BLAZAR DEMOGRAPHICS AND PHYSICS
eds. Paolo Padovani and C. Megan Urry
ISBN: 1-58381-059-5

Vol. CS-228 DYNAMICS OF STAR CLUSTERS AND THE MILKY WAY
eds. S. Deiters, B. Fuchs, A. Just, R. Spurzem, and R. Wielen
ISBN: 1-58381-060-9

Vol. CS-229 EVOLUTION OF BINARY AND MULTIPLE STAR SYSTEMS
A Meeting in Celebration of Peter Eggleton's 60th Birthday
eds. Ph. Podsiadlowski, S. Rappaport, A. R. King, F. D'Antona, and L. Burderi
IBSN: 1-58381-061-7

Vol. CS-230 GALAXY DISKS AND DISK GALAXIES
eds. Jose G. Funes, S. J. and Enrico Maria Corsini
ISBN: 1-58381-063-3

Vol. CS-231 TETONS 4: GALACTIC STRUCTURE, STARS, AND THE INTERSTELLAR MEDIUM
eds. Charles E. Woodward, Michael D. Bicay, and J. Michael Shull
ISBN: 1-58381-064-1

Vol. CS-232 THE NEW ERA OF WIDE FIELD ASTRONOMY
eds. Roger Clowes, Andrew Adamson, and Gordon Bromage
ISBN: 1-58381-065-X

Vol. CS-233 P CYGNI 2000: 400 YEARS OF PROGRESS
eds. Mart de Groot and Christiaan Sterken
ISBN: 1-58381-070-6

Vol. CS-234 X-RAY ASTRONOMY 2000
eds. R. Giacconi, S. Serio, and L. Stella
ISBN: 1-58381-071-4

Vol. CS-235 SCIENCE WITH THE ATACAMA LARGE MILLIMETER ARRAY (ALMA)
ed. Alwyn Wootten
ISBN: 1-58381-072-2

Vol. CS-236 ADVANCED SOLAR POLARIMETRY: THEORY, OBSERVATION, AND INSTRUMENTATION, The 20th Sacramento Peak Summer Workshop
ed. M. Sigwarth
ISBN: 1-58381-073-0

Vol. CS-237 GRAVITATIONAL LENSING: RECENT PROGRESS AND FUTURE GOALS
eds. Tereasa G. Brainerd and Christopher S. Kochanek
ISBN: 1-58381-074-9

Vol. CS-238 ASTRONOMICAL DATA ANALYSIS SOFTWARE AND SYSTEMS X
eds. F. R. Harnden, Jr., Francis A. Primini, and Harry E. Payne
ISBN: 1-58381-075-7

Vol. CS-239 MICROLENSING 2000: A NEW ERA OF MICROLENSING ASTROPHYSICS
ed. John Menzies and Penny D. Sackett
ISBN: 1-58381-076-5

ASP CONFERENCE SERIES VOLUMES

Published by the Astronomical Society of the Pacific

PUBLISHED: 2001 (* asterisk means OUT OF STOCK)

Vol. CS-240 GAS AND GALAXY EVOLUTION,
A Conference in Honor of the 20th Anniversary of the VLA
eds. J. E. Hibbard, M. P. Rupen, and J. H. van Gorkom
ISBN: 1-58381-077-3

Vol. CS-241 CS-241 THE 7TH TAIPEI ASTROPHYSICS WORKSHOP ON
COSMIC RAYS IN THE UNIVERSE
ed. Chung-Ming Ko
ISBN: 1-58381-079-X

Vol. CS-242 ETA CARINAE AND OTHER MYSTERIOUS STARS:
THE HIDDEN OPPORTUNITIES OF EMISSION SPECTROSCOPY
eds. Theodore R. Gull, Sveneric Johannson, and Kris Davidson
ISBN: 1-58381-080-3

Vol. CS-243 FROM DARKNESS TO LIGHT:
ORIGIN AND EVOLUTION OF YOUNG STELLAR CLUSTERS
eds. Thierry Montmerle and Philippe André
ISBN: 1-58381-081-1

Vol. CS-244 YOUNG STARS NEAR EARTH: PROGRESS AND PROSPECTS
eds. Ray Jayawardhana and Thomas P. Greene
ISBN: 1-58381-082-X

Vol. CS-245 ASTROPHYSICAL AGES AND TIME SCALES
eds. Ted von Hippel, Chris Simpson, and Nadine Manset
ISBN: 1-58381-083-8

Vol. CS-246 SMALL TELESCOPE ASTRONOMY ON GLOBAL SCALES, IAU Colloquium 183
eds. Wen-Ping Chen, Claudia Lemme, and Bohdan Paczyński
ISBN: 1-58381-084-6

Vol. CS-247 SPECTROSCOPIC CHALLENGES OF PHOTOIONIZED PLASMAS
eds. Gary Ferland and Daniel Wolf Savin
ISBN: 1-58381-085-4

Vol. CS-248 MAGNETIC FIELDS ACROSS THE HERTZSPRUNG-RUSSELL DIAGRAM
eds. G. Mathys, S. K. Solanki, and D. T. Wickramasinghe
ISBN: 1-58381-088-9

Vol. CS-249 THE CENTRAL KILOPARSEC OF STARBURSTS AND AGN:
THE LA PALMA CONNECTION
eds. J. H. Knapen, J. E. Beckman, I. Shlosman, and T. J. Mahoney
ISBN: 1-58381-089-7

Vol. CS-250 PARTICLES AND FIELDS IN RADIO GALAXIES CONFERENCE
eds. Robert A. Laing and Katherine M. Blundell
ISBN: 1-58381-090-0

Vol. CS-251 NEW CENTURY OF X-RAY ASTRONOMY
eds. H. Inoue and H. Kunieda
ISBN: 1-58381-091-9

Vol. CS-252 HISTORICAL DEVELOPMENT OF MODERN COSMOLOGY
eds. Vicent J. Martínez, Virginia Trimble, and María Jesús Pons-Bordería
ISBN: 1-58381-092-7

PUBLISHED: 2002

Vol. CS-253 CHEMICAL ENRICHMENT OF INTRACLUSTER AND INTERGALACTIC MEDIUM
eds. Roberto Fusco-Femiano and Francesca Matteucci
ISBN: 1-58381-093-5

ASP CONFERENCE SERIES VOLUMES

Published by the Astronomical Society of the Pacific

PUBLISHED: 2002 (* asterisk means OUT OF STOCK)

Vol. CS-254 EXTRAGALACTIC GAS AT LOW REDSHIFT
eds. John S. Mulchaey and John T. Stocke
ISBN: 1-58381-094-3

Vol. CS-255 MASS OUTFLOW IN ACTIVE GALACTIC NUCLEI: NEW PERSPECTIVES
eds. D. M. Crenshaw, S. B. Kraemer, and I. M. George
ISBN: 1-58381-095-1

LISTINGS OF IAU VOLUMES MAY BE FOUND ON THE NEXT PAGE

INTERNATIONAL ASTRONOMICAL UNION (IAU) VOLUMES

Published by the Astronomical Society of the Pacific

PUBLISHED: 1999

Vol. No. 190 NEW VIEWS OF THE MAGELLANIC CLOUDS
eds. You-Hua Chu, Nicholas B. Suntzeff, James E. Hesser, and David A. Bohlender
ISBN: 1-58381-021-8

Vol. No. 191 ASYMPTOTIC GIANT BRANCH STARS
eds. T. Le Bertre, A. Lèbre, and C. Waelkens
ISBN: 1-886733-90-2

Vol. No. 192 THE STELLAR CONTENT OF LOCAL GROUP GALAXIES
eds. Patricia Whitelock and Russell Cannon
ISBN: 1-886733-82-1

Vol. No. 193 WOLF-RAYET PHENOMENA IN MASSIVE STARS AND STARBURST GALAXIES
eds. Karel A. van der Hucht, Gloria Koenigsberger, and Philippe R. J. Eenens
ISBN: 1-58381-004-8

Vol. No. 194 ACTIVE GALACTIC NUCLEI AND RELATED PHENOMENA
eds. Yervant Terzian, Daniel Weedman, and Edward Khachikian
ISBN: 1-58381-008-0

PUBLISHED: 2000

Vol. XXIVA TRANSACTIONS OF THE INTERNATIONAL ASTRONOMICAL UNION REPORTS ON ASTRONOMY 1996-1999
ed. Johannes Andersen
ISBN: 1-58381-035-8

Vol. No. 195 HIGHLY ENERGETIC PHYSICAL PROCESSES AND MECHANISMS FOR EMISSION FROM ASTROPHYSICAL PLASMAS
eds. P. C. H. Martens, S. Tsuruta, and M. A. Weber
ISBN: 1-58381-038-2

Vol. No. 197 ASTROCHEMISTRY: FROM MOLECULAR CLOUDS TO PLANETARY SYSTEMS
eds. Y. C. Minh and E. F. van Dishoeck
ISBN: 1-58381-034-X

Vol. No. 198 THE LIGHT ELEMENTS AND THEIR EVOLUTION
eds. L. da Silva, M. Spite, and J. R. de Medeiros
ISBN: 1-58381-048-X

PUBLISHED: 2001

IAU SPS ASTRONOMY FOR DEVELOPING COUNTRIES
Special Session of the XXIV General Assembly of the IAU
ed. Alan H. Batten
ISBN: 1-58381-067-6

Vol. No. 196 PRESERVING THE ASTRONOMICAL SKY
eds. R. J. Cohen and W. T. Sullivan, III
ISBN: 1-58381-078-1

Vol. No. 200 THE FORMATION OF BINARY STARS
eds. Hans Zinnecker and Robert D. Mathieu
ISBN: 1-58381-068-4

Vol. No. 203 RECENT INSIGHTS INTO THE PHYSICS OF THE SUN AND HELIOSPHERE: HIGHLIGHTS FROM SOHO AND OTHER SPACE MISSIONS
eds. Pål Brekke, Bernhard Fleck, and Joseph B. Gurman
ISBN: 1-58381-069-2

INTERNATIONAL ASTRONOMICAL UNION (IAU) VOLUMES

Published by the Astronomical Society of the Pacific

PUBLISHED: 2001

Vol. No. 204 THE EXTRAGALACTIC INFRARED BACKGROUND AND ITS COSMOLOGICAL IMPLICATIONS
eds. Martin Harwit and Michael G. Hauser
ISBN: 1-58381-062-5

Vol. No. 205 GALAXIES AND THEIR CONSTITUENTS AT THE HIGHEST ANGULAR RESOLUTIONS
eds. Richard T. Schilizzi, Stuart N. Vogel, Francesco Paresce, and Martin S. Elvis
ISBN: 1-58381-066-8

Vol. XXIVB TRANSACTIONS OF THE INTERNATIONAL ASTRONOMICAL UNION REPORTS ON ASTRONOMY
ed. Hans Rickman
ISBN: 1-58381-087-0

Complete lists of proceedings of past IAU Meetings are maintained at the IAU Web site at the URL: http://www.iau.org/publicat.html

Volumes 32 - 189 in the IAU Symposia Series may be ordered from:

Kluwer Academic Publishers
P. O. Box 117
NL 3300 AA Dordrecht
The Netherlands

Kluwer@wKap.com

All book orders or inquiries concerning ASP or IAU volumes listed should be directed to the:

The Astronomical Society of the Pacific Conference Series
390 Ashton Avenue
San Francisco CA 94112-1722 USA

Phone: 415-337-2126
Fax: 415-337-5205

E-mail: catalog@astrosociety.org
Web Site: http://www.astrosociety.org